Analysis of Pile Foundations Subject to Static and Dynamic Loading

Analysis of Pile Foundations Subject to Static and Dynamic Loading

Edited by Amir M. Kaynia

NORWEGIAN GEOTECHNICAL INSTITUTE, OSLO, NORWAY
NORWEGIAN UNIVERSITY OF SCIENCE AND TECHNOLOGY (NTNU),
TRONDHEIM, NORWAY

CRC Press
Taylor & Francis Group
Boca Raton   London   New York   Leiden

CRC Press is an imprint of the
Taylor & Francis Group, an **informa** business

A BALKEMA BOOK

CRC Press/Balkema is an imprint of the Taylor & Francis Group, an informa business

Typeset by codeMantra

Library of Congress Cataloging-in-Publication Data Applied for

Published by   CRC Press/Balkema
                Schipholweg 107C, 2316 XC Leiden, The Netherlands
                e-mail: Pub.NL@taylorandfrancis.com
                www.routledge.com – www.taylorandfrancis.com

ISBN: 978-0-367-37416-7 (Hbk)
ISBN: 978-1-032-01547-7 (Pbk)
ISBN: 978-0-429-35428-1 (eBook)

DOI: 10.1201/9780429354281

Contents

Preface

Amir M. Kaynia
Norwegian Geotechnical Institute, Oslo, Norway
Norwegian University of Science and Technology, Trondheim, Norway

Pile foundations are often used when good soil conditions required to ensure satisfactory performance of the foundations are not available at shallow depths. Performance could be related to bearing capacity, settlement, or differential settlement. In soils that are prone to liquefaction during earthquake, piles are sometimes used to transfer the loads through the liquefiable layer to the stable ground at depth. In other cases, performance might be in terms of stiffness and vibration criteria such as in machine foundations under high frequency loads or in earthquake engineering under a wide range of loading frequencies.

While piles of different materials and cross sections have traditionally been used in the building industry and other onshore structures, tubular steel piles have been the dominant type of foundations in offshore structures both for providing capacity and for simpler construction. Piles have more recently been the main foundation type for offshore wind structures using steel monopiles with diameters as large as 10 m.

This book aims at reviewing and presenting available computational tools and design principles for piles used in a host of applications and for different loading conditions. The chapters provide a combination of basic engineering solutions and latest research findings in a balanced and coherent manner. The main motivation has been to present these methods as belonging to the family of numerical methods. The author has experienced that some practicing engineers have the notion that there exist several hardly related analysis methods which depend on pile type, loading, and application. Even among researchers, there are sometimes a misunderstanding that one should expect different answers because the analysis methods are different. The fact that there are many books and articles that each promote one method of analysis for one loading condition or application has not been helpful in this regard. The chapters in this book are written by world-renowned experts in the field, and each addresses one aspect of the subject in detail. However, the materials are presented in a unified manner based on rigorous numerical and analytical methods and in the end they are simplified to give engineering solutions. The chapters illustrate how the different analysis methods and tools are related and how simplifications emerge from them.

STRUCTURE OF THE BOOK

The book is composed of ten chapters. The first four chapters are devoted to static and cyclic loads, and the remaining chapters deal with dynamic and earthquake loadings. Chapter 1 presents the basic elements and steps in pile analysis under static loads together with clear references to the appropriate design regulations in Eurocode 7. It discusses both the conventional analysis tools and simple methods based on pile-soil interaction (subgrade) springs in the lateral direction, the so-called p-y curves, and axial direction, the so-called t-z curves. Chapter 2 shows how to rigorously derive the pile-soil springs presented in Chapter 1 using advanced finite element models and measured stress-strain response of the soil. It also explains how one could use the same analyses to compute hysteretic damping in the springs. Chapter 3 presents application of finite element (FE) methods in 3D analyses of piles, especially large-diameter monopiles used for offshore wind structures. It additionally demonstrates how one could use these analyses to calibrate project-specific p-y springs. As part of an application example, it presents the principles of design of offshore wind monopiles for the applicable limit states. Chapter 4 presents the basic principles of design of suction piles which are often used for anchoring floating offshore structures in deep water. It is shown that while the loading conditions are different in these piles, the principles of the analyses are the same as those in Chapters 1–3.

Chapters 5–10 are devoted to dynamic and earthquake analyses. Chapters 5 and 6 present development of approximate springs and dashpots for dynamic analyses in vertical and horizontal directions using advanced FE and analytical methods. The two chapters also discuss dynamic pile-soil-pile interaction and develop simple tools for computing dynamic interaction factors. Chapter 7 extends the concept of interaction curves to inelastic soil response and shows how the pile-soil-pile interaction effect is substantially reduced due to soil and interface nonlinearities. Chapter 8 extends the methods described in Chapter 6 to earthquake loading. It also discusses design issues including the effect of seismic waves on kinematic interaction forces in piles. Chapter 9 presents issues related to response of piles in liquefiable soil. It discusses both rigorous numerical methods and simplified methods of design. Chapter 10 presents special tools for analyzing foundations comprising large number of piles as encountered in machine foundations, foundations for tall structures and nuclear power plants. In addition to demonstrating some of the basic features of the dynamic stiffnesses, as presented in Chapters 5–7, it discusses several practical issues encountered in the dynamic analysis of large pile foundations.

Editor

Amir M. Kaynia is Technical Expert in Vibration and Earthquake Engineering at NGI, and Professor of Structural Engineering at Norwegian University of Science and Technology (NTNU). He has received his BSc degree from Tehran University and his MSc and PhD degrees in structural engineering from Massachusetts Institute of Technology (MIT). His areas of research and engineering practice include earthquake engineering, soil dynamics, and soil-structure interaction. He has published more than 200 papers in peer reviewed journals and international conference proceedings, has authored six book chapters, and has held numerous keynote lectures worldwide. He has led major international design projects onshore and offshore and has coordinated several research projects funded by the European Commission and the Norwegian Research Council related to earthquake engineering, offshore wind structures and dynamics of high-speed lines. He is a member of several national and international scientific committees dealing with earthquake engineering and soil-structure interaction. He is also chairman of the earthquake engineering committee for Norwegian National Standards and is a member of PT4 for revision of Part 5 of Eurocode 8.

Contributors

Knut H. Andersen Marine Geotechnics, Norwegian Geotechnical Institute, Oslo, Norway

Jamie J. Crispin Department of Civil Engineering, University of Bristol, United Kingdom

Raffaele Di Laora Department of Engineering, Università degli Studi della Campania "Luigi Vanvitelli", Aversa (CE), Italy

Evangelia Garini National Technical University of Athens, Greece

George Gazetas National Technical University of Athens, Greece

S.P. Gopal Madabhushi Department of Engineering, University of Cambridge, Cambridge, United Kingdom

Hans Petter Jostad Advanced Modelling, Norwegian Geotechnical Institute, Oslo, Norway

Amir M. Kaynia Geohazards and Dynamics, Norwegian Geotechnical Institute, Oslo, Norway and Department of Structural Engineering, Norwegian University of Science and Technology (NTNU), Trondheim, Norway

Rasmus T. Klinkvort Integrated Geotechnology, Norwegian Geotechnical Institute, Oslo, Norway

George E. Mylonakis Department of Civil Engineering, University of Bristol, United Kingdom, Department of Civil Engineering, University of California at Los Angeles, USA and Department of Civil and Infrastructure Engineering, Khalifa University, United Arab Emirates

Joani Radhima National Technical University of Athens, Greece

Emmanouil Rovithis Institute of Engineering Seismology and Earthquake Engineering EPPO-ITSAK, Thessaloniki, Greece

Nallathamby Sivasithamparam Advanced Modelling, Norwegian Geotechnical Institute, Oslo, Norway

Christos Vrettos Division of Soil Mechanics and Foundation Engineering, Technical University of Kaiserslautern, Kaiserslautern, Germany

Youhu Zhang School of Civil Engineering, Southeast University, Nanjing, China

Chapter 1

Design of piles for static loads

Christos Vrettos
Technical University of Kaiserslautern

CONTENTS

1.1   INTRODUCTION

Piles are used in the foundation of high-rise buildings, towers, or bridges for transferring and distributing structural loads into deeper, load-bearing layers. Pile foundations are relatively expensive, but they are often an economical alternative to raft foundations. They are often installed in relatively competent ground primarily aiming at fulfilling strict serviceability criteria in terms of settlement or tilt. Design considerations shall take into account the functional significance of the structure, the level of confidence in the soil parameters necessary for proper pile design, the adequacy of the analysis tools to assess pile-soil-structure interaction and load transfer mechanisms, and the extent of construction controls. Although pile load tests are expensive and time-consuming, they are indispensable for confirming and optimizing a pile foundation design in terms of number, pile length, and layout of piles. Especially for uncertain subsoil conditions, a carefully planned pile load test program prior to construction will usually reduce overall foundation costs by allowing for a lower factor of safety in terms of load capacity and a more reliable estimation of differential settlements under service loads. Equally important is the execution in the field by specialized contractors, in particular when soft soil layers of largely unknown characteristics are involved or when alternative, unconventional construction proposals are evaluated for reducing costs. Despite the advanced state-of-practice, many critical issues are still under research, in particular with novel applications of existing technologies, e.g., offshore wind developments.

Numerous studies during the last decades have addressed the behavior of single piles, and much knowledge has been gained mostly from instrumented pile load tests. However, piled foundations comprise mostly pile groups of various sizes that respond to loading through interaction with the surrounding soil along their shaft and their base. Due to this complex interaction, the behavior of a pile group is different from that of a single pile dependent on the pile layout (diameter and length of individual piles, pile spacing), the soil properties, and the load characteristics. The foundation behavior becomes more intricate when the topsoil layer has sufficient strength to participate in the bearing mechanism transforming the entire foundation to a piled raft foundation. In that case, the performance of the foundation in terms of serviceability is the main concern, which requires more advanced design concepts and analysis tools. The state of the art has been presented in the overview articles by Poulos (1989), Randolph (2003), and Mandolini et al. (2005) and in more detail in Kempfert et al. (2003), and in the books by Reese & Van Impe (2011), Fleming et al. (2009), Viggiani et al. (2012), Guo (2013), and Tomlinson & Woodward (2015).

While most of the solutions and design principles presented in this chapter are universal, more focus is placed on European practice in connection with the limit state design concepts of the Structural Eurocodes that are also adopted in many countries outside the European Union. EN 1997 (Eurocode 7 or EC7 in short form) is the relevant code for geotechnical design, cf. Frank et al. (2005), Bond & Harris (2008). National Annexes to EC7 provide the link between EC 7 and the national standards. In Germany, the National Annex to EN 1997-1 (EC7-1) and DIN 1054:2010-12 with Supplementary

Regulations to DIN EN 1997-1 have been merged in a user-friendly Handbook for Eurocode 7, Volume 1 (2011) that is abbreviated herein as Handbook EC7-1.

According to EC7, a limit state is defined as any limiting condition beyond which the structure ceases to fulfil its intended function. Limit state design considers the performance of a structure, or structural elements, at each limit state. Typical limit states are strength, serviceability, stability, and fatigue. Different factors are applied to loads and material strengths to account for their different uncertainties. Verification that ultimate limit states (ULSs) are not exceeded is performed by applying partial factors to the main variables of the calculation model, and by requesting that the fundamental inequality $E_d \leq R_d$ is fulfilled, where E_d denotes the design value of the action effects, and R_d denotes the design value of the resistance.

Currently, the verification against exceedance of the ULS is a straightforward well-established procedure allowing robust designs. However, because pile foundations are often composed of floating files or piles with limited base resistance, the focus in the design has gradually shifted to the assessment of the displacements. The increasing popularity of piled raft foundations has accelerated this trend. Methods to predict displacements are available at different levels of sophistication, but their accuracies are often restricted by the nonlinearity of the soil behavior. Despite the tendency to employ three-dimensional continuum analyses (such as finite element analyses), simplified design procedures remain an integral part of preliminary foundation design and a useful means for high-level control of the advanced analyses.

In the next sections, we first give a brief overview of common pile types and will outline the basic design principles. We then address the axial bearing capacity and deformation of single piles and pile groups. We elaborate on the interaction mechanisms in piled rafts and the available analysis tools with emphasis on settlements. Methods to assess the lateral response of single piles and pile groups are treated next.

The present chapter covers only routine design aspects for common structures, whereas piles under seismic and strong cyclic loading are presented in other chapters in the book. We skip lengthy details and focus on the presentation of the principles underlying the various calculation models and design procedures. Reference to comprehensive books and state-of-the-art papers is provided for further reading. Wherever possible, the notation used by each method considered is translated to a common one. This is, however, not always the case, and it is thus advisable to consult the original source when implementing a particular method. Due to space limitations, correlations for the various soil parameters entering the calculation models are not included. Several textbooks and most of the references listed herein contain a wealth of information on this topic.

1.2 TYPES OF PILES

A wide range of pile types is available, with the selection depending on the soil conditions, groundwater elevations, site conditions, and the length of the piles to satisfy specific structural requirements. Common piles may be divided into two main categories, namely, bored piles (cast-in-place piles) and displacement piles (driven piles), as described in the following.

i Bored piles (cast-in-place piles) whereby the soil is removed either by installing a steel casing or without casing in stable cohesive soil often using a bentonite slurry to support the borehole. Alternatively, continuous flight auger piles are drilled uncased by the use of a continuous hollow auger stem that provides support for the borehole by the soil resting on the auger blades. The pressure and volume of the concrete are controlled to construct a continuous pile without defects. Advantages over precast piles include the ease of changing lengths and diameters, the relatively low material cost, and the possibility to evaluate soil conditions during the drilling process and confirm or revise the pile foundation design. Requirements on the execution are given, for example, in the European execution code EN 1536:2015.

ii   Displacement piles (driven piles) whereby the soil is being displaced by driving the pile or the casing in the ground. This increases the lateral stresses in the surrounding soil leading to soil densification and prestressing in cohesionless and unsaturated cohesive soils and excess pore water pressures in saturated cohesive soils. They are installed by driving, vibrating, jacking, or screwing; therefore, the pile spacing must be sufficiently large and the driving sequence optimized in order to avoid damage to adjacent piles or structures. Guidelines for the installation are provided for example in execution code EN 12699:2015.

Typical examples of displacement piles are:

a   Prefabricated reinforced or nonreinforced concrete piles with solid or hollow cross-sections, with or without prestressing. The pile length is adapted to the local soil conditions using different kinds of splicing systems. This type of pile is advantageous for situations where the required pile length is not exactly known before installation. Risk of damage is primarily associated with transport and tension forces that may develop during driving, especially in layered soils;

b   Prefabricated steel piles of different cross-sections. Among them, H-piles have the advantage of a wide variety of sizes and lengths, easy handling and driving, but have the disadvantage of high material cost;

c   Cast-in-place concrete displacement piles of the Franki type where a steel pipe that is sealed at the base by a concrete plug is driven into the ground by a drop hammer. The widened pile base induced by compaction yields high tip resistances. Proper concreting to provide adequate compressive strength is essential when crossing soft soil layers.

A different type of piles are micropiles, which are small-diameter cast-in-place piles used in places with restricted access and low headroom interiors for underpinning foundations, for extension or reinforcement of existing infrastructure, for stabilization of slopes, or as tension elements to provide uplift safety for ground slabs in deep excavations. Construction of micropiles involves drilling of pile shaft, placing steel reinforcement (typically an all-thread bar), and pumping high-strength cement grout that ensures the load transfer to the soil. Due to their slenderness, the lateral load capacity of micropiles is often small compared with traditional piles. Specifications for the proper execution are given in the execution code EN 14199:2015 and in the system-specific provisions of the authority approvals.

Merits of the various pile types, recommendations for the selection of an appropriate pile type, rules of thumb for the layout, and aspects relevant to the proper execution are summarized in the Recommendations for Piling EAP issued by the German Geotechnical Society (DGGT, 2012). They complement classical references such as Tomlinson & Woodward (2015). For micropiles, one may additionally consult the report of the respective French national project FOREVER (Cyna et al., 2004) and the Reference Manual by FHWA (Sabatini et al., 2005).

1.3 DESIGN PRINCIPLES

Foundation piles are predominantly axially loaded, often accompanied by lateral loading. Usually, these two components of action are considered decoupled and the resulting sectional forces superimposed.

In the terminology of the Eurocodes, the basis for the design are the characteristic values for actions and resistances denoted by the index 'k'. They are obtained from experiments, measurements, calculations, or experience by considering the type of structure and design method, the quality and size of samples, and the sensitivity of the particular parameter. The characteristic values of actions are multiplied by partial load factors, while those of the resistances are divided by appropriate partial safety factors. The resulting values are called design values denoted by the index 'd'.

Two limit states are distinguished: Ultimate limit state ULS defining states associated with collapse, structural failure, excessive deformation, or loss of stability of the whole structure or any part of it; serviceability limit state SLS that corresponds to conditions beyond which specified service requirements are no longer met. Due to the novelty of limit state design in most European countries, the wide variety of soil conditions and soil testing techniques, and the different design traditions, EC 7 allows for three different so-called Design Approaches (DA 1, DA 2, and DA 3) when assessing the design values R_d and E_d for persistent or transient situations. The selection of the particular design approach and the values of the partial factors is left to national determination. Discussion and comparisons of results using the three design approaches may be found in Frank et al. (2005), Bond & Harris (2008).

Regarding pile design, distinction is made between internal and external bearing capacity. The former refers to verification against failure of the pile material (concrete, steel, etc.), while the latter refers to the bearing capacity verification of the ground supporting the pile. In the Eurocode terminology, this corresponds to the ultimate limit state STR for the internal pile structural capacity and GEO for the external bearing capacity. For piles in tension, the limit state UPL (uplift) has to be verified as well. Limit state GEO, which is the focus in this chapter, can be combined with any of the Design Approaches depending on the design philosophy and tradition in each country.

Concerning the actions on piles, it is distinguished between (i) foundation loads, e.g., from the superstructure; (ii) soil-specific loads (e.g. from downdrag, soil movement, bending due to settlement, and lateral earth pressure); and (iii) dynamic and cyclic loads. From these characteristic actions, denoted by E_k, one obtains the characteristic loads on the piles, denoted by $F_{G, k}$ and $F_{Q, k}$, and, $H_{G, k}$ and $H_{Q, k}$, from permanent loads (indexed 'G') and variable loads (indexed 'Q'), which act on the pile head in axial and transverse horizontal direction, as well as bending moments $M_{G, k}$ and $M_{Q, k}$. If variable actions are determined using combination factors, the index 'rep' is added to the load designations. All these loads have to be transferred to the ground fulfilling the internal and external capacity checks mentioned above. The partial safety factors are applied both to actions and to resistances, and depend on the load case by distinguishing between persistent, transient, and accidental design situation.

Possible limit states for pile foundations include geotechnical failure in compression, tension, and under transverse loading, as well as structural failure by buckling, shear, and bending. According to EN 1997-1 (Part 1 of EC7), the design of piles against geotechnical failure shall be based on one of the following approaches:

- Results of static load tests, which have been demonstrated, by means of calculations or otherwise, to be consistent with other relevant experience,
- Empirical or analytical calculation methods whose validity has been demonstrated by static load tests in comparable situations,
- Results of dynamic load tests whose validity has been demonstrated by static load tests in comparable situations,
- Observed performance of a comparable pile foundation, provided that this approach is supported by the results of site investigation and ground testing.

SLS is verified by calculating the pile settlements due to axial loading or pile head deflections due to lateral loading, respectively, and comparing the results against the specified requirements. Although settlement is generally not a concern if the pile is driven to refusal, it may become a design issue if no stiff substratum is encountered or if the length of the pile is reduced to optimize construction cost. It should always be born in mind that such calculations, even using sophisticated constitutive soil models, are subject to uncertainties due to the inherent variability of soil strata and the complexity of nonlinear soil behavior. Hence, results of such analyses will provide merely estimates of the expected response. For small-diameter piles, bearing in medium-dense to dense soils, or for tension piles, the verification that ULS is not exceeded is usually sufficient to fulfil the serviceability criteria in the supported structure.

In this overview, when addressing the bearing capacity of piles, we mainly present the state-of-practice currently followed in Germany. It is based on the long tradition in DIN standards for pile design as summarized for geotechnical design in the Handbook EC7-1 (2011). The Recommendations on Piling — EAP issued by the German Geotechnical Society (DGGT, 2012) as a companion to EC7 implemented its concepts and philosophy and cover the whole spectrum of provisions necessary for the design, testing, and construction of pile foundations. The original version of EAP in German was published in 2012 and was later translated into English in 2013. We will refer to it simply as EAP. Of course, reference to other codes and international practice is given as well throughout the text.

EAP excludes the application of calculation models or numerical analyses of soil continua to assess bearing capacity and suggests empirical values of resistance dependent on the pile type and the in-situ soil properties. The respective tables and formulae were derived over the years from the back-analysis of a large number of instrumented full-scale pile tests and have been recently updated. A similar concept has been traditionally followed in France (Frank, 2017) and in several other countries. Use of such code-based, empirical relationships inevitably leads to conservative designs. Project-specific pile load tests yield mostly higher bearing capacities, thus justifying the necessary additional expense.

Overviews on the design practice in Europe related to the introduction of Eurocode 7 are provided among others by Møller et al. (2016) for Denmark, Burlon et al. (2016) for France, Moormann (2016) for Germany, and Bond & Simpson (2010) for the UK.

1.4   SINGLE PILES AND PILE GROUPS UNDER VERTICAL LOADING

1.4.1   Pile resistance from tests

1.4.1.1   Static load tests

The purpose of a pile load test is the determination of a characteristic resistance vs. settlement or resistance vs. heave curve in order to derive the pile resistances for compression or tension, $R_{c,k}$ and $R_{t,k}$, respectively, for the verification of the ULS and the SLS. Adequate instrumentation allows the determination of both base resistance and shaft resistance. Failure in piles under compression is defined either by a suitable creep criterion or when the settlement of the pile top equals 10% of the pile base diameter. For noncircular base shapes, an equivalent diameter may be used. Tension piles shall always be tested to failure.

Regarding the necessary instrumentation, separate monitoring of shaft and base resistance is required when soil is strongly nonhomogeneous, shaft resistance after reaching peak value drops to a much lower residual value, and when strict criteria are imposed on limiting displacement values. In most cases, recording of pile head settlement is considered sufficient.

When determining the characteristic resistance $R_{c,k}$ from static pile load tests according to Eurocode 7, the dispersion of the data is taken into account by two correlation factors ξ_1 and ξ_2. Both factors depend on the number of piles tested and are applied to the mean and minimum measured resistances according to the following equation:

$$R_{c,k} = \mathrm{Min}\left\{\frac{\overline{R}_{c,m}}{\xi_1} ; \frac{R_{c,m,min}}{\xi_2}\right\} \tag{1.1}$$

where $\overline{R}_{c,m}$ and $R_{c,m,min}$ are the average and minimum values of measured resistance, respectively. Values for the correlation factors are given in Table 1.1.

The correlation factors are only slightly different from those recommended in the EC7-1 in combination with the National Annex, cf. Frank et al. (2005). One can identify the advantage of carrying out more load tests since the values of correlation factors reduce as the number of load tests increases. For structures which have sufficient rigidity to transfer loads from 'weak' to 'strong' piles, the ξ values may be divided by 1.1. In any case, ξ_1 should not be less than 1.0. The background for the implementation in France is outlined in Burlon et al. (2014) and Frank (2017).

For tension piles, the characteristic pile resistance $R_{t,k}$ must be determined by static pile tests. For a large number of tension piles, Handbook EC7-1 (2011) requires that at least two piles or 3% of all piles should be tested. The characteristic resistance is determined in accordance with Eq. (1.1).

Table 1.1   Correlation factors ξ depending on the number of pile load tests n (Handbook EC7-1, 2011)

n	1	2	3	4	≥ 5
ξ_1	1.35	1.25	1.15	1.05	1.00
ξ_2	1.35	1.15	1.00	1.00	1.00

1.4.1.2 Impact and dynamic load tests

Under certain circumstances, the pile resistances may be determined from dynamic load tests. In general, it is distinguished between impact (drop hammer) and dynamic tests. For impact tests, the evaluation procedure is based on pile driving formulae or wave equation methods. In dynamic tests, one measures the time-dependent force and movement at the pile head. Evaluation by means of the direct method (CASE and TNO method) yields the axial pile bearing resistance, while in the extended method (CAPWAP and TNOWAVE), the entire system is modelled yielding the settlement vs. resistance curve.

The characteristic compressive resistance is obtained in accordance with Eq. (1.1) where ξ_1 is replaced by ξ_5, and ξ_2 by ξ_6, and the correlation factors ξ_i are defined as follows:

$$\xi_i = \left(\xi_{0,i} + \Delta\xi\right) \cdot \eta_D \tag{1.2}$$

where $\xi_{0,i}$ is the basic value dependent on the number of piles tested, n, $\Delta\xi$ is an increase factor dependent on the calibration method used, and η_D is a model factor to account for the back analysis method. The factors $\xi_{0,5}/\xi_{0,6}$ vary between 1.60/1.50 for $n=2$ to 1.40/1.25 for $n \geq 20$. The factor $\Delta\xi$ ranges from 0 for calibration on static pile test results on the same site to 0.40 for calibration based on documented or common empirical data for pile resistances. The model factor η_D ranges between 1.05 and 1.20 for impact tests and between 0.85 and 1 for dynamic tests. Details are given in the EAP.

1.4.2 Pile resistance from empirical correlations

The external bearing capacity R of a single pile under axial compression loading is made up of two components, the base or tip resistance, R_b, and the shaft resistance, R_s, which both depend on the actual settlement of the pile, s. In terms of their characteristic values and for layered soil with distinct layers $i = 1,\ldots, n$ the total resistance reads:

$$R_{c,k}(s) = R_{b,k}(s) + R_{s,k}(s) = q_{b,k} \cdot A_b + \sum_i q_{s,k,i} \cdot A_{s,i} \tag{1.3}$$

where A_b is the pile base area, and $A_{s,i}$ is the pile shaft surface in layer i. Characteristic values for base resistance $q_{b,k}$ and shaft resistance (skin friction) in each layer $q_{s,k,i}$ may directly be determined from values of soil parameters using appropriate charts and tables. Figure 1.1 shows the different shapes of the load vs. settlement curve for base resistance q_b and for shaft resistance q_s. Note that q_b is almost parabolic, while q_s is bi-linear and, most importantly, attains its maximum value at much smaller settlement. Thus, during loading, shaft resistance is fully mobilized much earlier than base resistance. Figure 1.1 indicates that the base resistance is fully mobilized at settlement s_g with

$$s_g = 0.10 \cdot d_b$$

Pile resistance R_k

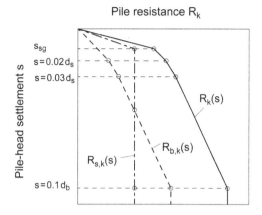

Figure 1.1 Elements of characteristic resistance vs. settlement curves for bored piles (EAP).

where d_b is the pile base diameter. For the pile settlement at the SLS, the relevant parameter is the pile shaft diameter d_s.

The settlement for the full mobilization of shaft resistance is defined by the following equation in which settlement is given in [cm] and resistance in [MN]:

$$s_{sg} = 0.50 \cdot R_{s,k}\left(s_{sg}\right) + 0.50 \leq 3.00 \tag{1.4}$$

When determining the characteristic axial pile resistance vs. heave curve for bored piles it is set

$$s_{sg,tension} = 1.30 \cdot s_{sg} \tag{1.5}$$

Empirical correlations for shaft and base resistances for various pile types are given in the EAP. They have been derived from a statistical analysis of a large number of well-documented static pile load tests in sites with sufficient knowledge of the prevailing soil conditions (Kempfert & Becker, 2010). Correlation factors ξ were not applied. Tables 1.2–1.5 summarize ranges of values for bored piles dependent on (i) the CPT tip cone resistance q_c for cohesionless soils and (ii) the characteristic value of undrained shear strength $c_{u,k}$ for cohesive soils. They are valid for pile diameter d_b or d_s from 0.30 to 3.00 m of piles embedded at least 2.50 m in the bearing layer. Below the pile base, the bearing layer must have a thickness of $3 \cdot d_b$ or at least 1.50 m. In this region, the bearing layer must exhibit an average CPT tip cone resistance $q_c \geq 7.5$ MN/m^2 or an undrained strength $c_u \geq 0.1$ MN/m^2. When specifying the base resistance from q_c and $c_{u,k}$, the relevant zone extends $1 \cdot d_b$ above and $4 \cdot d_b$ below the pile base level for pile diameters up to $d_b = 0.6$ m and $1 \cdot d_b$ above and $3 \cdot d_b$ below the base for $d_b > 0.6$ m. For pile shaft resistance, the average value over a layer with small scatter in the data shall be taken. The empirical values for bored piles may also be used for diaphragm wall elements (barrettes) and bored pile walls. Values for rocks are reproduced in Table 1.6. For all these tables, intermediate values may be obtained by linear interpolation.

Table 1.2 Range of values for the pile base resistance of bored piles in cohesionless soil according to EAP

	Pile base resistance $q_{b,k}$ [kN/m²]		
Relative pile head settlement s/d_s and s/d_b	Mean CPT cone tip resistance q_c [MN/m²]		
	7.5	15	25
0.02	550–800	1050–1400	1750–2300
0.03	700–1050	1350–1800	2250–2950
0.10 (= s_g)	1600–2300	3000–4000	4000–5300

For bored piles with widened base, values shall be reduced to 75%.

Table 1.3 Range of values for the pile base resistance of bored piles in cohesive soil according to EAP

	Pile base resistance $q_{b,k}$ [kN/m²]		
Relative pile head settlement s/d_s and s/d_b	Shear strength of the undrained soil $c_{u,k}$ [kN/m²]		
	100	150	250
0.02	350–450	600–750	950–1200
0.03	450–550	700–900	1200–1450
0.10 (= s_g)	800–1000	1200–1500	1600–2000

For bored piles with widened base values shall be reduced to 75%.

Table 1.4 Range of values for pile shaft resistance of bored piles in cohesionless soil according to EAP

Mean CPT cone tip resistance [MN/m²]	Pile shaft resistance $q_{s,k}$ [kN/m²]
7.5	55–80
15	105–140
≥25	130–170

Table 1.5 Range of values for pile shaft resistance of bored piles in cohesive soil according to EAP

Shear strength of the undrained soil $c_{u,k}$ [kN/m²]	Pile shaft resistance $q_{s,k}$ [kN/m²]
60	30–40
150	50–65
≥250	65–85

A similar methodology is followed in France (Frank, 2017). The empirical values initially derived from pile load tests in terms of field values of the Ménard pressuremeter (MPM) and the CPT have been enhanced by additional results, re-evaluated, updated, and currently constitute the basis for the French standard NFP 94-262 for the application of Eurocode 7 to deep foundations (AFNOR, 2012).

Several empirical formulae have been developed in the last decades directly correlating pile shaft and pile base resistance to CPT-data, mostly to the tip cone resistance. Some of them utilize also the friction ratio readings of the CPT. The rationale behind it is that the cone penetrometer can be regarded as a mini pile and that CPT provides a continuous profile of soil stratification. Both instrumented and noninstrumented pile tests have been used for the calibration. Methods that provide design values directly for the pile capacity, in a manner analogous to the EAP-values presented above, have been reviewed and categorized by Niazi & Mayne (2013). Only a few of them provide values for both shaft and base resistance. One may mention the NGI-BRE method applicable to driven/jacked piles in clays (Almeida et al., 1996; Powell et al., 2001), the UniCone method for all types of pile and soil (Eslami & Fellenius, 1997), and the Fugro V-K method for offshore driven piles in clays (Van Dijk & Kolk, 2011). The KTRI method (Takesue et al., 1998) for all pile and soil types, and the UWA'13 method (Lehane et al., 2013) for driven/jacked piles in clays provide information only for the shaft resistance. The incorporation of such empirical formulae into code-based routine design still requires additional verification and assessment of their reliability.

For piles in rock, limited data are available. Furthermore, parameters to classify rock were incomplete and the unconfined compressive strength q_u typically used for this is often determined on intact rock. Recommended range of values from EAP are reproduced in Table 1.6 and are valid under the condition of a minimum embedment into rock of 0.5 m for $q_{u,k} \geq 5$ MN/m^2 and 2.50 m for $q_{u,k} \leq 0.50$ MN/m^2. That reference also contains suggested values for piles in weathered rock as well as in siltstone, mudstone, and sandstone by considering the degree of weathering and the discontinuities spacing.

For grouted displacement piles and micropiles with $d_s \leq 0.30$ m, the general rule is to perform pile load tests, and empirical values should be used only in exceptional cases. Resistance in compression and tension is provided only by the shaft resistance:

$$R_k(s) = R_{s,k}(s) = \sum_i q_{s,k,i} \cdot A_{s,i} \tag{1.6}$$

Empirical values for shaft resistance specifically for grouted micropiles from EAP are reproduced in Tables 1.7 and 1.8. For grout body lengths larger than 12 m, the execution of pile load tests is strongly recommended, as this limit corresponds to the current state of experience.

Table 1.6 Range of values for pile base and pile shaft resistance for bored piles in rock according to EAP

Uniaxial compression strength $q_{u,k}$ [MN/m^2]	Pile base resistance $q_{b,k}$ [kN/m^2]	Pile shaft resistance $q_{s,k}$ [MN/m^2]
0.5	1500–2500	70–250
5.0	5000–10,000	500–1000
20.0	10,000–20,000	500–2000

Table 1.7 Range of values for pile shaft resistance of grouted micropiles ($d_s \leq 0.30$ m) in cohesionless soil according to EAP

Mean CPT cone tip resistance [MN/m²]	Pile shaft resistance $q_{s,k}$ [kN/m²]
7.5	135–175
15	215–280
≥25	255–315

Table 1.8 Range of values for pile shaft resistance of grouted micropiles ($d_s \leq 0.30$ m) in cohesive soil according to EAP

Shear strength of the undrained soil $c_{u,k}$ [kN/m²]	Pile shaft resistance $q_{s,k}$ [kN/m²]
60	55–65
150	95–105
≥250	115–125

1.4.3   Code-based verification of limit states

1.4.3.1   Bearing capacity verification

In the ULS verification, the characteristic values of the loads imposed by the super-structure on the piles are multiplied by the partial factors to obtain the design values:

$$F_{c,d} = F_{c,G,k} \cdot \gamma_G + F_{c,Q,k} \cdot \gamma_Q \tag{1.7}$$

where γ_G and γ_Q are the partial safety factors for unfavorable permanent and variable actions, respectively. For persistent design situations, for example, Handbook EC7-1 (2011) specifies $\gamma_G = 1.35$ and $\gamma_Q = 1.50$.

The design value of the pile compressive resistance $R_{c,d}$ may be obtained either separately for the base and the shaft resistances or as a total (combined) pile resistance. In the first case, the relevant partial factors γ_b and γ_s are applied to the characteristic base and shaft resistances, $R_{b,k}$ and $R_{s,k}$; in the latter, the relevant partial factor γ_t is applied to the total resistance $R_{c,k}$. EC7-1 suggests different sets of values for the individual partial safety factors dependent on the type of pile. One of the sets corresponds to identical values for γ_b, γ_s and γ_t, a procedure that is often suggested in the codes. The shaft resistance partial factor for tension is always higher than for compression. We then have

$$R_{c,d} = R_{b,d} + R_{s,d} = \frac{R_{b,k}}{\gamma_b} + \frac{R_{s,k}}{\gamma_s} = \frac{R_{c,k}}{\gamma_t} \tag{1.8}$$

$$R_{t,d} = R_{s,d} = \frac{R_{s,k}}{\gamma_{s,t}} = \frac{R_{t,k}}{\gamma_{s,t}} \tag{1.9}$$

for compression and tension, respectively. The partial safety factors suggested by EC7-1 Handbook (2011) are summarized in Table 1.9.

Table 1.9 Partial safety factors for pile resistances for all design situations (Handbook EC7-1, 2011)

Pile resistance	Factor	Value
In compression from pile load test	$\gamma_b, \gamma_s, \gamma_t$	1.10
In tension from pile load test	$\gamma_{s,t}$	1.15
From empirical correlations for pile in compression	$\gamma_b, \gamma_s, \gamma_t$	1.40
From empirical correlations for pile in tension (only in exceptional cases)	$\gamma_{s,t}$	1.50

Finally, the bearing capacity in compression is checked:

$$F_{c,d} \le R_{c,d} \tag{1.10}$$

For piles in tension, we proceed analogously. Note that for persistent design situation, the average partial factor for actions is $\gamma_{G,\,Q} \approx 1.4$, and the partial factor for resistance is $\gamma_t = 1.4$ when pile resistance is determined from empirical correlations. This yields $1.4 \cdot 1.4 \approx 2.0$ that corresponds to the value of the traditional global safety factor.

Partially embedded piles (e.g., in waterfront structures) and piles in soft soil with an undrained shear strength $c_{u,k} \le 10 \, \text{kN/m}^2$ have to be checked against buckling. It should be kept in mind that micropiles may be prone to buckling even when $c_{u,k} > 10 \, \text{kN/m}^2$.

1.4.3.2 Serviceability verification

The verification of serviceability is performed by setting a limitation on the deformation, mostly at the pile head. This requires the definition of a suitable load vs. settlement curve derived either from a pile load test or from an empirical curve suitable for the soil conditions prevailing at the site. Alternatively, the deformation may be estimated from an analysis of the pile response utilizing analytical or numerical models.

Crucial to the serviceability are the differential settlements between the individual piles within a pile group or between supports consisting of single piles or pile groups. The amount of the actual differential settlement Δs depends mainly on the pile type and the absolute settlement. Typical values for $\Delta s/s$ range from 1/3 for bored piles to 1/4 for displacement pile foundations. Displacements associated with serviceability requirements usually range from approximately 2 cm to 0.035 times the pile diameter. If pile load tests are used to define the curve for characteristic load vs. displacement, correlation factors ξ are omitted as they refer only to the bearing capacity. If curves derived from the empirical values summarized above are adopted, it is assumed that they are identical for both the ULS and the SLS.

The verification is formally carried out either by comparing the characteristic load to the load corresponding to a specified deformation criterion or by comparing the calculated characteristic deformation with the allowable one.

Differential settlements are likely to occur even when pile interaction effects are not significant. They may result from the inherent heterogeneity of the ground, the pile installation procedure, or the position of the pile within the pile group. For an estimate, the simplified procedure suggested by the EAP assumes that

$$\Delta s_k = \kappa \cdot s_k \tag{1.11}$$

with κ being an empirical factor. The load vs. settlement curve is then shifted by this amount. In the absence of specific data, EAP suggests as a first approximation κ = 0.15.

1.4.4   Settlement of single piles

1.4.4.1   Approximate closed-form solutions

In routine design, pile reaction is modelled by nonlinear springs. In the absence of pile load test results, an initial assessment of the spring stiffness may be accomplished utilizing relationships derived from continuum-based solutions using numerical methods in combination with soil models of different complexity. Although nonlinearity of soil behavior is undisputed, solutions based on elasticity theory are still the first choice due to the relatively small deformations occurring under service loading of structures. The required two elastic constants are assessed as equivalent values that reflect the anticipated strain level. A versatile closed-form solution, derived by Randolph & Wroth (1978) that extended the work by Frank (1974), combines expressions for shaft and base resistance. The crucial assumptions are: (i) pile loading is transferred to the soil along the shaft via shearing on concentric cylinders up to a distance r_m from the pile, (ii) the base response corresponds to that of a rigid footing with spring stiffness K_b, and (iii) shaft and base responses are decoupled. The resulting expression for axial stiffness, i.e., the pile-top settlement w due to an applied load P, for a pile of diameter d and length l is reproduced, e.g., in Fleming et al. (1992). We present here a more concise form utilizing the dimensionless base stiffness coefficient Ω introduced by Mylonakis & Gazetas (1998). The expression for the single pile stiffness then becomes

$$\frac{P}{w} = E_p \cdot A_p \cdot \mu \cdot \frac{\Omega + \chi \cdot \tanh(\mu \cdot l)}{1 + \Omega \cdot \tanh(\mu \cdot l)} \tag{1.12}$$

$$\mu = \sqrt{\frac{2 \cdot \pi \cdot G_l}{\zeta \cdot E_p \cdot A_p}} \tag{1.13}$$

$$\Omega = \frac{K_b}{E_p \cdot A_p \cdot \mu} \tag{1.14}$$

$$K_b = \frac{2 \cdot G_b \cdot d}{(1-\nu)} \cdot \eta \tag{1.15}$$

$$\zeta = \ln(2 \cdot r_m/d) = \ln\left\{\left[0.25 + (2.5 \cdot \chi \cdot (1-\nu) - 0.25) \cdot (G_l/G_b)\right] \cdot 2 \cdot l/d\right\} \tag{1.16}$$

$$\chi = \frac{\bar{G}}{G_l} \leq 1 \tag{1.17}$$

where E_p is the Young's modulus of the pile material, A_p is the cross sectional area of the pile, G_l is the shear modulus of the soil at the level of the pile base ($z = l$), $\bar{G}$ is the mean value of the shear modulus over the pile length, G_b is the shear modulus of the bearing stratum at the base for end-bearing piles, v is the Poisson's ratio of the soil, and η is the ratio of underream for underreamed piles. Typically, $\eta = 1$.

The situation for a linear depth variation of the soil modulus is depicted in Figure 1.2. For layered soil profiles, it is sufficiently accurate to compute a weighted average of the modulus along the pile length and use the solution for a homogeneous soil.

Note that in the above Eqs. (1.12)–(1.17), the soil stiffness is expressed in terms of the shear modulus G, instead of the Young's modulus E or the oedometric modulus (constrained modulus) E_{oed} that are customarily used in foundation engineering. Since the load transfer is primarily due to shearing, use of G is more appropriate and simplifies the expressions. G can be determined from E_{oed} or E using the well-known relations of linear elasticity:

$$G = \frac{(1 - 2 \cdot v)}{2 \cdot (1 - v)} E_{oed} = \frac{E}{2 \cdot (1 + v)} \tag{1.18}$$

Like the Young's modulus E, Poisson's ratio v varies with strain level, and its value is uncertain. For drained conditions, values $v = 0.2$–0.3 are often suggested (Lo Presti, 1994).

The parameter ζ, defined by Eq. (1.16), is a measure of the radius of influence of the pile and has been estimated from finite element analyses. As elasticity theory overestimates the extent of the domain affected by the spread of stresses, Eq. (1.16) increases monotonically with increasing length of the pile, which contradicts intuition. To obtain a realistic estimate, an upper bound must be imposed on ζ, e.g., presuming that beyond a radius of ten times the pile diameter the displacements in the soil become negligibly small. This would yield an upper bound for ζ equal to $\ln(20) = 3$. The parameter ζ may be assigned a higher value, e.g., 5, to approximately capture nonlinear effects in the soil zone next to the pile.

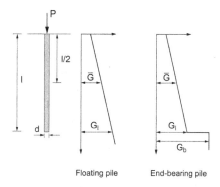

Figure 1.2 Pile embedded in soil with depth-dependent modulus.

From Eq. (1.12) one may identify the conditions for two limiting cases: (i) for $l/d < 0.25$ $\sqrt{E_p/G_l}$ the pile may be assumed approximately rigid, and (ii) for $l/d > 1.5\sqrt{E_p/G_l}$ the pile is considered very long with no load reaching its lower end, cf. Fleming et al. (2009).

Of interest is also the distribution of the displacement w with depth along the pile:

$$\frac{w}{w_b} = \cosh\big(\mu \cdot (l - z)\big) \tag{1.19}$$

where w_b is the displacement at the pile base.

The ability of Eq. (1.12) to reproduce elastic continuum solutions has been re-examined by Guo & Randolph (1997), and an enhanced version of the equation for the shaft load transfer factor ζ has been proposed to improve modelling fidelity in vertically nonhomogeneous soils.

Equation (1.12) has been developed for floating piles or piles with limited base resistance. For end-bearing piles, one has to merely increase the value of G_b accordingly to yield a perfect rigid base, namely

$$\xi \to 0: \quad \frac{P}{w} = K_1 = \frac{E_p \cdot A_p \cdot \mu}{\tanh(\mu \cdot l)} \tag{1.20}$$

As known, the major difficulty in applying elasticity solutions in practice is the selection of an appropriate value for the elastic modulus describing the strain-dependent soil stiffness. The use of values determined from soil laboratory tests does not yield reliable estimates for the pile response. It is thus recommended to calibrate such predictive equations from load vs. settlement curves obtained from a reference pile load test in similar ground conditions and then to extrapolate to other pile geometries. The curves that correspond to the tabulated values in the EAP are suitable as first estimates.

For a rough assessment of the effects of the nonlinearity, the simplified method proposed by Mayne & Schneider (2001) may be utilized. The representative soil shear modulus is adjusted to the mobilization of the overall bearing capacity of the pile by adopting the approximate nonlinear load-deformation relationship as suggested by Fahey & Carter (1993)

$$\frac{G}{G_{max}} = 1 - f \cdot \left(\frac{q}{q_u}\right)^g \tag{1.21}$$

with values $f = 1$ and $g = 0.3 \pm 0.1$, where G_{max} is the small-strain shear modulus, G is the shear modulus corresponding to the load q, and q_u is the ultimate bearing resistance of the pile as determined, e.g., from pile load tests or code specifications. This means, for example, that if the service load of a pile is increased from 50% of the ultimate bearing resistance to 60%, the effective stiffness of the embedded pile is reduced by 24%.

An alternative means to assemble a nonlinear load vs. displacement curve in conjunction with Eq. (1.12) has been proposed by Berardi & Bovolenta (2005) by defining the equivalent secant modulus dependent on the ratio between the actual pile settlement w and the pile diameter d as follows:

$$\frac{G}{G_{max}} = \frac{1}{5.81 \cdot 10^2 \cdot (w/d) + 0.80} \tag{1.22}$$

The ratio w/d is a measure of the induced shear strain in the soil. The small-strain modulus G_{max} in the above equations may be estimated from the back-analysis of site-specific field tests using established correlations or directly using one of the empirical equations available in the literature.

In an alternative approach considered by the same authors, the modulus reduction with shear strain is taken from published curves, e.g., as a function of the plasticity index as compiled by Vucetic & Dobry (1991) or Vardanega & Bolton (2013). This concept has been later applied by Niazi et al. (2014) to derive appropriate modulus reduction curves for different types of pile (bored, driven, augured, or jacked) and ground. The curve fitting was based on the back analysis of a large number of pile load tests at sites with sufficient information on the ground properties, mostly from CPT soundings. The following equation has been proposed:

$$\frac{G}{G_{max}} = \frac{1}{1 + 3.634 \cdot \alpha_1 \cdot \alpha_2 \cdot 10^2 (w/d)^{0.942 \cdot \beta_1 \cdot \beta_2}} \tag{1.23}$$

$$\alpha_2 = c_1 - c_2 \cdot \tanh(c_3 \cdot PI - c_4) \tag{1.24}$$

$$\beta_2 = d_1 - d_2 \cdot \tanh(d_3 \cdot PI - d_4) \tag{1.25}$$

with the plasticity index PI given in [%]. Values of the parameters in the above equations are given in Table 1.10.

1.4.4.2 Load-transfer approach

This approach is based on an idealization of the soil reaction at distinct points along the pile-soil interface by means of discrete independent (Winkler) bi-linear or nonlinear springs, e.g., Kraft et al. (1981). These so-called 't-z' curves relate load transfer t to pile displacement z at various points along the pile. The reaction at the pile base is defined analogously by means of 'Q-z' curves. The load-displacement relationship at the pile head, together with the distribution of load and displacement down

Table 1.10 Parameters for Eqs. (1.23)–(1.25) (Niazi et al., 2014)

Pile type	α_1	β_1	c_1	c_2	c_3	c_4	d_1	d_2	d_3	d_4
Driven	0.837	1.068	2.1	1.8	0.03	0.01	1.08	0.3	0.025	0.95
Jacked	0.648	1.247	1.4	1.3	0.015	0.2	1.05	0.21	0.03	1.2
Auger	1.176	1.013	1.8	1.5	0.03	0.23	1.05	0.17	0.037	1.5
Bored	1.912	0.971	1.4	1.3	0.02	0.07	1.16	0.3	0.025	1.0

the pile, can be calculated using a step-by-step approach, as implemented in several commercial software codes. This semi-empirical analysis method is adopted for long piles, especially where nonlinear soil behavior is essential and/or soil stratification is highly variable.

A major drawback of the method is that Winkler springs do not constitute an intrinsic soil property, and further depend on the pile geometry. Hence, their proper assessment from basic soil properties for strength and stiffness is difficult, requiring engineering judgement when site conditions considerably deviate from those of the reference field test.

Relations for the nonlinear springs for pile shaft and base may be developed from theoretical considerations, but in practice, they are obtained by back analysis of an instrumented pile test, thus taking also into account the effects of pile installation. The American Petroleum Institute API (2000) guideline provides recommendations to calculate ultimate shaft resistance, ultimate base resistance, and load transfer curves for shaft resistance (t–z) and end bearing (Q–z) for driven piles in clays and sands.

For clays, the ultimate unit skin friction $q_{s,u}$ is obtained by multiplying the undrained shear strength c_u by a reduction factor $\alpha \le 1$,

$$q_{s,u} = \alpha \cdot c_u \tag{1.26}$$

$$\alpha = 0.5 \cdot \left(\frac{c_u}{\sigma'_v}\right)^{-0.5} \quad \text{for } c_u/\sigma'_v \le 1$$

$$\alpha = 0.5 \cdot \left(\frac{c_u}{\sigma'_v}\right)^{-0.25} \quad \text{for } c_u/\sigma'_v > 1 \tag{1.27}$$

where σ'_v is the effective overburden stress at the considered depth. The ultimate base resistance is given by

$$q_{b,u} = 9 \cdot c_u \tag{1.28}$$

For cohesionless soils, the respective relations are

$$q_{s,u} = K \cdot \sigma'_v \cdot \tan\delta \tag{1.29}$$

where K is the coefficient of lateral earth pressure, and δ is the soil-pile friction angle. The ultimate base resistance is given by

$$q_{b,u} = \sigma'_v \cdot N_q \tag{1.30}$$

where N_q is a dimensionless bearing capacity factor. API (2000) provides recommendations for estimating N_q, δ, and a limiting end bearing pressure $q_{b,u,lim}$ dependent on the soil type and the in-situ relative density.

Recommended load-transfer (t–z) curves for clays exhibit a nonlinear function reaching $q_{s,u}$ at local pile deflection ratio $z/d = 0.01$ followed by a reduction to a residual

Table 1.11 Mobilized shaft and base resistance according to API (2000)

Clays		Sands		Clays and sands	
z/d	$q_s/q_{s,u}$	z [cm]	$q_s/q_{s,u}$	z/d	$q_b/q_{b,u}$
0.0	0.0	0.0	0.0	0.0	0.0
0.0016	0.30	0.254	1.0	0.002	0.25
0.0031	0.50	∞	1.0	0.013	0.50
0.0057	0.75			0.042	0.75
0.0080	0.90			0.073	0.90
0.0100	1.00			0.100	1.00
0.0200	0.70–0.90				
∞	0.70–0.90				

value in the range of $(0.7–0.9)\cdot q_{s,u}$. For sands, the curve increases linearly reaching $q_{s,u}$ at local pile deflection $z = 0.254\,\text{cm}$ and remaining constant afterwards. For the base resistance, the $Q–z$ curves for clays and cohesionless soils are identical; they increase nonlinearly to the respective $q_{b,u}$ at $z/d = 0.1$. The respective numerical values for the mobilized shaft and base resistance are given in Table 1.11.

Unified nonlinear load transfer curves have been derived from instrumented tests by Bohn et al. (2017) by assuming either a cubic-root or a hyperbolic variation with the pile displacement s. The hyperbolic load transfer functions for pile shaft and base are given by

$$q_s = \frac{q_{s,u}}{1 + M_t \cdot d/s}; \quad q_b = \frac{q_{b,u}}{1 + M_q \cdot d/s} \qquad (1.31)$$

with $M_t = 0.0038$ for the pile shaft and $M_q = 0.01$ for the pile base. As in all other relationships, the key parameter is the ultimate values for shaft and base resistance that could be inferred using CPT- or MPM-based empirical correlations.

Numerous relations have been developed over the years to estimate the shaft and base bearing capacity by applying soil mechanics theory and/or by linking theory to results of in-situ tests, mostly CPT. In addition to those selected from the review by Niazi & Mayne (2013) in Section 1.4.4.1 above, one should mention the studies by Jardine et al. (2005), White & Bolton (2005), Togliani (2008), Fleming et al. (2009), Doherty & Gavin (2011).

1.4.4.3   Continuum-based solutions

In *Boundary Element Methods* (BEMs), the soil is idealized as a continuum and elastic theory is utilized to compute the load-settlement characteristics. The effects of slippage at the pile-soil interface and end-bearing failure are considered in an approximate manner.

The basis is Mindlin's solution for a point load within a homogeneous, isotropic elastic half space. Soil layering can be considered by suitable averaging procedures for the soil moduli. Yielding of the soil is incorporated by imposing limiting stresses along the shaft and at the base, e.g., by adopting expressions similar to those given above for the load-transfer analyses, namely, for cohesive soils in terms of the undrained shear strength and for cohesionless soils in terms of the effective overburden stress, the interface friction angle and the angle of friction of the soil. Yielding at the base is assessed by appropriate bearing capacity factors.

Nonlinear soil behavior can be considered approximately, e.g., by assuming that the soil stiffness at the pile-soil interface varies according to the convenient hyperbolic model, cf. Poulos (1989) and Randolph (1994):

$$E_t = E_i \left(1 - R_f \frac{q}{q_u}\right)^2 \tag{1.32}$$

where E_t is the tangent modulus of the soil, E_i is the initial tangent modulus, R_f is the hyperbolic curve-fitting constant, q is the pile-soil stress, and q_u is the ultimate pile-soil stress. R_f defines the degree of nonlinearity of the stress-strain response and ranges between 0 and 1. Different values should be assigned to the response of pile shaft, in the range 0–0.5, and greater than 0.9 for the base. The pile response is then determined incrementally in an equivalent linear analysis.

More advanced approaches rely on the application of *Finite Element Methods* (FEMs) in conjunction with a suitable constitutive model for the soil (e.g. Jardine et al., 1986; Trochanis et al., 1991; Potts & Zdravkovic, 2001). Mostly elastoplastic models have been utilized including among others Mohr-Coulomb (MC) models with associated or nonassociated flow rules, Drucker-Prager cap models, and more recently the Hardening-Soil (HS) model (Schanz et al., 1999). Despite the acknowledged accuracy of such analyses, the calculated pile response strongly depends on the adopted constitutive soil model and in particular on the discretization of the interface pile-to-soil and the domain surrounding the pile. Different strategies have been proposed for that, including the adjustment of the size of the elements in direct contact with the pile. The suitability of such algorithms shall be checked against the results of instrumented pile load tests.

In the recent years, numerous pile foundation projects have been realized in several metropolitan cities, and in many of them, pile load tests have been executed to assess the ground response. Often, a set of parameters for common soil models is available to the designer, at least for preliminary design. From the numerous studies on this subject, we portray in Figure 1.3 the results from the back-analysis of a pile load test in Frankfurt clay, cf. Wehnert & Vermeer (2004). It compares the MC and the HS soil

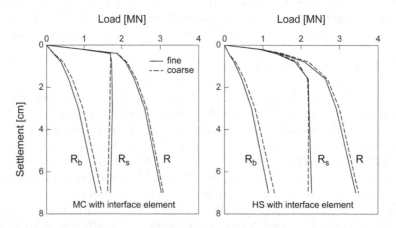

Figure 1.3 Example for base, shaft, and total pile resistance dependent on mesh fineness and soil model (Wehnert & Vermeer, 2004).

model at two mesh finenesses using interface elements. Without interface elements the dependence on mesh geometry is significant.

The strong nonlinearity of the single-pile response seldom governs the design of large pile foundations, as these are dominated by the interaction effects within the pile group. The settlement limitations imposed by the superstructure usually keep the mobilization of pile resistance at moderate levels, and the assessment of an appropriate soil stiffness is the crucial part of the design.

1.4.5 Bearing capacity of pile groups

1.4.5.1 Piles in compression

The verification of the pile group capacity is carried out by considering two ULSs: (i) failure of the individual piles and (ii) failure of the entire soil block encompassing all piles, which is more likely to govern when piles are closely spaced. In the verification, the ultimate vertical compressive resistance is the lesser of the two resistances, i.e., sum of individual pile resistances and block resistance. In clays, the resistance of the individual pile is set equal to that of an isolated pile. For the group resistance, the adhesion at the periphery is assumed equal to the undrained shear strength of the soil. In sands, densification due to pile driving may increase the bearing capacity of the individual piles. In general, block failure is more likely in clay than in sand.

The EAP suggests for the verification of the ULSs a more refined procedure. For the verification against individual pile failure, each pile is assigned a resistance $R_{G,\,i}$ depending on its position within the pile group by distinguishing between central, corner, and edge piles as

$$R_{G,i} = R_{c,k} \cdot G_{R,i} \tag{1.33}$$

where $R_{c,k}$ is the resistance of a single, isolated pile, and $G_{R,i}$ is a group factor being the product of three factors λ_i that account for the effects of soil type and pile group geometry (λ_1), group size (λ_2), and pile type (λ_3). For bored piles, $\lambda_3 = 1$, while λ_2 is given in charts for the three different pile positions dependent on the pile settlement. For the application in Eq. (1.33), the curves for the limiting settlement $s_g = 0.1\,d$ are suggested. The total resistance is then the sum of the individual pile resistances. This method is suitable for smaller groups with no more than 9 piles per side. As for the verification of the ULS of block failure according to the EAP, the area to be considered for the total base resistance is the sum of the individual pile footprint areas and not that of the entire block. It should be mentioned that Eq. (1.33) is the outcome of an analysis of the nonlinear pile-soil interaction under the assumption of a constitutive model (MC) for the soil. It shall thus be regarded as an approximation. In most instances in practice, the piles along the perimeter will be made shorter than the central ones to avoid large contrasts in pile stiffness. Hence, the application of group factors will be restricted to the case where for construction convenience all piles have the same length. Further, it should be noted that when piles in the group are connected by a pile cap of sufficient rigidity to redistribute the loads, the verification of the ULS for each pile may be omitted.

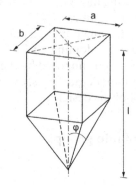

Figure 1.4 Soil wedge attached to a pile within a group subjected to axial tension.

1.4.5.2   Piles in tension

The same principles apply for tension piles, as typically encountered when base slabs of deep excavations are anchored in the ground using micropiles to prevent uplift due to high groundwater pressures. The following two ULSs are considered: (i) pull-out of individual piles assuming that all piles carry the same load and (ii) uplift of the entire soil block encompassing all piles. According to the provisions of Eurocode 7, case (i) is assigned to the limit state GEO (bearing failure), whereas case (ii) is assigned to the limit state UPL (uplift, stability). This may lead to a gap in the safety level for the two cases. In most instances, block uplift is the relevant verification and is performed first. The geometry of the soil block anchored to a single pile is displayed in Figure 1.4 with φ denoting the friction angle of the soil. The total block weight is then

$$G_{E,k} = n \cdot \left[a \cdot b \cdot \left(l - \frac{1}{3}\sqrt{a^2 + b^2} \cdot \cot\varphi \right) \right] \cdot \eta \cdot \gamma \tag{1.34}$$

where n is the number of piles in the group, γ is the weight density of the attached block of soil, and $\eta = 0.8$ is a reduction factor to cover uncertainties in the calculation model. Background information can be found in Quarg-Vonscheidt & Walz (2001).

1.4.6   Settlement of pile groups

1.4.6.1   Methods of analysis

Due to the interaction of piles in a pile group, the resistance vs. settlement curves of the individual piles are different from those of the single isolated pile, additionally depending on the position within the group, as exemplified in Figure 1.5. It may also happen that for large settlements, the resistance of the piles in a group is larger than that of the isolated single pile. At working load conditions, however, a pile group will settle more than the single one for the same average load. Although the application of computational methods for the assessment of resistance vs. settlement is not allowed in design practice by Handbook EC7-1 (2011), a realistic prediction of group behavior is

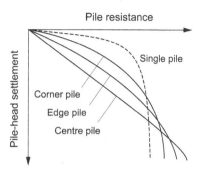

Figure 1.5 Effect of pile position on the resistance vs. settlement curve for a 5 × 5 pile group (Kempfert et al. 2003).

feasible only by employing an appropriate mechanical model. This, in turn, is difficult to be established due to the influence of the installation method and the prevailing soil conditions. A practical solution consists in using the resistance curve for a single pile and applying a theoretical model that describes as closely as possible the group effects. In this sense, a distinction between bored and displacement piles is necessary. While the installation of displacement piles makes the reliable prediction of group effects almost impossible due to the induced soil displacement, the careful installation of bored piles alters the surrounding soil to a much lesser extent. In general, due to the prestressing of the soil between the piles and the influence of the induced driving energy, the expected settlement of displacement piles will be smaller than that of bored piles. The distance, beyond which the interaction between two neighboring piles becomes negligible, is often set to 6–8 times the pile diameter. As shown above in the numerical simulation example for the pile response in Figure 1.3, the nonlinear part is strongly affected by the interface strength and stiffness properties and to a lesser extent by the details of the immediate and far field. It should also be noted that linear elasticity predicts a very large zone of influence that is not realistic under static loading conditions. Despite these shortcomings, linear elasticity constitutes the basis for assessing interaction effects between piles in a group. A large number of computational methods are available in the literature which have to be used with engineering judgement. Any calculation method should be validated using documented pile load tests and case histories. A range of values, rather than a single predictive value, should always be provided, as no computational method can capture the inherent variability of the ground and the complex interaction phenomena prevailing within a pile group.

Available analysis methods are divided into (i) simplified methods that involve some form of approximation for the response of the soil and the embedded pile group as a unit, and (ii) more refined analyses with the settlements being determined from explicit consideration of pile-soil-pile interaction effects using linear or nonlinear continuum models for the soil. Depending on the degree of the modelling fidelity, it is then distinguished between the interaction factor method and the numerical continuum-based methods, Poulos (2006). Over the years, simplified closed-form solutions have been derived from complete solutions for the interaction factors that are suitable for preliminary design and are included herein.

1.4.6.2   Simplified methods

For pile foundations comprising piles founded on a stratum of limited resistance or/ and underlain by a soft layer, the total settlement may be calculated by adding the settlement of an equivalent deep seated raft foundation at the level of the pile base and the settlement of the individual piles. For floating pile groups, the position of the equivalent raft is usually set at a depth of 2/3 of the pile length, and the load spread angle is assumed 1:4, cf. Tomlinson (1986). Van Impe (1991) suggests that the equivalent raft method should be limited to cases in which the pile cross-sectional area is larger than about 10% of the plan area of the group.

Another simplified approach refers to the equivalent pier method where the pile group is replaced by an equivalent pier consisting of the piles and the soil between them. The stiffness of the pier is defined as the area-weighted average value for the pile-soil block. The equivalent pier approach is advantageous for estimating the response of driven pile groups where installation effects on pile interaction are difficult to quantify, and it also allows the assessment of the rate of settlement of pile groups in clays (Poulos, 2001a).

Pile group settlement w_G may be related to the settlement w_{av} of a single pile at the average load of a pile in the group through the settlement ratio R_S

$$w_G = R_S \cdot w_{av} \tag{1.35}$$

The settlement ratio may be determined by a variety of analysis methods. An empirical relationship has been derived from a large number of documented case histories with a variety of pile configurations by Mandolini et al. (2005). They suggested the following equations for the maximum expected and most probable value of the settlement ratio

$$R_{S,\max} = \frac{0.5}{R} \cdot \left(1 + \frac{1}{3 \cdot R}\right) \cdot n \tag{1.36}$$

$$R_S = 0.29 \cdot n \cdot R^{-1.35} \tag{1.37}$$

where R denotes the aspect ratio introduced by Clancy & Randolph (1993) to characterize the pile group

$$R = \left(n \cdot s/l\right)^{0.5} \tag{1.38}$$

and n is the number of piles, s the pile spacing, and l the pile length.

1.4.6.3   Interaction factor method

The interaction factor α_v is defined as the displacement induced in a pile due to the load on a neighboring pile divided by the displacement of the loaded pile. It refers to two identical piles and offers a simple means to analyze pile groups. Interaction factors have been initially derived for a homogeneous linear elastic half-space (Poulos, 1968; Poulos & Davis, 1980) and were later extended to cover soils with stiffness increasing with depth. For the latter case, interaction factors are smaller. The settlement of a pile within a group of identical piles is then obtained by superposition of the two-pile interaction factors. The main approximation of this method is that the interaction factors do not consider the presence of the other piles. Imposition of the condition of all piles being connected by a rigid cap and thus having the same settlement yields a system of linear equations for the unknown pile head forces and the settlement of the group. For piles with an appreciable shaft resistance, pile-to-pile interaction effects along the shaft will dominate the group response compared to base-to-base interaction. Interaction factors have been computed for a wide range of situations, including corrections to account for the effects of a finite layer depth or soil modulus increasing with depth, and the Poisson's ratio of the soil (Butterfield & Banerjee, 1971; Poulos & Davis, 1980).

Based on the assumptions underlying their solution for a single pile, Eq. (1.12), Randolph & Wroth (1979) derived the following expressions for the interaction factors for shaft, $\alpha_{v,\,s}$, and base, $\alpha_{v,\,b}$:

$$\alpha_{v,s} = \ln\left(r_m/s\right)/\zeta = 1 - \ln\left(2 \cdot s/d\right)/\zeta \tag{1.39}$$

$$a_{v,b} = d/\left(\pi \cdot s\right) \tag{1.40}$$

where ζ is given by Eq. (1.16). Due to the stronger interaction between shafts than between bases, the load proportion transmitted to the base is larger in a group than in a single pile. Therefore, the base interaction is often neglected.

From an extensive parametric study by varying the length-to-diameter ratio l/d, the spacing-to-diameter ratio s/d, and the pile to soil stiffness ratio E_p/G_l, a simple equation has been derived by Randolph (Fleming et al., 1992) for the settlement ratio defined by Eq. (1.35):

$$R_S = n^{(1-e)} \tag{1.41}$$

The exponent e is generally lying between 0.4 and 0.6 for most pile groups. As an approximate rule of thumb, Poulos (1989) suggested $e = 0.5$ for piles in clay and $e = 0.67$ for piles in sand. The actual value of e will depend on l/d, s/d, and E_p/G_l as well as on the Poisson's ratio v of the soil and the nonhomogeneity parameter of the soil χ as defined by Eq. (1.17). Fleming et al. (1992) provide graphs for determining the exponent e. These graphs have been parametrized by the author (Vrettos, 2012) using linear regression:

$$e = e_1\left(l/d\right) \cdot c_1\left(E_p/G_l\right) \cdot c_2\left(s/d\right) \cdot c_3\left(\rho\right) \cdot c_4\left(v\right) \tag{1.42}$$

$$e_1 = \left(l/d\right)^3 \cdot 4.5 \cdot 10^{-7} - \left(l/d\right)^2 \cdot 9.0 \cdot 10^{-5} + \left(l/d\right) \cdot 4.9 \cdot 10^{-3} + 0.47 \tag{1.43}$$

$$c_1 = -\left[\lg\left(E_p/G_l\right)\right]^3 \cdot 3.8 \cdot 10^{-2} + \left[\lg\left(E_p/G_l\right)\right]^2 \cdot 0.28 - \lg\left(E_p/G_l\right) \cdot 0.45 + 0.85 \tag{1.44}$$

$$c_2 = -(s/d)^3 \cdot 2.5 \cdot 10^{-4} + (s/d)^2 \cdot 9.6 \cdot 10^{-3} - (s/d) \cdot 0.13 + 1.32 \tag{1.45}$$

$$c_3 = -\chi^2 \cdot 0.20 + \chi \cdot 0.56 + 0.69 \tag{1.46}$$

$$c_4 = -v^2 \cdot 0.14 - v \cdot 0.20 + 1.07 \tag{1.47}$$

where s is the average pile spacing within the pile group, and lg(.) indicates the decadic logarithm. The above equations are valid for $10 \leq l/d \leq 100$, $2 \leq \lg(E_p/G_l) \leq 4$, $0.5 \leq \rho \leq 1$, $0 \leq v \leq 0.5$. We observe that piles with $d = 1.20$ m and $l = 12$ m, typically used for piled rafts in sandy soil, are already at the lower limit of the slenderness ratio, indicating that the above curves were developed primarily for long and slender piles.

Application of Eqs. (1.41)–(1.47) is valid only for the average settlement of a pile group composed of identical piles without taking into account the exact geometry of the pile group nor the location of the individual piles within the group. It should be noted that in a small group, the proportion of piles along the periphery is higher than in a large group. Hence, these equations should be used primarily for preliminary design purposes.

The above method does not consider the reinforcing effect of the 'receiver' pile. Mylonakis & Gazetas (1998) suggested an enhanced model to capture this effect. Their method assumes a Winkler model for the single pile response and an attenuation function for the soil settlement variation with radial distance from the 'source' pile according to Randolph & Wroth (1978), i.e., independent of depth. The following closed-form solution has been derived for piles in homogenous soil. First, the single pile response is calculated from Eq. (1.12) by setting $G_l = G_b = \bar{G} = G$ and replacing in Eq. (1.13)

$$2 \cdot \pi \cdot G / \zeta = k \tag{1.48}$$

where k is the Winkler modulus, cf. Mylonakis (2001).

The interaction factor is then determined as the product of two terms, representing the logarithmic decay inherent to the Randolph & Wroth (1979) approach and a so-called diffraction factor ξ:

$$\alpha_v = \frac{\ln(r_m/s)}{\zeta} \xi \tag{1.49}$$

Neglecting the minor effects of base-to-base interaction yields (Mylonakis & Gazetas, 1998)

$$\xi = \frac{1}{2} - \frac{\mu \cdot l \cdot (\Omega^2 - 1) + \Omega}{(\Omega^2 + 1) \cdot \sinh(2 \cdot \mu \cdot l) + 2 \cdot \Omega \cdot \cosh(2 \cdot \mu \cdot l)} \tag{1.50}$$

It can be seen that for common cases of floating piles, the diffraction factor varies between 0.5 and 1 and attains 0.5 at the long pile limit. The above methodology has been extended to a Winkler modulus increasing with depth by Crispin et al. (2018).

The settlement of the pile group is obtained by superposition of the two-pile interaction factors and imposing the appropriate boundary condition of a rigid or flexible pile cap. Since interaction factors are customarily defined only for two identical piles, some adjustment is necessary in order to include dissimilar piles. Some approximations are proposed by Poulos (2006).

Over the years, several simplified analytical and empirical models have been proposed to capture nonlinear vertical pile-soil-pile interaction. The increased level of shear strain in the close vicinity of a loaded 'receiver' pile yields a locally reduced soil stiffness that inevitably affects the value of the applicable interaction factor. A review of the different approaches has recently been provided by Sheil et al. (2019). In routine design using interaction factors, the nonlinearity is typically considered merely in the response of the 'source' pile.

The result of the interaction between piles is that, depending on the stiffness of the pile cap, central piles will exhibit a considerably lower stiffness compared to piles along the periphery and corner piles carrying the greatest proportion of the load. This is the consequence of the assumed linear-elastic soil behavior for the soil. For the corner piles, the computed pile stiffness is unrealistically high, thus calling for an adjustment. Ideally, all piles should have the same utilization degree. Piles at the edge are accordingly shortened in order to redirect part of the load to the central piles, thus achieving a uniform load distribution among the piles. This again requires the application of interaction factors for dissimilar piles of dissimilar lengths for which explicit expressions do not exist. Matters are further complicated when considering nonhomogenous soil profiles or irregular soil layering. Such conditions impose limitations on the interaction factor methods and can be tackled only in an approximate manner, cf. Basile (2003) or Poulos (2006). In these two references, one may also find an overview of the main features of available software, DEFPIG (Poulos, 1990) and PIGLET (Randolph, 1996).

1.4.6.4   Continuum-based numerical methods

The proper consideration of pile-soil-pile interaction effects requires the use of rigorous continuum-based solutions. Initially, BEMs for homogeneous soil were used for this purpose (Butterfield & Banerjee, 1971). Since these methods are restricted to linear-elastic material behavior, they had to be combined with nonlinear algorithms to capture the nonlinear pile response, for example, by assuming a hyperbolic stress-strain relationship (Chow, 1986; Basile, 1999). A yield criterion is applied to limit shaft and base resistance. Further assumptions are needed to capture the nonhomogeneity/layering of the soil by using some weighted average of the soil moduli when calculating the influence functions between 'source' and 'receiver' pile elements, e.g., Chow (1987), Yamashita et al. (1987), Poulos (1989). A software code incorporating these features is REPUTE (Geocentrix, 2018).

Advances in computing power during the past decades promoted the widespread use of the FEM in pile-soil-pile interaction problems. It enables the incorporation of a variety of nonlinear constitutive laws for the soil, arbitrary pile group geometries, and a direct coupling with the superstructure. So, the focus has gradually shifted to the selection/validation of appropriate site-specific soil models and to the discretization details. Several general-purpose software codes are currently used in design work and parametric studies, including ABAQUS, FLAC, or PLAXIS, c.f. Katzenbach et al. (2009), Comodromos (2004), McCabe & Sheil (2015). Moreover, to facilitate efficient modelling of large pile groups, PLAXIS has introduced the concept of the embedded pile (Engin & Brinkgreve, 2009).

1.4.7 Piled raft foundations

Piled raft is a new foundation concept for high-rise buildings and has extensively been used in particular in Germany since the beginning of the 1990s (Katzenbach et al., 2000). This foundation type is a viable alternative to conventional pile or raft foundations in competent ground. The combined foundation can support the applied axial loading with an appropriate factor of safety at a tolerable level of settlement under working loads. The implementation of this foundation type has led to abolition of complicated settlement-correction techniques. In recent years, the available computational methods in combination with data from measurements in real projects have allowed a realistic modelling of the complicated bearing behavior of this composite foundation system.

The overall bearing behavior is described by the piled raft coefficient α_{pr} that defines the proportion of load carried by the piles. Due to the strong nonlinearity of the pile bearing behavior, the piled raft coefficient depends on the stress level and, accordingly, on the amount of settlement of the piled raft foundation. The piles can be loaded up to their ultimate bearing capacity and can be spaced strategically to achieve a more uniform settlement and reduce internal forces in the raft, thus providing a more economical solution. Verification of the ULS for the individual piles is not necessary, as long as the ULS of the composite foundation is not exceeded. The latter may be determined by neglecting the pile resistances.

Code-based design for this type of deep foundation has been introduced in Handbook EC7-1 (2011). Reference is made to the corresponding guideline (Hanisch et al., 2001) and that a project-specific approval by the building control authority may be required. The guideline requires the applied computational model to be able to simulate (i) the bearing behavior of a single pile taking into account the shear forces along the pile shaft and the compression at the pile base, and (ii) the interaction effects between individual piles and between pile and raft, as schematically shown in Figure 1.6. The principles of this guideline have been adopted by the ISSMGE (Katzenbach and Choudhury, 2013). Back analysis of a static pile test is commonly used to verify the assumptions of the model and the values of the parameters. Figure 1.7 shows for an isolated piled raft unit the variation of the contact pressure as the applied load increases.

The design process aims at optimizing the position and the geometry of the piles. The solution of the structure-foundation-soil interaction problem is obtained by employing a pseudo-coupled procedure that requires an interaction between the designers of the superstructure and the foundation system. The interface in this design procedure is commonly defined in terms of the modulus of subgrade reaction for the raft and the spring coefficients for the individual piles.

The geotechnical design part consists in estimating the deformation of the composite system and the distribution of the load into its two components, namely, the pile group and the raft. The available methods may be divided into (i) approximate analytical,

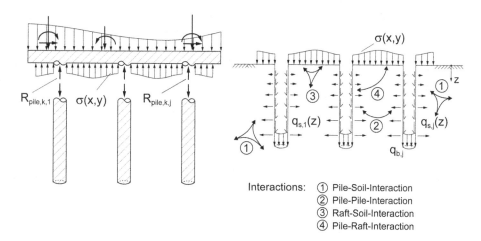

Interactions: ① Pile-Soil-Interaction
② Pile-Pile-Interaction
③ Raft-Soil-Interaction
④ Pile-Raft-Interaction

Figure 1.6 Interactions in a piled raft foundation. (Adapted from ISSMGE Guideline (2013).)

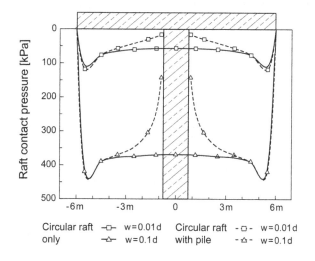

Circular raft –□– w=0.01d   Circular raft - □ - w=0.01d
only   –△– w=0.1d   with pile   - △ - w=0.1d

Figure 1.7 Variation of contact pressure with increasing load and settlement level based on the Frankfurt Clay model. (Adapted from Hanisch et al. (2001).)

(ii) approximate numerical, and (iii) refined numerical; the choice is dictated by the importance of the project. A method belonging to the first category is that by Randolph (1983, 1994), which is presented herein in more detail. Methods of the second category model the structural elements with finite-elements and apply approximate methods of elastic continua for calculating the interaction between the structural elements; e.g., Clancy & Randolph (1993), Horikoshi & Randolph (1998), Kitiyodom & Matsumoto (2003), Poulos (1994), Russo (1998), Ta & Small (1996), Yamashita et al. (1998). The third category includes boundary element methods, as applied by Butterfield and Banerjee (1971), Franke et al. (1994), as well as finite element methods, as implemented by Katzenbach & Reul (1997), Reul (2004), Comodromos et al. (2016). The latter methods progressively dominate final design since they offer the possibility of capturing soil behavior by appropriate nonlinear constitutive models. Overviews are presented by, among others, Katzenbach et al. (2000), Poulos (2001b), Mandolini et al. (2005), Katzenbach et al. (2009). A comparison of the methods applied to a well-documented case study involving a pile group with irregular pile configuration is presented in Richter & Lutz (2010).

For the preliminary design, where different foundation alternatives are compared, a flexible yet simple method is required to assess the influence of the pile group configuration and the soil parameters. The approximate method outlined by Randolph (1994) is most suitable for this purpose. It estimates the stiffness of the system from a combination of the stiffnesses of the individual components as

$$K_{pr} = \frac{K_p + K_c \cdot \left(1 - 2 \cdot \alpha_{cp}\right)}{1 - \alpha_{cp}^2 \cdot K_c / K_p} \tag{1.51}$$

where K_{pr} is the overall stiffness of the piled raft, K_p is the stiffness of the free-standing pile group, K_c is the stiffness of the free-standing raft, and α_{cp} is the average interaction factor between the piles and the pile cap.

The proportion of load carried by the piles is

$$\alpha_{pr} = \frac{P_p}{P} = \frac{K_c \cdot \left(1 - \alpha_{cp}\right)}{K_p + K_c \cdot \left(1 - 2 \cdot \alpha_{cp}\right)} \tag{1.52}$$

where P_p is the load carried by the piles, and P is the total load borne by the piled raft foundation.

The interaction factor α_{cp}, which is the key parameter in the above expressions, is obtained by assuming that the settlement of the surrounding ground decays logarithmically with distance:

$$\alpha_{cp} = \frac{\ln\left(r_m / r_c\right)}{\ln\left(r_m / r_0\right)} = 1 - \frac{\ln\left(r_c / r_0\right)}{\zeta} \tag{1.53}$$

where r_c is the equivalent radius of the raft area associated with each single pile, and ζ from Eq. (1.16) reflects the zone of influence of a single pile.

The stiffness of the pile group K_p is obtained by the weighted superposition of the stiffness of each pile according to Eqs. (1.41)–(1.47) assuming that all n piles are identical and uniformly arranged within the pile group:

$$K_p = n^{(1-e)} \cdot K_1 \tag{1.54}$$

where K_1 is the vertical stiffness of a single isolated pile. The latter can be calculated by, for example, Eq. (1.12).

The raft stiffness K_c is obtained via elastic continuum theory, taking into account the actual variation of the soil stiffness with depth. Available solutions for rigid rectangular plates with dimensions $2b \times 2a$ $(b > a)$ on homogeneous or inhomogeneous soil have been presented by various authors, with the solutions taking the general form

$$K_c = \frac{G \cdot a}{(1 - v)} \cdot I_\delta \qquad (1.55)$$

where I_δ is a shape factor and G is the effective shear modulus of the soil. I_δ can be approximated by the following equation (Pais & Kausel, 1988):

$$I_\delta = 1.6 + 3.1 \cdot (b/a)^{0.75} \qquad (1.56)$$

An inhomogeneous/layered soil can be replaced by an equivalent homogeneous soil with modulus computed by adjusting the settlement of a perfectly flexible plate resting on the actual inhomogeneous soil to be equal to that of the same plate on the equivalent homogeneous soil. The Steinbrenner approximation (Poulos & Davis, 1980) may be used for this purpose. To account for the effects of a finite raft rigidity, K_c for a perfectly rigid raft may be multiplied by a reduction factor ranging between 1.0 for a rigid raft to 0.75 for a flexible raft, typically around 0.85. One should bear in mind that due to the different stress distribution patterns, the effective shear modulus for the raft stiffness is different from the shear modulus used for the piles and the pile group. When removing the piles, the shear modulus for the raft should yield a realistic value of the raft settlement.

The application of the above simple method has some limitations. To guarantee a positive valued overall stiffness K_{pr} in Eq. (1.51), it is required that

$$K_{pr} \geq K_c \cdot \alpha_{cp} \qquad (1.57)$$

For high-rise building foundations, this condition is violated in situations where the reduction of the differential settlements requires only a small number of piles underneath the rigid core of the building. In this case, the foundation can be modelled as a pier-supported raft foundation, with the equivalent pier corresponding to a squat pile with slenderness ratio $l/d < 1$. From Eq. (1.57), it follows that for small values of the slenderness ratio, this method becomes inaccurate.

The method outlined above estimates the overall stiffness, the average settlement, and the proportion of load carried by each system component, which is sufficient for a preliminary design. Each pile in the group is assumed to have the same stiffness, i.e., it makes no difference if a particular pile is located in the center, at the corner, or along the periphery of the group. Hence, the pile length used in the calculation shall be a weighted average of the individual pile lengths. The approximate formulae have been derived from numerical solutions for large pile groups. With a decreasing ratio between footprint area and perimeter of the pile group, the proportion of peripheral piles within the group increases, and the method becomes less accurate. An extension of this method that allows the accommodation of variable pile distance and length as well as different stiffness for central, peripheral, and corner piles has been suggested by the author (Vrettos, 2012).

The *Messeturm* in Frankfurt may serve both as an example and as a case study to demonstrate the suitability of Randolph's method to reflect the basic features of a piled raft foundation. The square raft (58.8×58.8 m) is supported by 64 piles with diameter 1.30 m and variable length. There are 16 piles of length 34.9 m in the central area, 20 piles of length 30.9 along the inner periphery and 28 piles of 26.9 m along the outer periphery (Katzenbach et al., 2000; Reul & Randolph, 2003). The average pile length is thus 30.15 m. The Young's modulus of the piles E_p is set equal to $22 \cdot 10^3$ MN/m^2. For the over-consolidated Frankfurt clay, a two-layer profile is usually adopted in design that comprises a constant modulus in the top layer and a linearly increasing one underneath. The modulus profile has been back-calculated from several case studies, mostly raft foundations, and is considered representative at the location. Assuming a Poisson's ratio $v = 0.33$, it is composed by a top layer with Young's modulus 62.4 MN/m^2 (reloading modulus) down to a depth of $z = 25.7$ m below the surface of the clay, underlain by a layer with linearly increasing Young's modulus with $E(z) = 6.24 + 2.183 \cdot z$. The bottom of the raft lies 4.83 m underneath the clay surface which gives at the average level of the pile tip a Young's modulus $E_l = 82.43$ MN/m^2. Hence, according to Eq. (1.17), we have $\chi = 0.793$. With these values, we obtain from Eq. (1.12) for the single pile stiffness $P/w = 829.60$ MN/m. The stiffness of the pile group K_p is determined from Eqs. (1.41)–(1.47) equal to $K_p = 8333$ MN/m and the pile cap to pile interaction factor $\alpha_{cp} = 0.551$. The raft stiffness is determined using the Steinbrenner approximation for a perfect flexible raft (rectangular loaded area) by neglecting contributions from depths larger than the limiting depth where additional stress becomes equal to 20% of the overburden stress. The corresponding Young's modulus amounts to 104.6 MN/m^2. The rigid raft stiffness, as calculated for this Young's modulus from Eqs. (1.55) and (1.56), is reduced by 15% to consider the finite flexibility of the raft. This yields $K_r = 6.894$ MN/m. The overall stiffness of the piled raft from Eq. (1.51) is $K_{pr} = 10,191$ MN/m and the load taken by the piles leads to $\alpha_{pr} = 0.594$ that compares well with the measurements that indicated 55% of the load carried by the piles (Katzenbach et al., 2000). For the settlement estimation, the effective load is set equal to 1600 MN corresponding to a pressure of 0.463 MN/m^2. Division by the overall stiffness K_{pr} yields an average settlement of 15.7 cm that agrees reasonably well with the observed settlement of 14.4 cm. The main reason for the good performance of the simplified method is the adequacy of the soil modulus profile that dominates the response, as in any elasticity-based approach.

1.5   SINGLE PILES AND PILE GROUPS UNDER LATERAL LOADING

1.5.1   General

Typical lateral loading of pile foundations originates from wind and earthquake actions, which are generally small compared to static vertical loading. There are, however, structure types, such as offshore structures and wind turbines, where lateral loading dominates the response. These are treated in other chapters of this book. Concerning the type of loading, one distinguishes between (i) direct external loading applied at the level of the pile head and (ii) transverse ground load caused by moving ground, typically encountered in bridge abutments or when piles are installed within a slope for stabilization (Reese & Van Impe, 2011). We will focus on the first category.

For a single laterally loaded pile, two failure modes may occur: (i) yielding of the soil along the pile, typically in case of short piles ($l/d < 6$), or (ii) bending failure and local yielding for longer piles. In practice, the majority of piles are long, flexible piles. The design is often controlled by serviceability rather than ULSs, requiring an elaborate analysis taking into account the stratification details of the ground. In assessing the pile response, the fixity condition of the pile head to the pile cap shall be properly considered. Due to the fast degradation of soil stiffness with increasing pile head displacement, soil models with nonlinear stiffness are indispensable when large movements are expected.

For pile groups, the effects of pile-soil-pile interaction and possible shadow effects of closely spaced piles should be considered when assessing the lateral resistance by means of calculations or from pile load tests.

1.5.2   Lateral resistance of single piles

Field load tests are often performed for convenience on a pair of piles by loading the piles against each other, thus avoiding the installation of costly reaction systems. However, such tests provide information only on the pile response in the SLS, as it is not possible to reach failure condition in long, flexible piles. Hence, the ultimate lateral resistance is specified for the soil in conjunction with some model for the embedded pile, usually by adopting a beam on Winkler springs approach combined with an appropriate modulus of subgrade reaction at any point along the pile. The depth distribution may be determined from the deflection curve of the pile as measured by inclinometer along the pile shaft at the various loading levels. The capacity of the pile section to carry the induced bending moments usually governs the design.

The ultimate unit lateral resistance p_u for *purely cohesive soil* is assumed proportional to the undrained shear strength c_u. An expression widely used in practice derives from the work by Broms (1964a) that considers a constant resistance along the pile except in a zone of 1.5 pile diameters at the pile head with zero resistance to account for the low resistance near the soil surface; that is,

$$p_u = \begin{cases} 0 & \text{for} \quad 0 \le z \le 1.5 \cdot d \\ 9 \cdot c_u & \text{for} \quad z > 1.5 \cdot d \end{cases} \tag{1.58}$$

Fleming et al. (1992) suggest a bi-linear profile starting with $2 \cdot c_u$ at the surface, increasing to $9 \cdot c_u$ at $z = 3 \cdot d$ and remaining constant below this depth.

For piles in *purely cohesionless soil*, the suggestion by Broms (1964b) is widely used

$$p_u = 3 \cdot K_p \cdot \sigma_v' = 3 \cdot \tan^2\left(45^\circ + \varphi'/2\right) \cdot \sigma_v' \tag{1.59}$$

where K_p is the Rankine passive pressure coefficient, φ' is the effective friction angle of the soil, and σ_v' is the effective vertical stress at the point of interest. The proposal to replace $3\,K_p$ by K_p^2 is often adopted (Fleming et al., 1992), but it alters the underlying physics while the difference is small ($K_p = 3$ for a typical friction angle $\varphi' = 30^\circ$).

For *frictional-cohesive soil* with shear parameters φ and c in drained conditions, one of the earliest proposals derives from Brinch Hansen (1961):

$$p_u = K_q \cdot \sigma'_v + c \cdot K_c \tag{1.60}$$

where both coefficients are functions of the soil friction angle, the soil-pile interface roughness, and l/d (Poulos & Davis, 1980). This approach is discussed by Cecconi et al. (2019) who present their suggestion that is an extension of the original Broms' theory:

$$p_u = 3 \cdot K_p \cdot \sigma'_v + 9 \cdot c \cdot \sqrt{K_p} \tag{1.61}$$

Possible failure modes exhibit distinct features for the ultimate horizontal force H_u and the maximum bending moment M_{max}, which are derived from horizontal and moment equilibrium (Fleming et al., 1992; Viggiani et al., 2012; Cecconi et al., 2019). Lateral behavior is controlled by head fixity conditions at the cap. It is distinguished between two cases:

- *Free-head pile*: A 'short-pile' mode induces a rotation around a point in the lower part of the pile and bending moment M_{max} in the pile that is smaller than the yield moment M_y of the pile section; the ultimate horizontal force H_u depends only on the pile geometry, the eccentricity e of the load above the ground line, and the soil strength. When M_{max} exceeds M_y, a 'long-pile' mode prevails with a hinge at a certain depth, and with H_u depending on M_y but not on the slenderness l/d.
- *Fixed-head pile*: In a 'short-pile' mode (horizontal translation of pile and pile cap), the maximum moment occurs at the pile head. In a 'long pile' mode, one hinge develops at a certain depth and another one at the connection between pile head and pile cap. In an 'intermediate-length pile' mode, a single hinge occurs at the pile head and the solution depends on both the pile slenderness (l/d) and the yield bending moment M_y.

We elaborate only on the fixed-head pile since this is usually the case in practice. Distribution of soil resistance and sectional forces for the case of an immediate-length fixed-head pile in cohesive soil are exemplarily shown in Figure 1.8. Details on the statics are given by Viggiani et al. (2012), who set a priori the zone of zero reaction equal to $1.5 \cdot d$. We generalize herein the solution for an arbitrary depth z_0 for this zone.

In the 'short pile' mode, the maximum bending moment at the pile head and the associated ultimate lateral load are

$$\frac{M_{max}}{9 \cdot c_u \cdot d^3} = \frac{1}{2} \cdot \frac{l^2 - z_0^2}{d^2}; \quad \frac{H_u}{9 \cdot c_u \cdot d^2} = \frac{l - z_0}{d} \tag{1.62}$$

In the 'intermediate-length pile' mode, a hinge develops when M_{max} equals the plastic moment of the pile section M_y. The respective ultimate lateral load is then

$$\frac{H_u}{9 \cdot c_u \cdot d^2} = -\frac{l + z_0}{d} + \sqrt{2 \frac{l^2 + z_0^2}{d^2} + \frac{4 \cdot M_y}{9 \cdot c_u \cdot d^3}} \tag{1.63}$$

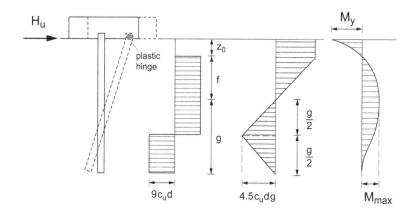

Figure 1.8 Deflection, soil resistance, shear force, and bending moment for intermediate-length pile in cohesive soil, fixed against rotation. (Adapted from Reese & Van Impe (2011).)

For the 'long pile' mode with two hinges, the expression for the ultimate load is

$$\frac{H_u}{9 \cdot c_u \cdot d^2} = -\frac{z_0}{d} + \sqrt{\frac{z_0^2}{d^2} + \frac{4 \cdot M_y}{9 \cdot c_u \cdot d^3}}$$

(1.64)

Figure 1.9 displays, for distinct values of M_y and for $z_0 = 1.5 \cdot d$, the variation of ultimate lateral resistance with pile slenderness for intermediate-length piles including the two bounds corresponding to short and long piles. The condition for long piles is given by

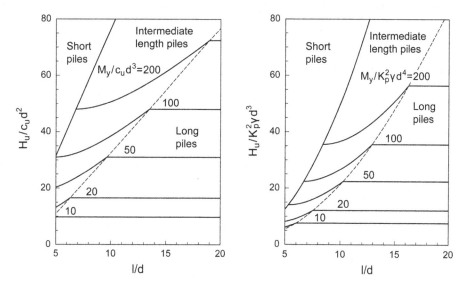

Figure 1.9 Ultimate lateral resistance for fixed-head piles in cohesive soil (a) and cohesionless soil (b). (Adapted from Viggiani et al. (2012).)

$$\text{long piles}: \frac{l}{d} \geq \sqrt{\frac{4 \cdot M_y}{9 \cdot c_u \cdot d^3}} + \sqrt{\frac{z_0^2}{d^2} + \frac{4 \cdot M_y}{9 \cdot c_u \cdot d^3}} \tag{1.65}$$

Similar charts may be derived for purely cohesionless soil, Viggiani et al. (2012). They are reproduced in Figure 1.9 for completeness. These authors adopted the term K_p^2 instead of $3 \cdot K_p$ in Eq. (1.59) for the lateral ultimate resistance. The condition for long piles is derived from the respective equations:

$$\text{long piles}: \frac{l}{d} \geq 2.784 \cdot \left(\frac{M_y}{K_p^2 \cdot \gamma \cdot d^4} \right)^{1/3} \tag{1.66}$$

Sign conventions are the same as those usually employed in the mechanics for beams. The pile-head bending moment is thus negative because of rotational restraint. In the graphs, the moment is plotted as positive for presentation convenience.

For the verification of the geotechnical ULS, EN 1997-1 provides two alternative solutions to determine the design value of the resistance: (i) GEO in conjunction with Design Approach DA2, abbreviated as GEO-2, where a partial safety factor is applied to the characteristic resistance force R_k, and (ii) GEO along with Design Approach DA3, abbreviated as GEO-3, where partial safety factors are applied to the characteristic shear strength parameters, i.e., angle of friction φ_k' and cohesion c_k' for drained conditions and undrained shear strength $c_{u,k}$ for undrained conditions.

For laterally loaded piles, Handbook EC7-1 (2011) and in accordance with the Piling Recommendations EAP specifies a particular type of GEO-2 verification that comprises the following steps: determination of subgrade modulus profile; calculation of sectional forces from characteristic loads and subgrade modulus; calculation of design values by the application of partial safety factors; verification that the characteristic normal stresses $\sigma_{h,k}$ between pile and soil (subgrade modulus multiplied by deflection) do not exceed at any point the characteristic passive earth pressure $e_{ph,k}$ determined under plane strain conditions; verification that the design value of the lateral soil reaction force is less than the design value of the three-dimensional earth resistance force in a zone extending down to the center of rotation of the pile.

In some countries, and also in offshore geotechnics, verification is often carried out in accordance with GEO-3, whereby the characteristic value of the ultimate load at failure is first calculated by assuming an appropriate model for the interaction between the embedded pile and the soil. The ultimate load will depend on the system characteristics (i.e. geometry and rigidity) and the shear strength of the soil. Suitable partial safety factors are then applied to the characteristic soil strength parameters to reduce them to the design values φ_d', c_d', or $c_{u,d}$. The load is gradually increased until no more load can be sustained. The corresponding load R_d is then compared to the design value of actions E_d. Details on this procedure and comparison with analyses following GEO-2 are given by Thieken et al. (2014).

1.5.3   Deflection of single piles

1.5.3.1   Methods of analysis

Depending on the load intensity, lateral pile response may be highly nonlinear. Deflection and head rotation are estimated by means of (i) the subgrade modulus approach with appropriate Winkler-type springs along the pile length, (ii) load-transfer curves, known as 'p-y' curves, where explicit, nonlinear relationships for the soil reaction as a function of the local pile deflection are employed, and (iii) solutions derived from continuum-based models for the pile-soil interaction that can accommodate both linear and nonlinear soil models.

Common to all methods is the strong dependency of the solutions on the soil stiffness at small to medium strain levels. Hence, the analysis should take account of potential variations of ground stiffness with depth. Moreover, for piles in layered soil, upper and lower values of soil stiffness in different layers should be combined in the most adverse manner.

1.5.3.2   Subgrade modulus approach

We adopt the concept of Winkler springs and assume a constant spring coefficient along the pile length, which is denoted by k with units of force per unit area. Under the action of a horizontal load H at the pile head ($z = 0$) and a line load $q(z)$ for the ground reaction, the lateral pile deflection $u(z)$ is described by the differential equation of beam bending. The pile length is l, and the pile bending stiffness is given by $E_p \cdot I_p$, where E_p is the Young's modulus of the pile material, and I_p is the area moment of inertia of the pile section. The simplest case refers to the free-headed pile, but we consider here the case of a fixed-head pile that is more common in practice. From the general solution of the differential equation and after inserting the appropriate boundary conditions, the following expressions are obtained for the pile head deflection $u_{0f} = u(z = 0)$ and the bending moment at the connection of pile to pile-cap $M_{0f} = M(z = 0)$, cf. Scott (1981):

$$u_{0f} = \frac{H \cdot \lambda}{k} \cdot \frac{\cosh 2l\lambda + \cos 2\lambda l + 2}{\sinh 2l\lambda + \sin 2l\lambda} \tag{1.67}$$

$$M_{0f} = -\frac{H}{2 \cdot \lambda} \cdot \frac{\cosh 2\lambda l - \cos 2\lambda l}{\sinh 2\lambda l + \sin 2\lambda l} \tag{1.68}$$

where

$$\lambda = \sqrt[4]{\frac{k}{4 \cdot E_p \cdot I_p}} \tag{1.69}$$

It can be deduced that from a value $\lambda \cdot l \approx 3$, the variation of the pile head deflection and the maximum bending moment becomes negligible, and thus, the following classification may be introduced (Hetenyi, 1946)

i   short piles for $\lambda \cdot l \leq \pi/4$
ii   intermediate-length piles for $\pi/4 < \lambda \cdot l \leq \pi$
iii long piles for $\lambda \cdot l > \pi$

For piles belonging to category (i), the deflection can be neglected because it is small in relation to the deformation of the surrounding soil. For piles in category (iii), the flexibility is so low that the spreading of the load at the bottom of the pile is not noticeable. Category (ii) requires a closer look. Since the majority of piles in practice are classified as fixed-head long piles, consideration here is confined to this category. For this limiting case, equations for pile-head displacement and moment simplify to

$$\text{long piles}: u_{0f} = \frac{H \cdot \lambda}{k}, \quad M_{0f} = -\frac{H}{2 \cdot \lambda} \tag{1.70}$$

From the above solutions, in conjunction with λ from Eq. (1.69), one may quantify the influence of the modulus of Winkler springs k. It can be seen that the deflection is proportional to $k^{-3/4}$, whereas the bending moment varies proportionally to $k^{-1/4}$. In order words, inaccuracies in the selection of k will strongly affect the deflection and, to a minor extent, the bending moment. This is the reason why according to the EAP the Winkler model is used solely to determine sectional forces in the pile.

It is postulated that k follows the same variation with depth as the soil Young's modulus E. Consequently, the assumption of a constant value is only a rough approximation since soil stiffness, even in uniformly layered soils, increases with effective confining stress. Solutions for various soil profiles have been derived for elasto-dynamic problems (Karatzia & Mylonakis, 2016). These solutions can be applied to static conditions by using the static counterpart of the soil modulus.

For the structural design of the piles only, EAP recommends the traditional empirical relationship widely used in Germany:

$$k \approx E_{oed} \tag{1.71}$$

In countries where the use of oedometric modulus E_{oed} is less common in design practice, the following relationship is used

$$k \approx \delta \cdot E \tag{1.72}$$

where δ is an empirical constant. The simple form of these equations has the advantage of being easy to recall. Their accuracy is investigated later in this chapter where drawbacks and limitations of the subgrade modulus approach become evident. An improvement in modelling fidelity is achieved by using nonlinear Winkler springs, the so-called p-y curves along the pile length. This method is addressed next.

1.5.3.3   Load-transfer approach

The load-transfer method using the so-called p-y curves is extensively applied in offshore foundations and is the preferred method in the respective standards, e.g., API (2000, 2011), ISO 19902. Recommendations are given for static, cyclic, and postcyclic loading.

We address herein only static loading of clays and sands in the presence of free water. For background information, the reader is referred to Reese & Van Impe (2011). In this method, the pile is modelled by a beam on elastoplastic supports representing the nonlinear local soil reaction. The distribution is obtained by enforcing the conditions of equilibrium and compatibility between pile deflection and soil resistance. The *p-y* curves were derived from lateral-load tests with measurement of bending moment in the pile as a function of depth. A method for numerically deriving *p-y* curves from the stress-strain characteristics of the soil is presented in Chapter 2.

For *soft clays*, the depth-dependent ultimate lateral resistance per unit length of the pile is given in Matlock (1970), API (2000):

$$p_u = 3 \cdot c_u \cdot d + \gamma' \cdot z \cdot d + J \cdot c_u \cdot z \le 9 \cdot c_u \cdot d \tag{1.73}$$

where z is the depth from ground surface to the depth under consideration, c_u is the undrained shear strength at depth z, γ' is the average effective unit weight of the soil down to the depth z, and J is an empirical factor ranging from 0.5 for soft clay (often used) to 0.25 for medium clay.

A further element of the *p-y* curve is the deflection y_c at which the lateral resistance attains its value p_u. This threshold deflection is expressed as a function of ε_{50} that denotes the axial strain at 50% of the peak deviatoric stress in an undrained compression test under the in-situ pressure:

$$y_c = 2.5 \cdot \varepsilon_{50} \cdot d \tag{1.74}$$

Finally, the relationship between lateral soil resistance p vs. lateral deflection y (*p-y* curve) is defined by a power-law function bounded by the value p_u:

$$p = \frac{p_u}{2} \left(\frac{y}{y_c} \right)^{1/3} \le p_u \tag{1.75}$$

For *stiff clays*, the concept by Reese et al. (1975) is similar, but the *p-y* curve is more complex in shape being composed of several parts. For details, see Reese & Van Impe (2011).

For *sands,* the method recommended in codes such as API (2011), ISO 19902 is based on the method by Reese et al. (1974). It considers two different failure mechanisms at shallow depths and larger depths with their associated ultimate pressures expressed by Eq. (1.76). The lesser of the two values at any depth is taken as the ultimate resistance:

$$p_u = \min \begin{cases} (C_1 \cdot z + C_2 \cdot d) \cdot \gamma' \cdot z \\ C_3 \cdot d \cdot \gamma' \cdot z \end{cases} \tag{1.76}$$

The coefficients C_1, C_2, and C_3 are defined in terms of the effective soil friction angle φ', either in analytical form (Reese & Van Impe, 2011), or in graphical form given in ISO 19902. For convenience, they are approximated herein for $25° \le \varphi' \le 40°$ by the following formulas:

$$C_1 = 0.3 + 6.9 \cdot (\tan\varphi')^{2.66}; \quad C_2 = 0.5 + 5.1 \cdot (\tan\varphi')^{1.56} \tag{1.77}$$

$$C_3 = 5.3 + 193 \cdot (\tan \varphi')^{3.85} \tag{1.78}$$

The theoretical p_u value is multiplied by an adjustment function $A = (3 - 0.8 \cdot z/d) \geq 0.9$, and the p-y relationship is computed as follows:

$$p = A \cdot p_u \cdot \tanh \left(\frac{k_i \cdot z}{A \cdot p_u} \cdot y \right) \tag{1.79}$$

where k_i is the rate of increase with depth of the initial modulus of subgrade reaction. ISO 19902 and API (2011) recommend k_i values dependent on the friction angle φ' as $k_i = 5.4/11/22/45$ MN/m^3 for $\varphi' = 25°/30°/35°/40°$.

Known limitations and drawbacks of the p-y approach include the decoupled soil-reaction springs, the accuracy of ultimate soil resistance, the effects of pile diameter and flexibility, and the initial stiffness. Such effects may be captured by numerical finite element analyses with appropriate constitutive soil models (e.g. Sørensen et al., 2010; Papadopoulou & Comodromos, 2014).

1.5.3.4   Continuum-based solutions

- Linear solutions

   Several explicit expressions derived by curve fitting to solutions obtained via the FEM are available. They mostly refer to an idealization of the soil as an elastic continuum. Solutions are provided in terms of lateral deflection and rotation at ground level for piles loaded by a horizontal force and a moment at the pile head. For fixed-head piles, the respective solution is derived by setting the pile head rotation equal to zero.

   If the pile head deformation is known, the influence of an increase in the pile length can be determined. As shown above for the subgrade modulus approach, beyond a certain threshold length, the pile length has little influence on the deformations. This so-called 'active' or 'critical' length l_c depends on the distribution of the modulus of elasticity of the soil along the depth, the pile-to-soil stiffness ratio, and the pile diameter. Formulae compiled in this section apply to long piles, that is, $l > l_c$; therefore, the dependence on the pile length is eliminated. Because in practice most cases refer to pile groups with the pile cap preventing free rotation, we confine our consideration to the particular case of long (flexible) piles fully restrained at the head.

   A distinction is made in the literature between two depth profiles: (i) constant modulus, which is typical for over-consolidated clays, and (ii) modulus increasing linearly with depth, which applies to normally consolidated clays and approximately to cohesionless soils. A normalized modulus is introduced relating the Young's modulus of the pile E_p to the soil Young's modulus $E(z)$ at a depth of one pile diameter ($z = d$) that is denoted by E_{zd}. For a uniform constant profile, $E_{zd} = E$, and for a profile with linearly increasing modulus

$$E = E_{zd} \cdot \frac{z}{d}; \quad \bar{E} = \frac{E_p}{E_{zd}} \tag{1.80}$$

For these two profiles, we summarize here published expressions for the critical length l_c, the pile deflection at ground surface u_{0f}, and the pile-head bending moment M_{0f} for *fixed-head, long piles* ($l > l_c$). For l_c, we adopt the suggestion by Gazetas (1991), while u_{0f} and M_{0f} are taken from Davies & Budhu (1986) for the constant profile and from Budhu & Davies (1987) for the linear profile.

Constant profile:

$$l_c = 2.0 \cdot d \cdot \bar{E}^{1/4} \tag{1.81}$$

$$u_{0f} = \frac{0.8 \cdot \bar{E}^{-2/11}}{E \cdot d} \cdot H; \quad M_{0f} = -0.24 \cdot \bar{E}^{3/11} \cdot H \cdot d \tag{1.82}$$

Linear profile:

$$l_c = 2.0 \cdot d \cdot \bar{E}^{1/5} \tag{1.83}$$

$$u_{0f} = \frac{1.4 \cdot \bar{E}^{-3/9}}{E_{zd} \cdot d} \cdot H; \quad M_{0f} = -0.4 \cdot \bar{E}^{2/9} \cdot H \cdot d \tag{1.84}$$

• Nonlinear solutions

The incorporation of a nonlinear constitutive model for the soil requires the use of specialized software. For fixed-head long piles, Davies & Budhu (1986), Budhu & Davies (1987, 1988) provide useful expressions for clay and sand assuming a constant or linear variation of soil modulus with depth. The expressions are given as reduction factors which are applied to the respective linear elastic solutions that are independent of the soil strength properties. For fixed-head piles,

$$u_{0f,y} = I_{u,y}^f \cdot u_{0f}, \quad M_{0f,y} = I_{M,y}^f \cdot M_{0f} \tag{1.85}$$

For clays, it is distinguished between heavily overconsolidated clay with a constant undrained shear strength c_u and soft clay with undrained shear strength c_u increasing linearly with depth, namely

$$c_u = c_{u,zd} \cdot z/d \tag{1.86}$$

For sand, the friction angle φ' is taken constant. The reduction factors are given in terms of a dimensionless load $\bar{H}$ that is defined as follows:

$$\text{Clay: } \bar{H} = \frac{H}{c_{u,zd} \cdot d^2} \tag{1.87}$$

$$\text{Sand: } \bar{H} = \frac{H}{K_p \cdot \gamma' \cdot d^3}, \quad K_p = \frac{1 + \sin\varphi'}{1 - \sin\varphi'} \tag{1.88}$$

The constant profile is recast by setting $c_{u,zd} = c_u$. For convenience, we retained the rigidity ratio as defined by Eq. (1.80), while in the original references, it is divided by 1000. This alters the numerical values of the constants in the equations.

$$\text{Clay, constant profile}: I_{u,y}^f = 1 + \frac{\bar{H} - 0.4 \cdot \bar{E}^{2/5}}{3.75 \cdot \bar{E}^{2/7}}, \quad I_{M,y}^f = 1 + \frac{\bar{H} - 0.71 \cdot \bar{E}^{1/4}}{12.3 \cdot \bar{E}^{2/9}} \tag{1.89}$$

$$\text{Clay, linear profile}: I_{u,y}^f = 1 + \frac{\bar{H} - 1.64 \cdot \bar{E}^{0.43}}{2.52 \cdot \bar{E}^{0.54}}, \quad I_{M,y}^f = 1 + \frac{\bar{H} - 0.77 \cdot \bar{E}^{0.53}}{6.52 \cdot \bar{E}^{0.56}} \tag{1.90}$$

$$\text{Sand, linear profile}: I_{u,y}^f = 1 + \frac{\bar{H} - 0.63 \cdot \bar{E}^{0.40}}{0.474 \cdot \left(\tilde{\varphi} \cdot \bar{E}\right)^{0.50}}, \quad I_{M,y}^f = 1 + \frac{\bar{H}}{8.91 \cdot \left(\tilde{\varphi} \cdot \bar{E}\right)^{0.35}} \tag{1.91}$$

$$\tilde{\varphi} = \exp\left(0.07 \cdot (\varphi - 30°)\right) \tag{1.92}$$

The application of these solutions in practice requires the selection of suitable pairs of stiffness and strength. For clays, several empirical relations are available linking soil modulus to undrained shear strength. For sands, the link may be established via the void ratio and the effective stress. It should be noted that friction angle increases at low stress levels. Nonlinearity is mainly noticeable close to the surface, and a typical value at a depth of about two pile diameters can be regarded as representative. The original references as well as Pender (1993) include numerical examples to quantify the effects of soil properties on the reduction factors.

1.5.3.5 Comparison between continuum-based solutions and subgrade modulus approach

A first comparison is based on the active or critical length and the slenderness of the pile for a constant soil stiffness profile. For the subgrade modulus approach, it was shown that the critical length corresponds to a slenderness $\lambda \cdot l \approx 3.14$, with λ from Eq. (1.69). Hence,

$$l_c = 3.14 \cdot \sqrt[4]{\frac{4 \cdot E_p \cdot I_p}{k}} \tag{1.93}$$

For the continuum-based solution, l_c is given by Eq. (1.81). For a cylindrical solid pile, $I_p = \pi \cdot d^4 / 64$. Equating the expressions for the critical length yields

$$k/E = 1.2 \tag{1.94}$$

that corresponds to Eq. (1.72) with $\delta = 1.2$, an approximation widely used in elastodynamics.

Next, we examine the response of fixed-head pile for a constant modulus soil profile. We compare Eq. (1.70) with Eq. (1.82) for the lateral deflection u_0 and Eq. (1.70) with Eq. (1.82) for the pile-head bending moment M_0. We obtain

equal deflection: $k/E = 2.317 \cdot \bar{E}^{-1/11}$; equal bending moment: $k/E = 3.698 \cdot \bar{E}^{-1/11}$

Finally, we compare solutions for the maximum moment along the pile for the case of a free-headed pile. From the subgrade modulus approach for long piles, $M_{max} = (H/\lambda) \cdot \sin(\pi/4) \cdot \exp(-\pi/4)$, and from the continuum solution by Davies and Budhu (1986), $M_{max} = 0.12 \, \bar{E}^{3/11} \, H \cdot d$. Equating in conjunction with Eq. (1.69) yields

equal maximum moment: $k/E = 10.23 \cdot \bar{E}^{-1/11}$

It can be seen that k/E strongly depends on the system considered and the quantity selected to match with. For a typical range of values for $\bar{E}$ between 300 and 800, the ratio k/E varies as follows: fixed-head, equal deflection: $k/E = 1.38$ to 1.26; fixed-head, equal moment: $k/E = 2.01$ to 2.20; free-head, equal maximum moment: $k/E = 5.56$ to 6.08.

1.5.4 Lateral bearing capacity of pile groups

The overall lateral capacity of a pile group is seldom of concern in nonseismic design situations, and the primary objective is thus the proper design of the piles to avoid structural failure of the piles. Verification of the ULS regarding the supporting ground follows the concepts outlined above in Section 1.5.2 for the case of a single, isolated pile. It must be ensured that the local normal pressure imposed by any pile can be sustained by the soil, i.e., it does not exceed the passive earth pressure at any depth along the pile length. For pile groups, an adjustment of that method is required to take into account the unequal distribution of pile-head loads within the group that is a consequence of (i) the cross-interaction of the piles in terms of stiffness enforced by the condition of a rigid pile cap (almost equal head deflection for all piles), and (ii) the so-called shadowing effect due to the overlapping of pile yield zones that results in a reduction of soil resistance for piles in a trailing row and an increase of the load carried by piles in a leading row. The following multistep procedure, based on the recommendations in the EAP, is outlined here.

Consider a rectangular group of equal piles with diameter d connected by a rigid cap and loaded at the pile-head level by a lateral force H_G acting perpendicular to one of the sides of the rectangle, cf. Figure 1.10. The pile center-to-center spacing is s_l in the direction of loading and s_t in the transverse direction. The individual pile-head loads H_i are determined from the following equations:

$$\frac{H_i}{H_G} = \frac{\alpha_i}{\sum \alpha_i} \tag{1.95}$$

where

$$\alpha_i = \alpha_l \cdot \alpha_t \tag{1.96}$$

The reduction factor α_i to be applied to each pile i is defined in Figure 1.10 dependent on the pile position within the group, i.e., different for the front row, the periphery, or the center. The respective expressions are

$$\alpha_l = 0.125 \cdot (s_l/d); \quad 2 \le s_l/d \le 6 \tag{1.97}$$

$$\alpha_{t,c} = 0.25 \cdot (1 + s_t/d), \quad \alpha_{t,e} = 0.1 \cdot (7 + s_t/d); \quad 2 \le s_t/d \le 3 \tag{1.98}$$

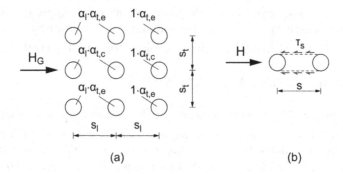

Figure 1.10 (a) Reduction factors α_i as a function of the pile position according to EAP;
(b) block failure under lateral load (Fleming et al., 1992).

Next, the characteristic value of the subgrade modulus profile $k_{s,k}(z)$ is established using the simplified relationship Eq. (1.71.)

For each pile, the subgrade modulus is adjusted to consider group effects utilizing the following expression:

$$k_{si,k}(z) = \alpha_i^n \cdot k_{s,k}(z) \tag{1.99}$$

with α_i from Eq. (1.96), and exponent n dependent on the depth profile of the subgrade modulus and the ratio between the flexural stiffness of the pile and the subgrade modulus. For the exponent n, the EAP provides relationships for subgrade modulus linearly increasing with depth (for use as an approximation for bored piles in normally consolidated cohesive soil or cohesionless soil) as well as for a constant profile (representative for over-consolidated clay). We reproduce herein merely the relationship for the constant profile:

$$n = 0.67 + 0.165 \cdot \left(l / \sqrt[4]{E_p I_p / k_{s,k}} \right) \quad \text{and} \quad 1 \le n \le 1.33 \tag{1.100}$$

The verification follows then the steps described in Section 1.5.2 for the single pile adopting the ULS GEO-2.

It should be mentioned that the range of application of the simplified relationship for the subgrade modulus Eq. (1.71) is limited to a maximum lateral deflection equal to the lesser value of 2.0 cm or 0.03 times the pile shaft diameter (Handbook EC7-1, 2011).

Another method, which is easier to implement, is that suggested by Fleming et al. (1992) where the ultimate lateral resistance for a pile located in the shadow zone of the front pile is calculated as the minimum of the ultimate resistances of an isolated single pile and that of the block of soil between two adjacent piles in the direction of loading, see right-hand side in Figure 1.10. The block resistance is $2 \cdot (s/d) \cdot \tau_s$, where s is the center-to-center pile spacing, d the pile diameter, and τ_s is the shear resistance on the side of the block. For cohesionless soil, τ_s may be taken as $\sigma_v' \cdot \tan\varphi'$ whereas for purely cohesive soil, τ_s equals c_u (Fleming et al., 1992).

For the application of the concept of p-y curves to pile group response, recourse to curves that have been derived to reproduce single pile behavior is not acceptable.

The significant interaction between the piles and the induced shadowing effects as well as the influence of structural restraint at the pile head can presently be assessed only by means of advanced continuum-based numerical methods. An attempt in this direction is described by Papadopoulou & Comodromos (2014) that present a relationship to transform single pile p-y curves into the so-called p_G-y_G curves for piles in the group.

1.5.5   Deflection of pile groups

In analogy to the case of vertical loading, pile group response to lateral loading is computed in routine design via the application of appropriate interaction factors that depend on the type of loading (force, moment) and the deformation mode (deflection, rotation), cf. Poulos & Davis (1980). Randolph (1981) presents useful approximations for the various required interaction factors. The combinations referring to moment and rotation as well as the cross-terms are in general of less importance to common foundation design, and hence we focus herein on that for the lateral deflection of fixed-head (restrained against rotation) piles. The respective factor α_{uf} depends on the position of the adjacent pile relative to the direction of loading:

$$\alpha_{uf} = 0.3 \cdot \chi \cdot \left(\frac{E_p}{G_c}\right)^{1/7} \cdot \frac{d}{s} \cdot \left(1 + \cos^2 \theta\right) \tag{1.101}$$

where s is the distance between adjacent piles (pile spacing), χ is the degree of inhomogeneity according to Eq. (1.17) ($\chi = 1$ for constant modulus, $\chi = 0.5$ for linear depth profile), and θ is the angle between the loading direction and the line connecting the two piles. When piles are in line with the load $\theta = 0$, G_c is an average modulus over the critical/active length of the pile l_c that is defined as follows:

$$G_c = \left(1 + 3 \cdot v/4\right) \frac{\int_0^{l_c} G(z)}{l_c} \tag{1.102}$$

$$l_c = d \cdot \left(\frac{E_p}{G_c}\right)^{2/7} \tag{1.103}$$

For closely spaced piles, when α_{uf} exceeds 0.33, it should be replaced by

$$\alpha_{uf} = 1 - \frac{2}{\sqrt{27 \cdot \alpha_{uf}}} \tag{1.104}$$

The determination of G_c requires an iteration due to the implicit nature of the equations involved. It can be seen that the interaction for $\theta = 90°$ is half the value for $\theta = 0°$, and that the interaction for a linear soil modulus profile is half of that for homogeneous soil with the same average modulus.

An alternative solution is to adopt the methodology suggested by the EAP for the ultimate resistance and then for the actions to use the characteristic values applicable to SLS verifications.

REFERENCES

AFNOR (2012) NF P94–262: Justification des ouvrages géotechniques Normes d'application nationale de l'Eurocode 7 – Fondations profondes.

Almeida, M.S.S., Danziger, F.A.B. & Lunne, T. (1996) Use of the piezocone test to predict the axial capacity of driven and jacked piles in clay. *Canadian Geotechnical Journal*, 33(1), 33–41.

API American Petroleum Institute (2000) *Recommended Practice 2A-WSD – Planning, Designing, and Constructing Fixed Offshore Platforms – Working Stress Design*, 21st Edition, Washington D.C., API Publishing Services.

API American Petroleum Institute (2011) *Recommended Practice RP2 GEO/ISO 19901-4, Geotechnical and Foundation Design Considerations*, 1st Edition, April 2011, Add. 1, October 2014, Washington D.C., API Publishing Services.

Basile, F. (1999) Non-linear analysis of pile groups. *Proceedings of the Institution of Civil Engineers: Geotechnical Engineering*, 137(2), 105–115.

Basile, F. (2003) Analysis and design of pile groups. In: Bull, J.W. (ed.) *Numerical Analysis and Modelling in Geomechanics*, London, UK, Spon Press, pp. 278–315.

Berardi, R. & Bovolenta, R. (2005) Pile-settlement evaluation using field stiffness non-linearity. *Proceedings of the Institution of Civil Engineers, Geotechnical Engineering*, 158(GE1), 35–44.

Bohn, C., Lopes dos Santos, A. & Frank, R. (2017) Development of axial pile load transfer curves based on instrumented load tests. *Journal of Geotechnical and Geoenvironmental Engineering*, 143(1): 04016081-1-04016081-15.

Bond, A. & Harris, A. (2008) *Decoding Eurocode 7*, London, UK, Taylor & Francis.

Bond, A.J. & Simpson, B. (2010) Pile design to Eurocode 7 and the UK National Annex. Part 2: UK National Annex. *Ground Engineering*, 43(1), 28–31.

Brinch Hansen, J. (1961) The ultimate resistance of rigid piles against transversal forces. *Bulletin No. 12*, Danish Geotechnical Institute.

Broms, B.B. (1964a) Lateral resistance of piles in cohesive soils. *Journal of the Soil Mechanics and Foundations Division*, ASCE, 90(2), 27–63.

Broms, B.B. (1964b) Lateral resistance of piles in cohesionless soils. *Journal of the Soil Mechanics and Foundations Division*, ASCE, 90(3), 123–156.

Budhu, M. & Davies, T.G. (1987) Nonlinear analysis of laterally loaded piles in cohesionless soils. *Canadian Geotechnical Journal*, 24(2), 289–296.

Budhu, M. & Davies, T.G. (1988) Analysis of laterally loaded piles in soft clays. *Journal of Geotechnical Engineering*, 114(1), 21–39.

Burlon, S., Frank, R., Baguelin, F., Habert, J. & Legrand, S. (2014) Model factor for the bearing capacity of piles from pressuremeter test results – Eurocode 7 approach. *Géotechnique*, 64(7), 513–525.

Burlon, S., Szymkiewicz, F., Le Kouby, A. & Volcke, J.-P. (2016) Design of piles French practice. In: *ISSMGE—ETC 3 International Symposium on Design of Piles in Europe*, Brussels.

Butterfield, R. & Banerjee, P. K. (1971) The elastic analysis of compressible piles and pile groups. *Géotechnique*, 21(1), 43–60.

Cecconi, M., Pane, V., Vecchietti, A. & Bellavita, D. (2019) Horizontal capacity of single piles: An extension of Broms' theory for c-φ soils. *Soils and Foundations*, 59, 840–856.

Chow, Y.K. (1986) Analysis of vertically loaded pile groups. *International Journal for Numerical and Analytical Methods in Geomechanics*, 10, 59–72.

Chow, Y.K. (1987) Axial and lateral response of pile groups embedded in nonhomogeneous soils. *International Journal for Numerical and Analytical Methods in Geomechanics*, 11, 621–638.

Clancy, P. & Randolph, M.F. (1993) An approximate analysis procedure for piled raft foundations. *International Journal for Numerical and Analytical Methods in Geomechanics*, 17, 849–869.

Comodromos, E.M. (2004) Response evaluation of axially loaded fixed-head pile groups using
3D nonlinear analysis. *Soils and Foundations*, 44(2), 31–39.

Comodromos, E.M., Papadopoulou, M.C. & Laloui, L. (2016) Contribution to the design
methodologies of piled raft foundations under combined loadings. *Canadian Geotechnical
Journal*, 53(4), 559–577.

Crispin, J.J., Leahy, C.P. & Mylonakis, G. (2018) Winkler model for axially loaded piles in inhomogeneous soil. *Géotechnique Letters*, 8(4), 290–297.

Cyna, H., Schlosser, F., Frank, R., Plumelle, C., Estephan, R., Altmayer, F., Goulesco, N.,
Juran, I., Maurel, C., Shahrour, I. & Vezole, P. (2004) Synthèse des résultats et recommanda-
tions du Projet national sur les micropieux FOREVER. Opération du Réseau Génie Civil et
Urbain, IREX, Paris, France, Presses de l'Ecole Nationale des Ponts et Chaussées.

Davies, T.G. & Budhu, M. (1986) Nonlinear analysis of laterally loaded piles in heavily overconsolidated clays. *Géotechnique*, 36(4), 527–538.

DGGT – German Geotechnical Society (2012) *Recommendations on Piling (EA-Pfähle)*, Berlin,
Germany, Ernst & Sohn.

DIN 1054:2010-12 Baugrund – Sicherheitsnachweise im Erd- und Grundbau – Ergänzenden
Regelungen zu DIN EN 1997-1. Berlin, Beuth Verlag.

DIN EN 1997-1/NA:2010-12: Nationaler Anhang – National festgelegte Parameter – Eurocode
7: Entwurf, Berechnung und Bemessung in der Geotechnik – Teil 1; Allgemeine Regeln. Berlin, Beuth Verlag.

Doherty, P. & Gavin, K. (2011) The shaft capacity of displacement piles in clay: A state of the
art review. *Geotechnical and Geological Engineering*, 29(4), 389–410.

EN 1997-1:2014. Eurocode 7 – Geotechnical design, Part 1: General rules. EN 1997-
1:2004 + AC:2009 + A1:2013, Brussels, Belgium, European Committee for Standardization.

EN 1536:2015. Execution of special geotechnical work – Bored piles. EN 1536:2010 + A1:2015,
Brussels, Belgium, European Committee for Standardization.

EN 12699:2015. Execution of special geotechnical work – Displacement piles, Brussels, Belgium,
European Committee for Standardization.

EN 14199:2015. Execution of special geotechnical works – Micropiles, Brussels, Belgium,
European Committee for Standardization.

Engin, H.K. & Brinkgreve, R.B.J. (2009) Investigation of pile behaviour using embedded piles.
In: *Proceedings of the 17th International Conference on Soil Mechanics and Geotechnical Engineering*, Alexandria, Egypt. Amsterdam, the Netherlands, IOS Press, pp. 1189–1192.

Eslami, A. & Fellenius, B.H. (1997) Pile capacity by direct CPT and CPTu methods applied to
102 case histories. *Canadian Geotechnical Journal*, 34(6), 880–898.

Fahey, M. & Carter, J.P. (1993) A finite element study of the pressuremeter test in sand using a
nonlinear elastic plastic model. *Canadian Geotechnical Journal*, 30(2), 348–362.

Fleming, W.G.K., Weltman, A.J., Randolph, M.F. & Elson, W.K. (1992) *Piling Engineering*, 2nd
Edition, Glasgow, Scotland, Blackie Academic and Professional.

Fleming, W.G.K., Weltman, A.J., Randolph, M.F. & Elson, W.K. (2009) *Piling Engineering*, 3rd
Edition, London & New York, Taylor & Francis.

Frank, R. (1974) Etude Théorique du Comportement des Pieux sous Charge Verticale.
Introduction de la Dilatance, Thèse de Docteur-ingénieur, Université Pierre et Marie Curie
(Paris VI).

Frank, R. (2017) Some aspects of research and practice for pile design in France. *Innovative
Infrastructure Solutions*, 2, 32. https://doi.org/10.1007/s41062-017-0085-4.

Frank, R., Bauduin, C., Driscoll, R., Kavvadas, M., Krebs Ovesen, N., Orr, T., Schuppener,
B. (2005) *Designers' Guide to EN 1997-1 Eurocode 7: Geotechnical Design – General Rules*,
London, UK, Thomas Telford Publishing.

Franke, E., Lutz, B. & El-Mossallamy, Y. (1994) Measurements and numerical modelling of high-rise building foundations on Frankfurt clay. In: Yeung, A. T. & Félio, G. Y. (eds.) *Vertical and Horizontal Deformation of Foundations and Embankments – Geotechnical Special Publication No. 40*, New York, NY, ASCE, Vol. 2, pp. 1325–1336.

Gazetas, G. (1991) Foundation vibrations. In: Fang, H.-Y. (ed.) *Foundation Engineering Handbook*, 2nd Edition, New York, NY, Van Nostrand Reinhold, pp. 553–593.

Geocentrix (2018) Repute 2.5 – Reference manual. https://www.geocentrix.co.uk/repute/manuals.html

Guo, W.D. (2013) *Theory and Practice of Pile Foundations*, Boca Raton, FL, CRC Press.

Guo, W.D. & Randolph, M.F. (1997) Vertically loaded piles in non-homogeneous media. *International Journal for Numerical and Analytical Methods in Geomechanics*, 21(8), 507–532.

Handbuch Eurocode 7 (2011) *Geotechnische Bemessung, Band 1: Allgemeine Regeln*, Berlin, Germany, Beuth Verlag.

Hanisch, J., Katzenbach, R. & König, G. (2001) *Kombinierte Pfahl-Plattengründungen*, Berlin, Germany, Ernst & Sohn.

Hetenyi, M. (1946) *Beams on Elastic Foundation*, Ann Arbor, MI, University of Michigan Press.

Horikoshi, K. & Randolph, M.F. (1998) A contribution to optimal design of piled rafts. *Géotechnique*, 48(3), 301–317.

ISO 19902:2007 + Amd 1:2013: Petroleum and natural gas industries – Fixed steel offshore structures, Geneva, Switzerland, International Standards Organize

Jardine, R., Chow, F., Overy, R. & Standing, J. (2005) *ICP Design Methods for Driven Piles in Sands and Clays*, London, UK, Thomas Telford Publishing.

Jardine, R.J., Potts, D.M., Fourie, A.B. & Burland, J.B. (1986) Studies of the influence of non-linear stress-strain characteristics in soil-structure interaction. *Géotechnique*, 36(3), 377–396.

Karatzia, X. & Mylonakis, G. (2016) Horizontal stiffness and damping of piles in inhomogeneous soil. *Journal of Geotechnical and Geoenvironmental Engineering*, 143(4), 04016113.

Katzenbach, R. & Choudhury, D. (2013) *ISSMGE Combined Pile-Raft Foundation Guideline*. International Society for Soil Mechanics and Geotechnical Engineering, Technical Committee TC 212. Technische Universität Darmstadt, Darmstadt, Germany, 1–23.

Katzenbach, R., Arslan, U. & Moorman, C. (2000) Piled raft foundation projects in Germany. In: Hemsley, J.A. (ed.) *Design Applications of Raft Foundations*, London, UK, Thomas Telford, pp. 323–391.

Katzenbach, R., Bachmann, G. & Gutberlet, C. (2009) Assessment of settlements of high-rise structures by numerical analysis. In: Bull, J.W. (ed.) *Linear and Non-linear Numerical Analysis of Foundations*, London, UK, CRC Press, pp. 390–419.

Katzenbach, R. & Reul, O. (1997) Design and performance of piled rafts. In: *Proceedings of the 14th International Conference on Soil Mechanics and Foundation Engineering*, Hamburg, Rotterdam, the Netherlands, Balkema, pp. 2253–2256.

Kempfert, H.-G. & Becker, P. (2010) Axial pile resistance of different pile types based on empirical values. In: *Proceedings of Geo-Shanghai 2010: Deep Foundations and Geotechnical In-Situ Testing (GSP 205)*. Reston, VA, ASCE, pp. 149–154.

Kempfert, H.-G., Eigenbrod, K. U. & Smoltczyk, U. (2003) Pile foundations. In: Smoltczyk, U. (ed.) *Geotechnical Engineering Handbook*, 6th Edition, Berlin, Germany, Ernst & Sohn, Vol. 3, pp. 83–227.

Kitiyodom, P. & Matsumoto, T. (2003) A simplified analysis method for piled raft foundations in non-homogeneous soils. *International Journal for Numerical and Analytical Methods in Geomechanics*, 27, 85–109.

Kraft, L.M., Ray, R.P. & Kagawa, T. (1981) Theoretical t-z curves. *Journal of Geotechnical Engineering*, 107(11), 1543–1561.

Lehane, B.M., Li, Y. & Williams, R. (2013) Shaft capacity of displacement piles in clay using the cone penetration test. *Journal of Geotechnical and Geoenvironmental Engineering*, 139(2), 253–266.

Lo Presti, D.C.F. (1994) General report: Measurement of shear deformation of geomaterials in the laboratory. In: Shibuya, S., Mitachi, T. & Miura, S. (eds.) *Pre-Failure Deformation of Geomaterials*, Rotterdam, the Netherlands, Balkema, pp. 1067–1088.

Mandolini, A., Russo & Viggiani, C. (2005) Pile foundations: Experimental investigations, analysis and design. In: *Proceedings of the 16th International Conference on Soil Mechanics and Geotechnical Engineering, Osaka*. Amsterdam, the Netherlands, IOS Press, pp. 177–213.

Matlock, H. (1970) Correlations for design of laterally loaded piles in soft clay. In: *Proceedings of the II Annual Offshore Technology Conference*, Houston, Texas, (OTC 1204), pp. 577–594.

Mayne, P.W. & Schneider, J.A. (2001) Evaluating axial drilled shaft response by seismic cone. In: *Foundations and Ground Improvement, Geotechnical Special Publication 113*. Reston, AV, ASCE, pp. 655–669.

McCabe, B.A. & Sheil, B.B. (2015) Pile group settlement estimation: Suitability of nonlinear interaction factors. *International Journal of Geomechanics*, 04014056.

Møller, O., Frederiksen, J.K., Augustesen, A.H., Okkels, N. & Sørensen, K.K. (2016) Design of piles – Danish practice. In: *ISSMGE – ETC 3 International Symposium on Design of Piles in Europe*.

Moormann, C. (2016) Design of piles – German practice. In: *ISSMGE – ETC 3 International Symposium on Design of Piles in Europe*.

Mylonakis, G. (2001) Winkler modulus for axially loaded piles. *Géotechnique*, 51(5), 455–461.

Mylonakis, G. & Gazetas, G. (1998) Settlement and additional internal forces of grouped piles in layered soil. *Géotechnique*, 48(1), 55–72.

Niazi, F.S. & Mayne, P.W. (2013) Cone Penetration Test based direct methods for evaluating static axial capacity of single piles: A state-of-the-art review. *Geotechnical and Geological Engineering*, 31(4), 979–1009.

Niazi, F.S., Mayne, P.W. & Woeller, D.J. (2014) Pile axial load-displacement response from geophysical component of seismic CPT. In: *CPT'14 – 3rd International Symposium on Cone Penetration Testing*, Las Vegas, NV, pp. 1103–1111.

Pais, A. & Kausel, E. (1988) Approximate formulas for dynamic stiffnesses of rigid foundations. *Soil Dynamics and Earthquake Engineering*, 7, 213–227.

Papadopoulou, M.C. & Comodromos, E.M. (2014) Explicit extension of the p-y method to pile groups in sandy soils. *Acta Geotechnica*, 9(3), 485–497.

Pender, M. (1993) Aseismic pile foundation design analysis. *Bulletin of the New Zealand National Society of Earthquake Engineering*, 26(1), 49–160.

Potts, D.M. & Zdravkovic, L. (2001) *Finite Element Analysis in Geotechnical Engineering*, Vol. 2, London, UK, Thomas Telford.

Poulos, H.G. (1968) Analysis of the settlement of pile groups. *Géotechnique*, 18(4), 449–471.

Poulos, H.G. (1989) Pile behaviour – Theory and application. *Géotechnique*, 39(3), 365–415.

Poulos, H.G. (1990) User's guide to program DEFPIG – Deformation Analysis of Pile Groups, Revision 6. School of Civil Engineering, University of Sidney.

Poulos, H.G. (1994) An approximate numerical analysis of pile-raft interaction. *International Journal for Numerical and Analytical Methods in Geomechanics*, 18, 73–92.

Poulos, H.G. (2001a) Pile foundations. In: Rowe, R.K. (ed.) *Geotechnical and Geoenvironmental Engineering Handbook*, Boston, MA, Klewer Academic, pp. 261–304.

Poulos, H.G. (2001b) Piled raft foundations – Design and applications, *Géotechnique*, 50(2), 95–113.

Poulos, H.G. (2006) Pile group settlement estimation — Research to practice. In: *Proceedings of Geo-Shanghai 2006: Foundation Analysis and Design: Innovative Methods (GSP 153)*, ASCE, Reston, VA, pp. 1–22.

Poulos, H.G. & Davis, E.H. (1980) *Pile Foundation Analysis and Design*, New York, NY, John Wiley & Sons.

Powell, J.J.M., Lunne, T. & Frank, R. (2001) Semi-empirical design for axial pile capacity in clays. In: *Proceedings of 15th International Conference on Soil Mechanics and Geotechnical Engineering*, Istanbul, Turkey. Rotterdam, the Netherlands, Balkema, pp. 991–994.

Quarg-Vonscheidt, J. & Walz, B. (2001) A new method for calculating axial capacity and group action of tension piles in dense sand. In: *Proceedings of 15th International Conference on Soil Mechanics and Geotechnical Engineering*, Istanbul, Turkey. Rotterdam, the Netherlands, Balkema, pp. 995–998.

Randolph, M.F. (1981) Response of flexible piles to lateral loading. *Géotechnique*, 31(2), 247–259.

Randolph, M.F. (1983) Design of piled raft foundations. In: *Proceedings of the International Symposium on Recent Developments in Laboratory and Field Tests and Analysis of Geotechnical Problems*, Bangkok. Rotterdam, the Netherlands, Balkema, pp. 525–537.

Randolph, M.F. (1994) Design methods for pile groups and piled rafts. In: Proceedings of the 13th International Conference on Soil Mechanics and Foundation Engineering, New Delhi, India. Rotterdam, the Netherlands, Balkema, Vol. 5, pp. 61–82.

Randolph, M.F. (1996) PIGLET – Analysis and design of pile groups. University of Western Australia.

Randolph, M.F. (2003) Science and empiricism in pile foundation design. *Géotechnique*, 53(10), 847–875.

Randolph, M.F. & Wroth, C.P. (1978) Analysis of deformation of vertically loaded piles. *Journal of the Geotechnical Engineering Division*, ASCE, 104(GT12), 1465–1488.

Randolph, M.F. & Wroth, C.P. (1979) An analysis of the vertical deformation of pile groups. *Géotechnique*, 29(4), 423–439.

Reese, L.C., Cox, W.R. & Koop, F.D. (1974) Analysis of laterally loaded piles in sand. In: *Proceedings of the VI Annual Offshore Technology Conference*, Houston, TX, 2(OTC 2080), pp. 473–485.

Reese, L.C., Cox, W.R. & Koop, F.D. (1975) Field testing and analysis of laterally loaded piles in stiff clay. In: *Proceedings of the VII Annual Offshore Technology Conference*, Houston, TX, 2(OTC 2312), pp. 672–690.

Reese, L.C. & Van Impe, W.F. (2011) *Single Piles and Pile Groups under Lateral Loading*, 2nd Edition, Boca Raton, FL, CRC Press.

Reul, O. (2004) Numerical study of the bearing behavior of piled rafts. *International Journal of Geomechanics*, 4(2), 59–68.

Reul, O. & Randolph, M.F. (2003) Piled rafts in overconsolidated clay: Comparison of in situ measurements and numerical analyses. *Géotechnique*, 53(3), 301–315.

Richter, T. & Lutz, B. (2010) Berechnung einer Kombinierten Pfahl-Plattengründung am Beispiel des Hochhauses 'Skyper' in Frankfurt/Main. *Bautechnik*, 87(4), 204–211.

Russo, G. (1998) Numerical analysis of piled rafts. *International Journal for Numerical and Analytical Methods in Geomechanics*, 22, 477–493.

Sabatini, P.J., Tanyu, B., Armour, T., Groneck, P. & Keeley, J. (2005) Micropile design and construction – Reference manual. Report No. FHWA NHI-05-039. Washington, DC, Federal Highway Administration.

Schanz, T., Vermeer, P.A. & Bonnier, P.G. (1999) The Hardening-Soil model: Formulation and verification. In: Brinkgreve, R.B.J. (ed.) *Beyond 2000 in Computational Geotechnics*, Rotterdam, the Netherlands, Balkema, pp. 281–290.

Scott, R.F. (1981) *Foundation Analysis*, Englewood Cliffs, NJ, Prentice Hall.

Sheil, B.B., McCabe, B.A., Comodromos, E.M. & Lehane, B.M. (2019) Pile groups under axial loading: An appraisal of simplified non-linear prediction models. *Géotechnique*, 69(7), 565–579.

Sørensen, S.P.H., Ibsen, L.B. & Augustesen, A.H. (2010) Effects of diameter on initial stiffness of p-y curves for large-diameter piles in sand. In: Benz, T. & Nordal, S. (eds.) *Numerical Methods in Geotechnical Engineering (NUMEG 2010)*, Boca Raton, FL, CRC Press, pp. 907–912.

Ta, L.D. & Small, J.C. (1996) Analysis of piled raft systems in layered soils. *International Journal for Numerical and Analytical Methods in Geomechanics*, 20, pp. 57–72.

Takesue, K., Sasao, H. & Matsumoto, T. (1998) Correlation between ultimate pile skin friction and CPT data. In: *Proceedings of the First International Conference on Site Characterization*, Atlanta, GA. Rotterdam, the Netherlands, Balkema, pp. 1177–1182.

Thieken, K., Achmus, M. & Schmoor, K.A. (2014) On the ultimate limit state design proof for laterally loaded piles. *Geotechnik*, 37(1), 19–31.

Togliani, G. (2008) Pile capacity prediction for in-situ tests. In: *Proceedings of the 3rd International Conference on Site Characterization (ISC'3)*, Taipei, Taiwan. London, UK, Taylor and Francis Group, pp. 1187–1192.

Tomlinson, M.J. (1986) *Foundation Design and Construction*, 5th Edition, Harlow, UK, Longman.

Tomlinson, M.J. & Woodward, J. (2015) *Pile Design and Construction Practice*, 6th Edition, Boca Raton, FL, CRC Press.

Trochanis, A.M., Bielak, J. & Christiano, P. (1991) Three-dimensional nonlinear study of piles. *Journal of Geotechnical Engineering*, ASCE, 117(3), 429–447.

Van Dijk, B.F.J. & Kolk, H.J. (2011) CPT-based design method for axial capacity of offshore piles in clays. In: *Proceedings of Frontiers in Offshore Geotechnics II*, Perth, Australia. London, UK, Taylor and Francis, pp. 555–560.

Van Impe, W.F. (1991) Deformations of deep foundations. In: Proceedings of the 10th European Conference on Soil Mechanics and Foundation Engineering, Florence, Italy, Vol. 3, pp. 1031–1062.

Vardanega, P.J. & Bolton, M.D. (2013) Stiffness of clays and silts: Normalizing shear modulus and shear strain. *Journal of Geotechnical and Geoenvironmental Engineering*, 139(9), 1575–1589.

Viggiani, C., Mandolini, A. & Russo, G. (2012) *Piles and Pile Foundations*, London, UK, CRC Press.

Vrettos, C. (2012) Simplified analysis of piled rafts with irregular geometry. In: *Proceedings of the 9th International Conference on Testing and Design Methods for Deep Foundations*, Kanazawa, Japan, pp. 453–460.

Vucetic, M. & Dobry, R. (1991) Effect of soil plasticity on cyclic response. *Journal of Geotechnical Engineering*, 117(1), 89–107.

Wehnert, M. & Vermeer, P.A. (2004) Numerical analyses of load tests on bored piles. In: *Proceedings of the 9th Symposium on Numerical Models in Geomechanics (NUMOG)*, Ottawa, Canada. Rotterdam, the Netherlands, Balkema, pp. 505–511.

White, D.J. & Bolton, M.D. (2005) Comparing CPT and pile base resistance in sand. *Proceedings of the Institution of Civil Engineers: Geotechnical Engineering*, 158(GE1), 3–14.

Yamashita, K., Tomono, M. & Kakurai, M. (1987) A method for estimating immediate settlement of piles and pile groups. *Soils and Foundations*, 27(1), 61–76.

Yamashita, K., Yamada, T. & Kakurai, M. (1998) Simplified method for analysing piled raft foundations. In: *Proceedings of the 3rd International Geotechnical Seminar on Deep Foundations on Bored and Auger Piles*, Ghent, Belgium. Rotterdam, the Netherlands, Balkema, pp. 457–464.

Chapter 2

Analysis of laterally loaded pile foundations using the subgrade reaction method

Youhu Zhang
Southeast University

CONTENTS

2.1 INTRODUCTION

Pile foundations are widely used in the offshore industry for supporting oil and gas platforms, wellheads (i.e., well conductors), wind turbines and substations, as well as for anchoring floating facilities (e.g. FPSO). Due to the nature of offshore environment, wind, waves, and current exert large horizontal load on the offshore structures, which must be withstood by the pile foundations supporting them. For a multi-pod jacket structure, the global over-turning moment induced by the environmental loading is mainly resisted by a "push-pull" mechanism, i.e., redistribution of the vertical load amongst the legs and to a less extent by the local pile head bending moment. For a

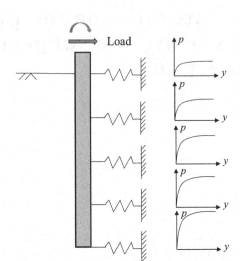

Figure 2.1 Schematic illustration of the subgrade reaction analysis of a laterally loaded pile.

monopod structure, such as a wind turbine supported by a monopile, the environmental load is translated into large overturning moment at the pile head level.

The pile response under environmental loading is commonly analysed using the subgrade reaction method. The pile-soil interaction is represented by distributed linear/non-linear springs along the pile, whilst the pile is modelled as an equivalent elastic beam. Figure 2.1 presents a schematic illustration of the analysis model for a pile foundation under lateral loading, where the lateral soil resistance is represented by the so-called *p-y* springs.

In the design of piled offshore structures, the following should be generally demonstrated:

1   The pile foundations have sufficient capacity to resist the extreme loading event (ultimate limit state design) and the accidental loading condition (accidental limit state design). Push-over analyses are commonly performed to understand the failure mechanism and weak points of the system. The modelling of the soil-pile interaction has a direct consequence on the sizing of the pile foundations. Cyclic loading effect typically needs to be considered.

2   The members of the offshore structure have sufficient fatigue life (i.e., fatigue limit state design). This includes the structure above the foundation and the pile foundations themselves. The soil-pile interaction springs influences the system dynamics (i.e., natural frequency) and therefore the amplification of the excitations (i.e., fatigue loads). The modelling of the soil-pile interaction also has a consequence on the distribution of the fatigue loads along the structure and therefore on which part of the structure should be strengthened. For wind turbines supported by monopiles, the fatigue limit state design can govern the pile sizing due to system natural frequency requirement, which must fall in-between the so-called 1P and 3P frequency band (Tempel & Molenaar, 2002).

3   The foundations do not accumulate excessive deformations that could jeopardize the normal operation of the structure. This is particularly relevant for bottom-fixed offshore wind turbine structures as the requirement of verticality is strict. A maximum permanent tilt of 0.5° is commonly adopted in design. In this regard, the evaluation of soil-pile interaction under repeated cyclic loading events (storms, typhoons) is critical to this assessment.

This chapter first reviews the industry practice for designing pile foundations under lateral loading in clay and sand. It then focusses on the current state of the art for modelling the lateral soil-pile interaction that is developed on the essential idea of linking the "micro" soil response observed at soil element level to "macro" soil-pile interaction behaviour. This chapter is devoted to the lateral pile response, and the vertical soil-pile interaction along the pile shaft and at the pile tip is outside the scope of this chapter.

2.2   CURRENT INDUSTRY PRACTICE

2.2.1   API p-y spring model for soft clay

The industry practice for design of laterally loaded piles in clay, such as the API RP-2GEO (American Petroleum Institute, 2014), is based on a limited database of field tests on relatively small diameter piles performed in the mid-1950s. Using the results of these field tests, empirical monotonic and cyclic p-y curves have been developed based on back analysis of the experimental results (Matlock, 1970). Figure 2.2 illustrates the construction of the monotonic and the cyclic p-y curves according to the API recommendation. In the normalised format, the shape of the monotonic p-y curve is defined by a cubic power law function:

$$\frac{p}{p_u} = 0.5 \left(\frac{y}{y_c} \right)^{\frac{1}{3}} \leq 1 \tag{2.1}$$

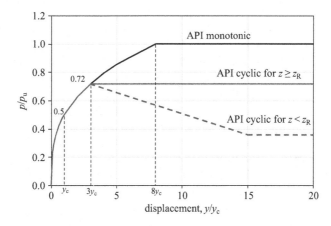

Figure 2.2 Illustration of the API (2014) p-y curves.

where p and p_u are the mobilised and the limiting lateral bearing capacity, respectively (force per unit pile length), y is the lateral pile deflection, and y_c is the pile deflection at half spring mobilisation (i.e., at $p/p_u = 0.5$). The overall spring stiffness is controlled by y_c, which is calculated by

$$y_c = 2.5\varepsilon_{50}D \tag{2.2}$$

where ε_{50} is defined as the axial strain at which 50% of maximum deviator stress is mobilised in an undrained compression test. The API document is not specific on the type of test. In practice, this is often chosen from unconsolidated undrained (UU) compression tests. A value in the range of 1%–2% is typically used in design. D is the diameter of the pile.

The ultimate strength of the p-y spring is calculated by

$$p_u = N_p s_u D \tag{2.3}$$

where s_u is the undrained shear strength of the soil at the depth of consideration, and N_p is the lateral bearing capacity factor and taken to be a linear function of soil depth z:

$$N_p = \left(3 + J\frac{z}{D}\right) + \frac{\gamma'z}{s_u} \tag{2.4}$$

where z is the soil depth in question, J is a dimensionless empirical factor and taken to be 0.5 for soft clay, and γ' is the effective unit weight of the soil. The first part of the equation represents the lateral resistance provided by the undrained shear strength of the soil, whilst the second part represents the contribution of effective soil weight.

The N_p factor calculated by Eq. 2.4 is limited to a maximum value of 9, which is taken to be the value mobilised by an in-plane flow-around mechanism. The onset depth (z_R) where the mechanism transits from a surface wedge mechanism to a flow-around mechanism is the depth at which N_p calculated by Eq. 2.4 first reaches the limiting value of 9.

As discussed by Zhang et al. (2016a), Eq. 2.4 implies assumption of a tension gap on the back side of the pile. The limiting N_p value of 9 also implies assumption of a fully smooth soil-pile interface according to the theoretical solution of Randolph and Houlsby (1984). Compared with results of upper bound plasticity analyses and finite element analyses, the rate of increase in N_p with normalised depth (z/D) suggested by API is considerably underestimated, thus leading to a too deep onset depth for flow-around mechanism. In conditions where suction can be mobilised (such as in soft clays), the flow-around mechanism is reached at even shallower depth due to mobilisation of both active and passive soil wedges. Jeanjean et al. (2017) presented a comprehensive comparison on the ultimate lateral capacity between API and recommendation based on upper bound and finite element analyses (Murff & Hamilton, 1993; Yu et al., 2015). The underestimation of the ultimate spring capacity is also confirmed by centrifuge model testing (Jeanjean, 2009) and field pile load test (Zhu et al., 2017).

For the cyclic p-y curve, the threshold for cyclic effects of the API model is set to be 72% of the ultimate monotonic capacity (i.e., $0.72p_u$). For cyclic mobilisation less than the threshold, the same monotonic curve is followed. At depths greater than the transition depth (z_R), at which soil is engaged in an in-plane flow-around mechanism, the peak lateral bearing capacity for cyclic loading is simply limited to $0.72p_u$. For depths shallower than the transition depth, a post-peak softening response is introduced that degrades the strength from $0.72p_u$ at $3y_c$ to a residual value at large deflection ($15y_c$). The magnitude of softening is linearly related to z/z_R.

Despite its widespread application of the API p-y model for soft clay, its limitations are clear. It is developed from a quite limited number of pile load tests in a specific soil and loading condition. The applicability of the model to other soil and loading conditions is questionable. As discussed in Zhang & Andersen (2017), the cubic power law format of the Matlock/API monotonic p-y curve is perhaps appropriate for the lightly over-consolidated Sabine River clay in which the tests were performed but may become inappropriate for other soil conditions, for example, heavily over-consolidated clay, where a different power law exponent is needed to fit the soil response. Furthermore, the ultimate strength of the p-y springs is considerably underestimated as discussed above. For the cyclic p-y response, the empirical models do not allow for consideration of neither site-specific soil response under cyclic loading nor the make-up of the cyclic loading. In this sense, it is rather arbitrary to apply the Matlock/API cyclic p-y curves for designing cyclically laterally loaded piles. In practice, one may also question whether the empirical cyclic p-y curves should be used along the entire pile length or just a certain section of the pile.

2.2.2 API p-y spring model for sand

The API p-y spring model for sand takes a hyperbolic tangent curve, proposed originally by O'Neil & Murchison (1983):

$$\frac{p}{p_u} = A \tanh\left(\frac{kz}{Ap_u} y\right) \tag{2.5}$$

where

A is a factor that depends on the type of loading: $A = 0.9$ for cyclic loading and $A = (3 - 0.8z/D) \geq 0.9$ for static loading. This implies that the cyclic loading only causes a reduction in capacity for soil depths less than $2.6D$.

Equation 2.5 implies an initial p-y stiffness (i.e., force per unit pile length versus lateral displacement y) at $y = 0$ that is equal to kz, i.e., a linearly increasing initial stiffness with depth. k therefore represents the gradient of the initial stiffness with depth. Its value is a function of the internal friction angle of the sand.

The ultimate strength of the p-y spring p_u (force per unit pile length) is the lesser of the resistance provided by the shallow mechanism (p_{us}) and by the deep mechanism (p_{ud}).

$$p_{us} = (C_1 z + C_2 D)\gamma' z \tag{2.6}$$

$$p_{ud} = C_3 D \gamma' z \tag{2.7}$$

where

C_1, C_2, and C_3 are functions of the internal friction angle of the sand, as illustrated in Figure 2.3.

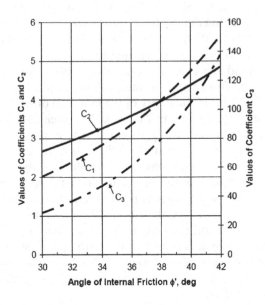

Figure 2.3 Variation of C_1, C_2, and C_3 with internal friction angle (API RP2GEO, 2014).

2.3   LINK BETWEEN SOIL-PILE INTERACTION AND THE STRESS-STRAIN RESPONSE OF THE SOIL

Lateral soil-pile interaction p-y curves/springs have been conventionally derived from field pile load tests (Matlock, 1970; Reese et al., 1975; O'Neil & Murchison, 1983; Georgiadis, 1983). However, a major weakness of the empirical p-y formulations is that their applicability is often limited to the soil conditions in which the pile load tests are performed. When they are applied to soil conditions that are different from the tested conditions, the applicability becomes uncertain. Further uncertainties exist with regard to the scalability between the model piles used in pile load tests and the piles being designed in reality. The loading condition may also present a limitation as the design loading condition may differ from that examined in the pile load tests.

The current state of the art in modelling the soil-pile interaction is to link the p-y curves with the stress-strain response of the soil measured at the soil element level using the conventional lab testing technique. In this way, site-specific p-y curves can be derived and used in design instead of using empirical p-y formulations derived from pile load tests which are performed with a model pile of a certain geometry under certain soil and loading conditions. The theoretical background of this approach is sound as the load-displacement response for soil-structure interaction is an integration of the stress-strain response of the soil over the volume that is mobilised by the structure. In the simplest form, the elastic stiffness coefficients are used to scale the soil stiffness to the foundation-soil interaction stiffness in ideally linearly elastic medium (e.g. Poulos & Davis, 1974). For the case of non-linear soil behaviour, it is observed that for many soil-structure interaction problems, the normalised load-displacement response is self-similar to the stress-strain response of the soil. Hence, it is possible to scale the soil-structure interaction response from the stress-strain response with certain scaling coefficients. The pioneering work by Osman & Bolton (2005) first applied this concept to derive the vertical load-settlement curve of a circular foundation resting on undrained clay. Osman et al. (2007) further extended the concept to derive the horizontal load-displacement and moment-rotation curves for a surface circular footing from the stress-strain response of the underlying soil. In recent years, this concept has been successfully applied to develop site-specific *p-y* curves for laterally loaded pile foundations. Scaling relationships between the *p-y* curve and the stress-strain response of the soil have been developed for different soil mechanisms (e.g. wedge mechanism and in-plane flow-around mechanism), different applications (e.g. conventional slender piles for offshore oil and gas platforms and monopiles for supporting offshore wind turbines), and different loading conditions (e.g. monotonic, cyclic loading). The rest of this chapter will introduce the current state of the art on modelling the soil-pile interaction under lateral loading that is fundamentally linked to soil response at element level.

2.4 SITE-SPECIFIC *P-Y* CURVES FOR SLENDER PILES

2.4.1 Monotonic *p-y* curves

The studies by Bransby (1999), Klar (2008), Klar & Osman (2008), and Yu et al. (2017) demonstrate from the numerical and theoretical perspectives that it is possible and rationale to link the pile-soil interaction *p-y* spring to the stress-strain response of the soil. Erbrich et al. (2010) presented a procedure that calculates the site-specific *p-y* curves by means of finite element/finite difference analyses using the stress-strain responses measured in laboratory tests and subsequently uses them in assessment of pile response under cyclic loading. However, a generalised scaling relationship is not developed. Zhang & Andersen (2017) proposed a framework which scales the *p-y* curve of a laterally loaded pile slice from the stress-strain response measured in direct simple shear (DSS) test using two scaling constants. Based on an extensive suite of parametric finite element analyses on a horizontal pile slice that is constrained to mobilise a plane strain failure mechanism, as illustrated in Figure 2.4, the scaling law is proved

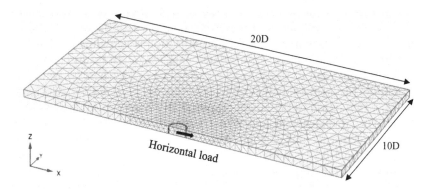

Figure 2.4 Illustration of the finite element model used to derive the *p-y* spring for a horizontal pile slice which is constrained to a plane strain mechanism (Zhang & Andersen, 2017).

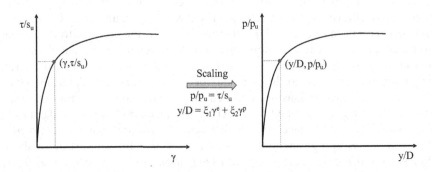

Figure 2.5 Schematic illustration of the *p-y* framework proposed by Zhang & Andersen (2017).

to be general and works for a wide range of stress-strain responses. The model is schematically illustrated in Figure 2.5. For a point on the normalised *p-y* curve with a mobilisation of p/p_u, the normalised lateral displacement y/D can be scaled from the strain evaluated at the same mobilisation ($\tau/s_u = p/p_u$) on the normalised stress-strain curve. ξ_1 and ξ_2 are two scaling coefficients: $\xi_1 = 2.8$ and $\xi_2 = 1.35 + 0.25\alpha$, where α is the interface roughness factor.

Jeanjean et al. (2017) performed an independent finite element study and proposed a similar scaling relationship. They concluded that the two scaling laws produce practically identical *p-y* curves.

Referring to Eq. 2.3, the lateral bearing capacity factor N_p for a flow-around mechanism is taken to be a linear function of the interface roughness factor α according to Randolph & Houlsby (1984):

$$N_p = 9 + 3\alpha \tag{2.8}$$

2.4.2   Reflection on API/Matlock (1970) *p-y* model

As already discussed in Section 2.2.1, the API/Matlock model expresses the lateral resistance mobilisation (p/p_u) as a power law relation with the normalised lateral displacement. The displacement used for normalisation is y_c, defined as lateral displacement at 50% mobilisation, and is calculated as $2.5 \cdot \varepsilon_{50} \cdot D$. In terms of shear strain in a UU test, it is equivalent to $y_c = 2.5 \cdot (2/3 \cdot \gamma_{50}) \cdot D = 1.67 \cdot \gamma_{50} D$, where γ_{50} is the shear strain corresponding to ε_{50} (in an undrained triaxial test, $\gamma = \varepsilon_a - \varepsilon_r = 1.5 \cdot \varepsilon_a$. Correspondingly, $\varepsilon_{50} = 2/3 \cdot \gamma_{50}$). In relation to the model proposed and Zhang & Andersen (2017), assuming the "elastic" displacement component (similarly the elastic shear strain) is negligible, the equivalent y_c can be calculated as $y_c = \xi_2 \cdot \gamma_{50} \cdot D$. The Zhang & Andersen model suggests $\xi_2 = 1.35 + 0.25\alpha$. With a fully rough interface, $\xi_2 = 1.6$, which is very close to 1.67 as implied by the Matlock *p-y* model.

It is noticed that the Matlock *p-y* expression (Eq. 2.1) is similar to the strength mobilisation framework proposed by Vardanega & Bolton (2011) for clay, which they argue that for most clays, the stress-strain response in shearing in the moderate stress range can be expressed reasonably by a power law:

$$\frac{\tau}{s_u} = 0.5\left(\frac{\gamma}{\gamma_{M=2}}\right)^b \quad \text{for } 0.2 \le \tau/s_u \le 0.8 \tag{2.9}$$

where $\gamma_{M=2}$ is the so-called mobilisation strain, which refers to shear strain at 50% strength mobilisation. Therefore, $\gamma_{M=2} = \gamma_{50}$. Parameter b is a material constant and is within a range of 0.3–1.2 with a mean value of 0.6. Vardanega et al. (2012) further found that $\gamma_{M=2}$ value increases with OCR for Kaolin clay, which leads to increasingly more ductile response. The data for Kaolin clay also suggest a slight increase of b value with OCR.

Essentially, the Matlock (1970) p-y model can be understood as a similar model to the Zhang & Andersen model that scales the soil stress-strain response to the p-y response, although the exact format of expressions used to describe the soil stress-strain responses is different. The scaling coefficient embedded in the definition of y_c for the Matlock model is even found to be similar to the Zhang & Andersen model for a rough pile-soil interface, as discussed previously. However, the Matlock model implies a constant b value of 0.33 for the power law description of the soil stress-strain response, which is probably appropriate for the Sabine River slightly over-consolidated clay on which the model was calibrated against. However, for other clays, when a different b value is required to provide a reasonable fit to the soil stress-strain response, the Matlock p-y model could result in inappropriate p-y curves.

2.4.3 Extension to cyclic loading

Combining the concept of equivalent number of cycles N_{eq}, which is defined as the number of cycles at the current load level that would have produced the same cyclic effects (changes in strength and stiffness) as by the actual previous cyclic load history, Zhang et al. (2016b, 2017) extended the monotonic framework proposed by Zhang & Andersen (2017) to the cyclic p-y response of piles. On a single pile element level, it is postulated that the cyclic effects due to lateral loading of a pile element under pressure p_{cy} for N number of cycles are analogous to the shearing of a DSS soil element under stress τ_{cy} for N number of cycles if $p_{cy}/p_u = \tau_{cy}/s_u$, as illustrated in Figure 2.6. By assuming so, a single N_{eq} can therefore be evaluated by the strain accumulation procedure

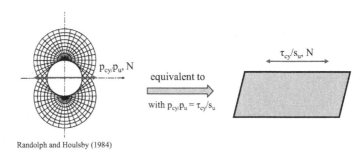

Figure 2.6 Analogy between loading of a pile element and shearing of a DSS soil element (Zhang et al., 2016b).

applicable for a DSS test, as described by Andersen (2015). For a pile element, if the previous loading history is equivalent to N_{eq} number of cycles at the current mobilisation level, the soil stress-strain response corresponding to N_{eq} can be derived from the cyclic strain contour diagram, which is established from cyclic soil element testing. By using the same scaling procedure used for monotonic p-y curves (i.e., Zhang & Andersen, 2017), the p-y curves for calculating the pile responses under the current cyclic loading can be derived.

Figure 2.7 provides a schematic illustration of the concept. The figure illustrates a pile element with a combination of average and cyclic mobilisations equal to 0.2 and 0.4 of the static capacity, respectively, and a $N_{eq} = 10$. Note that the cyclic mobilisation is defined as the mobilised soil pressure under the cyclic component of the pile load normalised by the ultimate static bearing pressure for the pile element in consideration. Similar definition applies to the average mobilisation level. The stress strain curves for average and cyclic components of loading can then be established by drawing a horizontal cross-section at $\tau_{cy}/s_u = 0.4$ and a vertical cross-section at $\tau_a/s_u = 0.2$, respectively. The cross-section lines intersect with the average or the cyclic strain contour lines, and the intersection points form the stress-strain curves.

Using these p-y curves, the pile response under the current load can be evaluated. Note that the total pile response (displacements and forces) is the sum of responses under average and cyclic components of the pile load. With the above assumptions, the pile response under a pile head load history can be analysed within the framework of a conventional beam-column p-y model, as schematically illustrated by Figure 2.8.

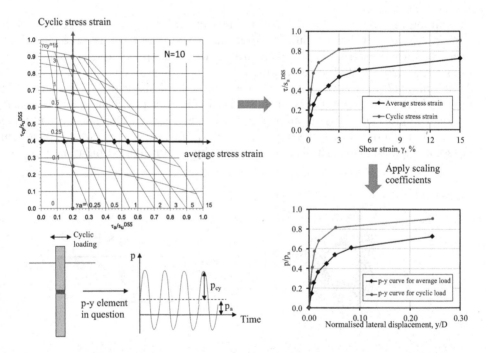

Figure 2.7 Schematic illustration on derivation of p-y curves for average and cyclic components of loading. (Modified from Zhang et al. (2017).)

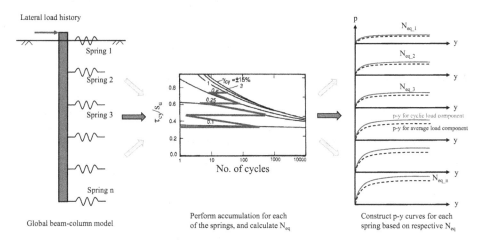

Figure 2.8 Schematic illustration of the general procedure to analyse an entire pile response (Zhang et al., 2017).

The design pile head load history is first sorted into load parcels with constant average and cyclic loads. The load history is then analysed in a parcel-by-parcel manner. The key is to keep track of the loading history for each of the *p-y* elements. By updating the global equilibrium and *p-y* springs at the beginning and at the end of each load parcel, the evolution of the pile response during the load history can be calculated. This procedure is explained in detail in Zhang et al. (2017) and implemented in a computer program called NGI-PILE.

2.4.4 Validation

The monotonic and cyclic *p-y* framework described above has been validated extensively by numerical analyses as well as by back-analysing physical model tests and field pile load tests. Jeanjean et al. (2017) presented a comprehensive validation of the monotonic framework through back-analysis of field pile load tests. Zhang et al. (2016b, 2017) presented validation of the cyclic framework by numerical analyses, firstly at a single pile element level (i.e., a horizontal pile slice) to verify the analogy illustrated in Figure 2.6 and then for a complete pile to verify the calculation procedure and redistribution of loads along the pile under cyclic loading. These were performed in finite element analyses with the UDCAM cyclic accumulation soil model (Jostad et al., 2014), which uses cyclic contour diagrams established from soil element tests as input. In the analyses, the stress history at each integration point of the entire finite element soil domain was kept track of, and the soil strength and stiffness at each point are constantly updated. Zhang et al. (2020a, b) present a comprehensive validation of the cyclic framework through back-analysis of four sets of pile load tests (centrifuge/field), covering different soil conditions (plasticity indices and overconsolidation ratios).

2.4.5 Simplified cyclic p-y curves for design of offshore structures

For a specific design case, pile analysis can be performed by considering site-specific cyclic soil properties and cyclic design loading history using the procedure described in Section 2.4.3. However, this presents a quite high demand on the competence of the design engineer as well as the input parameters. The design engineer must codify the procedure into a computer program. Furthermore, the cyclic contour diagrams and the cyclic pile head load history must be prepared beforehand, which may not always be possible. To address this challenge, Zhang et al. (2017) proposed a simplified procedure to evaluate the cyclic p-y curves, which is schematically illustrated in Figure 2.9.

The simplified cyclic p-y procedure involves four steps to calculate the pile response under peak storm load:

Step 1: Set-up the beam-column model with the monotonic p-y springs, which are derived by scaling the monotonic DSS stress-strain curves.

Step 2: Analyse the pile response under the peak storm load using the monotonic p-y springs. Calculate the spring mobilisations [*Mob*] and normalised depth [z/z_{rot}]. The mobilisation is defined as mobilised soil resistance (in pressure) divided by the ultimate static bearing pressure ($N_p \cdot s_u$). The rotation depth (z_{rot}) is defined as the first depth along the pile where the deflection reverses direction (Figure 2.9).

Step 3: Go into the look-up chart and find the p and y modifiers applicable to each of the p-y springs based on respective mobilisation level and normalised depth.

Step 4: Apply the modifiers to the monotonic p-y curves and recalculate the final pile response under the peak storm load. The equations shown in Figure 2.9 and the derivation of p, y modifiers will be explained below. Note that the modified p-y curves are not necessarily softer than the monotonic curves.

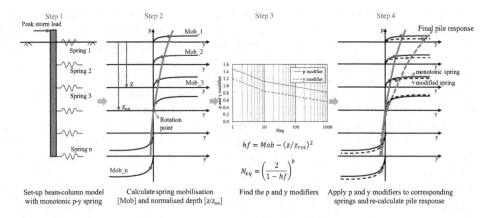

$$hf = Mob - (z/z_{rot})^2$$

$$N_{eq} = \left(\frac{2}{1-hf}\right)^b$$

Set-up beam-column model with monotonic p-y spring | Calculate spring mobilisation [Mob] and normalised depth [z/z_{rot}] | Find the p and y modifiers | Apply p and y modifiers to corresponding springs and re-calculate pile response

Figure 2.9 Simplified procedure to calculate cyclic pile response (Zhang et al., 2017).

The essential basis of the proposed simplified method is the assumption of

- a representative make-up of the storm load history for a specific type of structure in a specific geographic region, e.g., jacket pile in Gulf of Mexico (GoM);
- representative cyclic soil properties for a particular type of soil, e.g., normally consolidated clay in GoM.

The core of the simplified model is the relation between equivalent number of cycles (N_{eq}), the normalised depth (z/z_{rot}) and the mobilisation level (*Mob*) of the *p-y* springs. This relation needs to be established through parametric analyses beforehand using the full calculation procedure described above.

Zhang et al. (2019) presented recommendations of simplified cyclic *p-y* curves for practicing engineers following the simplified procedure described above. The simplified cyclic *p-y* curves allow for consideration of site-specific soil properties (shear strength and stress-strain response in monotonic and cyclic testing) and the magnitude of the environmental loading relative to the chosen pile dimensions. The approach undertaken was to apply the full cyclic procedure, as implemented in the beam-column design tool NGI-PILE, to carry out parametric analyses. For a specific type of pile foundation (e.g. piles for jacket structure) in a specific geographic region (e.g. GoM), representative make-up of the storm load history and cyclic soil properties are assumed. The parametric analyses investigate the effect of the global load scaling factor, the pile geometry (diameter, embedded length, and wall thickness), interface roughness, and pile head condition (free or constrained) on the distribution of N_{eq} at the end of the loading history. Based on the results from the parametric analyses, empirical relations between N_{eq}, the normalised depth z/z_{rot}, and spring mobilisation *Mob* are proposed. N_{eq} can be then used to find the respective *p* and *y* modifiers that account for the cyclic effect for soil springs at different depths (Figure 2.10). In Zhang et al. (2019), parametric analyses were carried out for the following:

1 Jacket pile in GoM clay
2 Spar anchor pile in GoM clay
3 Jacket pile in North Sea soft clay
4 Jacket pile in North Sea stiff clay

Table 2.1 summaries the recommendations for constructing simplified cyclic *p-y* curves based on the parametric analyses using the rigorous cyclic calculation procedure for four different applications. These simplified cyclic *p-y* curves are to be used in the four-step design process illustrated in Figure 2.9.

Zhang et al. (2019) demonstrated that for the jacket piles, the N_{eq} at the end of the cyclic loading history is typically less than 25. The maximum value is obtained at the surface, and it reduces rapidly with depth. This typically leads to a cyclic *p-y* curve that is stronger and stiffer than the monotonic *p-y* curve. On the soil-structure interaction level, this means that by considering of the cyclic effects, a stiffer global load-displacement response is obtained. The cyclic degradation, which does indeed occur under cyclic loading, is over-shadowed by rate effects because the reference static soil

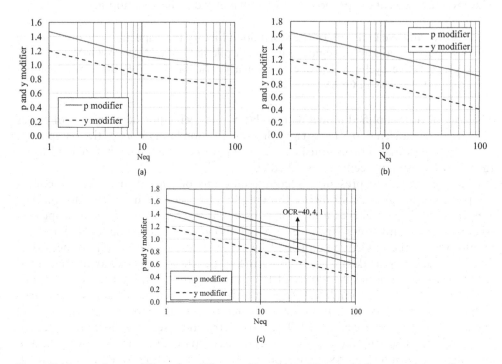

Figure 2.10 p and y modifiers for different N_{eq}: (a) wellhead platform on normally con-
solidated GoM clay; (b) wellhead platform on normally consolidated North
Sea soft clay; (c) wellhead platform on over-consolidated North Sea stiff clay
(Zhang et al., 2019).

properties are measured in standard laboratory DSS tests at a strain rate (typically
5%/hr) three orders of magnitude slower than that of the cyclic tests.

The above recommendations are a paradigm shift from the current method for cyclic
curves in the design code, e.g., API (2014), which recommends a cut-off at 72% of the
ultimate value of the monotonic p-y curve in the flow-around region and a post-peak
softening in the shallow wedge region, as illustrated in Figure 2.2. The Matlock cyclic
p-y model, which is the basis of the API recommendation, was primarily based on a
handful field tests carried out at the Sabine River test site in the 1950s (Matlock, 1970).
The model pile had a diameter of 0.324 m and a penetration depth of 12.8 m. There are
several important differences that have contributed to the significantly greater degradation suggested by the Matlock cyclic p-y model:

1   The monotonic pile tests at Sabine River were carried out under relatively fast
    loading, relative to the strain rate typically experienced in the laboratory soil ele-
    ment tests which measure the soil strength to be used in design, i.e., the reference
    point for cyclic degradation between Matlock p-y model and the current recom-
    mendation is different.
2   In the cyclic pile tests at Sabine River, four packages of constant amplitude cy-
    clic loading were consecutively applied to the pile head. The cyclic amplitude in-
    creased in ascending order with the package number. At the highest cyclic load

Table 2.1 Summary of recommendations in different applications

Application	Recommendation
Jacket pile in normally consolidated GoM clay	$hf = mob - (z/z_{rot})^2$ $N_{eq} = \left(\dfrac{2}{1-hf}\right)^{1.0} \leq 25$ Relation between N_{eq} with p and y modifiers as per Figure 2.10a
Spar anchor pile in normally consolidated GoM clay	$N_{eq} = 10$ along the entire pile length p modifier = 1.1 y modifier = 1.0 Alternatively, use monotonic p-y curves if cyclic effects to be neglected
Jacket pile in North Sea soft clay	$hf = mob - (z/z_{rot})^2$ $N_{eq} = \left(\dfrac{2}{1-hf}\right)^{1.25} \leq 25$ Relation between N_{eq} with p and y modifiers as per Figure 2.10b
Jacket pile in North Sea stiff clay	$hf = mob - (z/z_{rot})^2$ $N_{eq} = \left(\dfrac{2}{1-hf}\right)^{2.25} \leq 25$ Relation between N_{eq} with p and y modifiers as per Figure 2.10c

level, the peak horizontal load would have resulted in more than 10% pile diameter deflection at the ground surface if it was applied in a monotonic push-over test. Two hundred load cycles were applied in each load package, except the first package, where 400 cycles were applied. This differs significantly from the actual load history encountered by a jacket pile. Most of the load cycles are at very small load levels which do not induce much degradation of soil properties around the pile and there are only a small number of medium to large load cycles. The parametric analyses suggest that N_{eq} is typically limited to 25.

3 In the Sabine River tests, due to the presence of a surficial crust, a tension gap was developed. As the cyclic loading went on, the gap increased in size. Upon loading, soil resistance was not mobilised until the gap between the pile and the soil was closed. This phenomenon was concluded to be one of the main reasons that caused the degradation of global soil-pile interaction (Matlock & Tucker, 1961). However, this process is judged very unlikely to occur for jacket piles, which typically has a mudmat on the soil surface that is designed to support the jacket structure prior to piling. It acts as a seal and cuts off the seepage path for water getting into the pile-soil interface during cyclic loading, and soil is forced to adhere to the movement of the pile. The gapping mechanism, which led to the "apparent degradation" of soil-pile interaction stiffness in the Sabine River tests, is considered unlikely to occur in real life, even for piles in stiff clay.

2.5   SITE-SPECIFIC *P-Y* CURVES FOR MONOPILES

2.5.1   Difference in soil mechanism from slender piles

Large diameter steel pipe pile foundations, typically known as monopiles, are the predominant foundation solution for bottom fixed offshore wind turbines. Today's monopiles have a diameter (D) of 4–8 m and a penetration length (L) of 25–40 m. This leads to a penetration to diameter ratio typically around or less than 6. Monopile foundations for offshore wind turbines are predominantly loaded by a horizontal force and an over-turning moment. Similar to slender piles used for supporting offshore oil and gas platforms, design of the monopile foundation is typically performed using the subgrade reaction method. Monopiles are conventionally designed using *p-y* models borrowed from the offshore oil and gas industry. However, the question arises whether this is appropriate as monopiles differ in a few ways from slender pile foundations used in the oil and gas industry. Compared to piles used for supporting offshore jacket platforms, which have a typical pile diameter of 1.5–2.5 m and a penetration length of 40–100 m, monopiles are considerably stiffer due to smaller L/D ratio. As illustrated in Figure 2.11, under lateral loading, the soil around a monopile reacts predominantly in a wedge failure along the upper part of the pile and a rotational mechanism along the lower part and around the pile tip. In addition, the soil resistance at the pile tip may become an important component of soil reaction to lateral and overturning moment loading due to the large pile diameter. Furthermore, the vertical component of the pile-soil skin friction mobilised along a monopile may become a meaningful component of resistance to overturning moment loading. For a conventional slender pile,

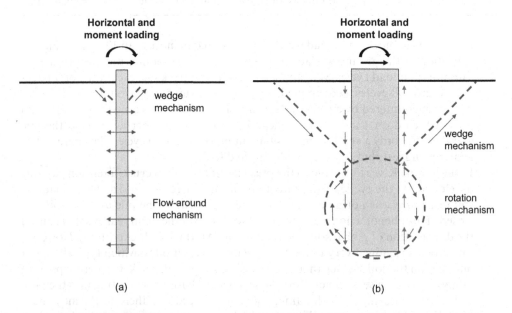

Figure 2.11 Comparison of failure mechanisms between (a) a slender pile and (b) a monopile.

the contribution of the pile tip is negligible, and sufficiently accurate response can be obtained by consideration of the lateral load-displacement springs (*p-y* curves) only.

Due to the differences mentioned above, the offshore wind industry has recognised that the *p-y* method used for design of conventional slender piles cannot be directly used for design of monopile foundations. The DNVGL standard for support structures for wind turbines DNVGL-ST-0126 recommends *p-y* curves to be calibrated by finite element analyses (DNVGL, 2018). A major joint industry research project (PISA) performed a comprehensive set of large-scale field testing of model monopile foundations in both sand and clay. Calibrated by the field experiments, finite element analyses were performed to develop new analysis model for monopiles (Byrne et al., 2017). A key feature of the new analysis model is the inclusion of the side shear rotational resistance and pile tip resistance, in addition to an updated lateral *p-y* formulation, as illustrated by Figure 2.12. However, as the model is calibrated against limited pile load test in a certain soil condition, site-specific calibration against finite element analyses is needed when applying the PISA model to wind farm locations. Zhang & Andersen (2019) and Fu et al. (2020) have further extended the concept of linking soil-pile interaction curves with the stress-strain response of the soil and developed site-specific models for monopile foundations, considering different soil types and different resistance components (i.e., *p-y*, base shear and rotational resistance due to vertical shaft friction). The subsections below will detail these developments.

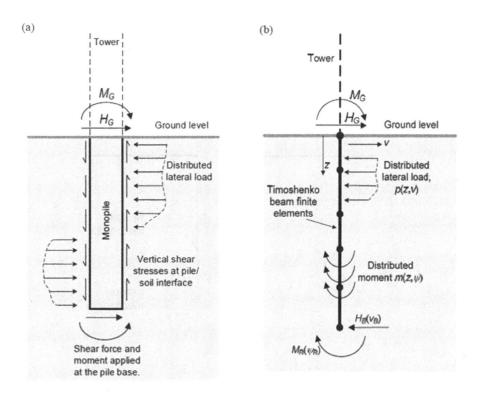

Figure 2.12 The PISA 1D analysis model: (a) components of soil resistance and (b) representation by soil reaction springs (Byrne et al., 2017).

2.5.2   Site-specific *p*-*y* model for the single-wedge mechanism in stiff clay

In a stiff over-consolidated clay site, a tension gap opens on the back side of a laterally loaded monopile. A conical wedge is mobilised in front of the pile, and a rotational mechanism is mobilised below the wedge. Zhang & Andersen (2019) performed a numerical study to investigate the *p*-*y* curves mobilised by the passive soil wedge and its link to the stress-strain response of the soil (Figure 2.13). Based on an assumption of isotropic strength and stiffness behaviour, they proposed a scaling relation of the *p*-*y* response for a passive wedge mechanism from the stress-strain response of the soil, as summarised in Table 2.2.

2.5.3   Site-specific *p*-*y* model for the double-wedge mechanism in soft clay

In clay sites which consist of normally consolidated to lightly over-consolidated soft clays or where a soft layer is present near the surface of the seabed, the surface material follows the movement of the pile, thus preventing the development of tension crack. In such soil conditions, the soil mechanism of a monopile consists of double conical wedges and a rational mechanism below, as illustrated by Figure 2.14a. Hong et al. (2017) revealed by centrifuge PIV tests that a flow-around region may exist between the top wedge mechanism and the bottom rotational mechanism for monopiles with higher *L*/*D* ratios.

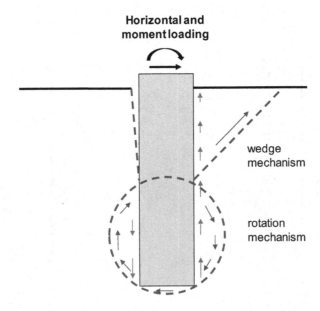

Figure 2.13 Soil mechanism of a monopile foundation under lateral and moment loading in stiff clay.

Table 2.2 Summary of equations for scaling p-y curve for a passive wedge mechanism from stress-strain response of the soil (Zhang & Andersen, 2019)

$$\frac{p}{p_u} = 1 - 2^{\left[-\left(y^p / y_{50}^p\right)^n\right]}$$

$$n = 0.55 - 0.05 (z/D) \geq 0.325$$

$$y = y^e + y^p; \quad y_{50} = y_{50}^e + y_{50}^p$$

$$\alpha = 1 \ (\text{rough}): \left(y_{50}^e / D\right) \cdot \left(G_{max}/s_u\right) = 0.6 + 0.28\left(\frac{z}{D}\right) - 0.029(z/D)^2$$

$$\alpha = 0 \ (\text{smooth}): \left(y_{50}^e / D\right) \cdot \left(G_{max}/s_u\right) = 0.6 + 0.41\left(\frac{z}{D}\right) - 0.039(z/D)^2$$

$$y^e / D = 2\left(y_{50}^e / D\right)\left(p/p_u\right)$$

$$y_{50}^p / D = \left[a + b(z/D)\right] \gamma_f^p$$

$$a = 0.01 + 0.015\alpha; \ b = 0.10 - 0.06\alpha$$

$$p_u = \left(N_{p0} s_u + \gamma' z\right) D$$

$$N_{p0} = N_1 - \left(N_1 - N_2\right)\left[1 - \left(\frac{z/D}{d}\right)^{0.6}\right]^{1.35} - (1-\alpha) \leq N_{pd}$$

$N_1 = 11.94; N_2 = 3.22$
$d = 14.5; N_{pd} = 9.14 + 2.8\alpha$
More general recommendation for N_{p0} factor in linearly increasing soil profiles can be found in Zhang et al. (2016a)

For the double-wedge mechanism, Fu et al. (2020) recommend that the shape of the p-y curves can be scaled from the stress-strain response using the same method and scaling coefficients as for the flow-around mechanism, despite the difference in the mechanism. This is primarily because soil on both sides of the monopile is mobilised, which is similar to the flow-around mechanism. However, the reduced ultimate capacity of the double-wedge mechanism as compared to the flow-around mechanism should be considered according to the recommendation by Zhang et al. (2016a). Fu et al. (2020) demonstrate that good modelling results are obtained following this approach.

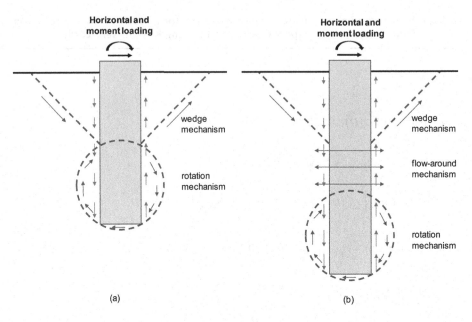

Figure 2.14 Soil mechanisms of monopile foundation in soft clay sites or where a surface soft clay is present: (a) without flow-around mechanism and (b) with flow-around mechanism (Fu et al., 2020).

2.5.4 Site-specific *s-y* model for pile tip base shear

An important learning from the PISA project (Byrne et al., 2017) is that for monopiles, it is essential to account for the soil resistance acting on the pile tip and the pile shaft in addition to the distributed lateral resistance along the pile for monopiles due to the smaller L/D ratio compared to slender piles used for supporting offshore oil and gas platforms. Zhang & Andersen (2019) examined the soil mechanism at the pile tip and found that it is sufficient to simplify the rotation scoop mechanism below the pile tip by a horizontal base shear (as illustrated in Figure 2.15). Even for a rotation centre as low as one pile diameter above the pile tip, the error involved in this simplification is less than 16% of the total soil resistance contributed by the pile tip. This simplification means that the moment resistance provided by the pile tip is negligible. Based on this simplification, Zhang & Andersen (2019) proposed site-specific base shear *s-y* spring, which is scaled from the DSS stress-strain response of the soil at the pile tip. The scaling model is similar to the flow-around *p-y* model, as illustrated by Figure 2.15. The scaling coefficients are $\xi_1 = 0.3$; $\xi_2 = 0.12$. The ultimate base shear resistance (s_{ult}, unit: force) is calculated by

$$s_{ult} = 0.25\pi D^2 s_u \tag{2.10}$$

Wang et al. (2020) proposed an alternative approach to account for the soil resistance contributed by the soil below the rotational point (RP). All the soil resistance below

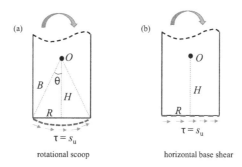

Figure 2.15 Simplification of the soil mechanism at pile tip (Zhang & Andersen, 2019).

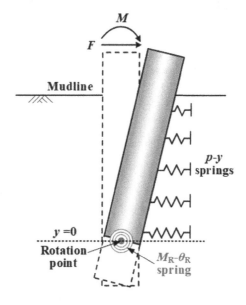

Figure 2.16 Lumped M_R-θ_R spring to capture the soil resistance below the rotation point
(Wang et al., 2020).

the RP, including the base shear and base moment, is lumped into a concentrated rotational spring, whilst the horizontal degree-of-freedom at the RP is constrained. In
application of this approach for monopile design, the soil resistance above the RP is
represented by distributed *p-y* springs, and the soil resistance below the RP is represented by the concentrated rotational spring, as illustrated in Figure 2.16. The depth
of the RP is determined by a trial analysis, which is generally not sensitive to the *p-y*
formulation (Zhang & Andersen, 2019). In Wang et al. (2020), the ultimate strength
and shape of the rotational spring were determined by finite element analyses, and
empirical formulations were developed from curve-fitting of the numerical results.

Lai et al. (2020) presented a further development of the model by means of finite
element parametric analyses and analytical study. Following the philosophy of linking

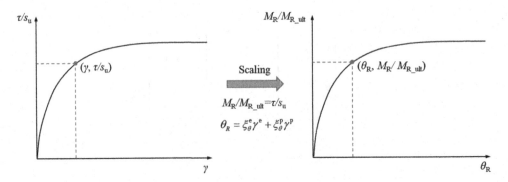

Figure 2.17 Scaling of the lumped M_R-θ_R spring from the stress-strain curve of the soil (Lai et al., 2020).

the pile-soil interaction response to the stress-strain response of the soil. Lai et al. (2020) proposed the following scaling framework to derive site-specific M_R-θ_R rotational spring, as illustrated in Figure 2.17:

$$\frac{M_R}{M_{R\_ult}} = \frac{\tau}{s_u} \tag{2.11}$$

$$\theta_R = \xi_\theta^e \gamma^e + \xi_\theta^p \gamma^p \tag{2.12}$$

$$\gamma^e = \frac{\tau}{G_{max}}; \tag{2.13}$$

$$\gamma^p = \gamma - \gamma^e \tag{2.14}$$

$$\xi_\theta^e = 0.63 + 0.32\left(\frac{H}{D}\right) \tag{2.15}$$

$$\xi_\theta^p = 0.19\left(\frac{H}{D}\right) + 0.34 \tag{2.16}$$

The ultimate strength of the lumped rotational spring is derived from upper bound analysis of a rotational scoop mechanism about the RP:

$$M_{R\_ult} = \left(\frac{1}{6}\pi D^3 s_{u0} + \pi s_{u0} DH^2\right) + k\left(\frac{1}{2}D^2 + 2H^2\right)^2\left[\frac{3}{8}t + \frac{1}{4}\sin(2t) + \frac{1}{32}\sin(4t)\right]$$
$$+ 0.73\left(\frac{2\pi}{3}s_{u0}H^3 + kH^4\right) \tag{2.17}$$

where $t = \arcsin\left(\dfrac{D}{\sqrt{D^2 + 4H^2}}\right)$

2.5.5 *m-θ* model for moment resistance due to vertical pile shaft friction

As found out by the PISA project, the rotational moment resistance due to the vertical friction acting in opposite directions along the pile shaft is an important component of soil resistance for short monopiles. Fu et al. (2020) presented a model that links the distributed moment spring (*m-θ*) with the DSS stress-strain response of the soil surrounding the monopile. The derivation of the scaling relationship is bridged by the relationship between the vertical shaft friction vs. relative pile-soil displacement (*t-z*) and the stress-strain response of the soil.

Based on the deformation mechanism proposed by Randolph & Wroth (1978), as illustrated in Figure 2.18, the *t-z* response can be solved analytically. The relative vertical soil-pile displacement (z) at a uniform interface shear stress (τ_0) is simply an integration of shear strain in the radial direction. Assuming that the stress-strain response is identical in the radial direction (i.e., neglecting the installation effect), and the stress-strain response described by the NGI-ADP model (Grimstad et al., 2012), a scaling relationship is derived between the stress-strain response and the *t-z* response:

$$\frac{z}{D} = \frac{z^e}{D} + \frac{z^p}{D} = \xi^e \gamma^e + \xi^p \gamma^p \qquad (2.18)$$

where γ^e and γ^p are the elastic and the plastic shear strain components, respectively, and can be calculated according to Eqs. 2.13 and 2.14, respectively. The elastic scaling factor $\xi^e = 1.15$. The plastic scaling factor ξ^p is not a constant but found to vary with the shear stress mobilisation, as shown in Figure 2.19. A constant value of 0.45 is recommended for simplicity.

Assuming that the stiffness for mobilisation of an interface shear stress due to rotation of the pile cross-section is similar to a uniform vertical displacement of the pile

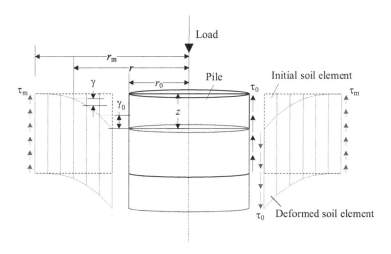

Figure 2.18 Deformation pattern and shear stress-strain distribution assumed for developing theoretical *t-z* response of the axially loaded pile (Fu et al., 2020).

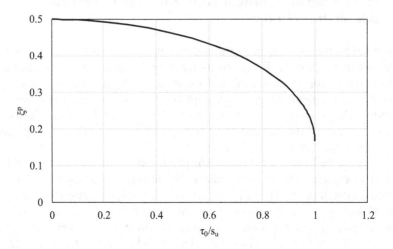

Figure 2.19 Variation of the plastic scaling coefficient with the mobilisation level (Fu et al., 2020).

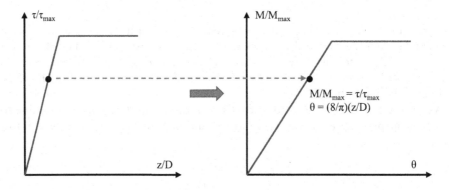

Figure 2.20 Scaling relationship between *t-z* spring and *m-θ* spring (Fu et al., 2020).

cross-section, the scaling between the normalised *t-z* response and the *m-θ* response is found to be a constant factor of $8/\pi$, as illustrated in Figure 2.20. This scaling relation is general to linear or non-linear *t-z* response. Using the relationship already established between *t-z* response and the stress-strain response, the scaling relation between the *m-θ* spring and the stress-strain response is finally established, as illustrated in Figure 2.21. The ultimate moment capacity per unit pile length is derived theoretically to be $D^2 \alpha s_u$ (α is the interface roughness factor). The scaling factors are $\zeta^e = 1.15$ and $\zeta^p = 0.45$.

Sensitivity analyses performed by Fu et al. (2020) suggest that the importance of *m-θ* spring is similar to the *s-y* spring for monopiles with *L/D* ratio equal to or less than 5. Neglecting base shear and the side shear can lead to significant over-estimation of the pile deflection and rotation.

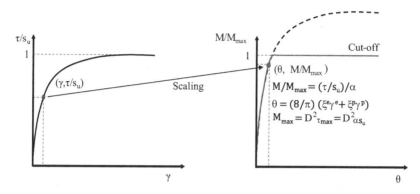

Figure 2.21 Scaling of distributed m-θ spring from the soil stress-strain response (Fu et al., 2020).

2.6 SOIL DAMPING FOR LATERALLY LOADED PILE FOUNDATIONS

As mentioned in the introduction, fatigue assessment is an important design aspect for offshore structures. The modelling of the soil-foundation interaction is critical as it influences the dynamic response of the structure and the environmental excitation, hence the loads on the structural members. Modelling of the soil-foundation interaction essentially includes two aspects: (i) soil-foundation interaction stiffness and (ii) the foundation damping due to the hysteretic behaviour of the soil under cyclic environmental excitation. The soil-foundation stiffness influences the system's natural frequency and therefore determines how close or far the system's natural frequency is to the environmental excitation frequencies. This stiffness can have a major impact on the system dynamics. The soil damping, together with other sources of damping in the system, influences the dynamic amplification for a given system stiffness and environmental excitation.

The previous sections of this chapter have introduced the current state of the art for modelling the stiffness of soil-pile interaction. However, despite its widely acknowledged significance, there is limited research on the modelling of soil damping in soil-pile interaction due to lack of established methods. Carswell et al. (2015) presented a case study on the effect of soil damping on the fatigue loads of a fictitious offshore wind turbine founded on monopile under extreme storm conditions. The soil damping was evaluated by a specialised finite element program and incorporated into the system as a lumped equivalent dashpot at the mudline. Inclusion of the soil damping is found to decrease the maximum and standard deviation of mudline moment by 7%–9%. Page et al. (2019) presented a case study where predicted and measured dynamic responses of an actual wind turbine founded on monopile in the North Sea are compared. The soil-pile interaction was modelled by a macrofoundation element formulated in multisurface plasticity theory. The model captures the different soil-pile stiffnesses in initial loading, unloading, and reloading, hence the energy dissipated (i.e., the soil damping) in hysteresis loops. It is found that by inclusion of the soil damping, the equivalent fatigue damage load is reduced by approximately 10% under idling condition at the base of the structure

The current design guidelines and standards, such as DNVGL-RP-C212 (DN-VGL, 2017), DNVGL-ST-0126 (DNVGL, 2018), and ISO 19901-4 (ISO, 2016), all stress the need to account for soil damping in modelling soil-structure interaction. However, there is little guidance on the evaluation of the soil damping. Kaynia & Andersen (2015) have shown how soil-foundation damping due to hysteretic soil response can be computed from a non-linear finite element analysis. They have also proposed a simple model to adapt the non-linear foundation response (backbone curve) to reproduce the computed soil-foundation damping during non-linear cyclic response. Use of this model has proven to provide a realistic representation of the non-linear foundation response in the response of Troll A platform (Kaynia et al., 2015). Alternative solutions preserving the shape of the backbone have been proposed by Hoogeveen et al. (2018) and Kaynia (2019). Zhang et al. (2021) developed a simplified model for quick estimate of foundation damping for soil-pile interaction under lateral loading. The model tackles the soil damping at a single p-y spring level, which can be either included as an equivalent dashpot at each spring level or a lumped dashpot at the mudline level for modelling the entire soil-pile interaction. The model follows the same philosophy that the soil-structure interaction is fundamentally linked to the soil response at the element level. Based on a comprehensive finite element parametric analysis, Zhang et al. (2021) found that the normalised lateral deflection (y/D) vs. damping ratio (d) for a horizontal pile slice engaged in a flow-around mechanism can be scaled from the cyclic shear strain (γ_{cy}) vs. damping ratio (d) relationship by a scaling factor (Figure 2.22). The scaling factor on the cyclic shear strain has a constant value of 3.6. The cyclic shear strain vs. damping ratio relation for the soil can be measured in the laboratory by resonant column tests and cyclic DSS tests or correlated from empirical databases, such as in Darendeli (2001). Once the site-specific γ_{cy} vs. d relation is established, the soil damping at each p-y spring level can be determined. Zhang et al. (2021) discussed methods on how the soil damping can be incorporated into structural analyses.

2.7 SUMMARY

Pile foundations supporting offshore structures (jackets, wind turbines, etc.) are often subjected to large lateral loading. In design practice, the pile-soil interaction is commonly dealt with using the subgrade reaction method (e.g., soil springs). This chapter presents the state of the art on the development of soil springs and damping for laterally loaded piles under static and cyclic loading for slender piles and for short stubby monopiles. It demonstrates that linking the soil-pile interaction to the stress-strain and damping response of the soil measured at the element level is an extremely powerful approach. By proper consideration of the pertinent soil failure mechanisms (e.g. wedge failure, flow-around failure, base shear, etc.) and the type of loading (static and cyclic), this method can produce site-specific soil springs that achieve good modelling results. This is a step change compared to existing design practice where soil-pile interaction curves are often based on empirical formulations curve-fitted from typically limited pile load tests which suffer uncertainties due to the differences in soil and loading conditions from the conditions examined in the pile load tests.

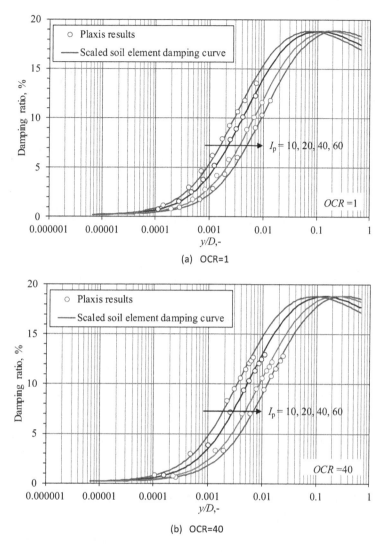

Figure 2.22 Example results: soil damping ratio (*d*) vs. normalised lateral displacement (*y/D*) for a laterally loaded pile slice and comparison with damping ratio (*d*) vs. scaled cyclic shear strain (γ_{cy}, multiplied by 3.6) (Zhang et al., 2021).

ACKNOWLEDGEMENTS

I would like to acknowledge the contribution by my former colleagues at the Norwegian Geotechnical Institute and collaborators and students on the various works summarised in this chapter. This includes Mr. Knut H. Andersen, Dr Philippe Jeanjean, Dr Lizhong Wang, Dr Yi Hong, Dr Dengfeng Fu, Mr. Yongqing Lai, and Mr. Kristoffer Aamodt.

REFERENCES

American Petroleum Institute (2014) API recommended practice 2GEO/ISO 19901-4. Geotechnical and foundation design considerations, 1st Edition, April 2011, Addendum 1, October 2014.

Andersen, K.H. (2015) Cyclic soil parameters for offshore foundation design. In Meyer (ed.) *Frontiers in Offshore Geotechnics III*, Taylor & Francis Group, London, pp. 5–82.

Bransby, M.F. (1999) Selection of p-y curves for the design of single laterally loaded piles. *International Journal for Numerical and Analytical Methods in Geomechanics* 23, 1909–1926.

Byrne, B.W., McAdam, R.A., Burd, H.J., Houlsby, G.T., Martin, C.M., Beuckelaers, W.J.A.P., Zdravkovic, L., Taborda, D.M.G., Potts, D.M., Jardine, R.J., Ushev, E., Liu, T., Abadias, D., Gavin, K., Igoe, D., Doherty, P., Gretlund, J.S., Andrade, M.P., Wood, A.M., Schroeder, F.C., Turner, S. & Plummer, M.A.L. (2017) PISA—New design methods for offshore wind turbine monopiles. *Proc. OSIG SUT Conference*, 2017, London.

Carswell, W., Johansson, J., Løvholt, F., Arwade, S.R., Madshus, C., DeGroot, D.J. & Myers, A.T. (2015) Foundation damping and the dynamics of offshore wind turbine monopiles. *Renewable Energy* 80, 724–736.

Darendeli, M.B. (2001) Development of a new family of normalised modulus reduction curves and material damping curves. PhD thesis, The University of Texas at Austin.

DNVGL (2017) DNVGL-RP-C212. Offshore soil mechanics and geotechnical engineering. Edition August 2017.

DNVGL (2018) DNVGL-ST-0126. Support structures for wind turbines. Edition July 2018.

Erbrich, C.T., O'Neill, M.P., Clancy, P. & Randolph, M.F. (2010) Axial and lateral pile design in carbonate soils. *Proc. 2nd International Symposium on Frontiers in Offshore Geotechnics (ISFOG II)*, pp. 125–154.

Fu, D., Zhang, Y., Aamodt, K.K. & Yan, Y. (2020). A multi-spring model for monopile analysis in soft clays. *Marine Structures* 72, 102768.

Georgiadis, M. (1983) Development of p-y curves for layered soils. *Proc. Conf. on Geotechnical Practice in Offshore Engineering*, Austin, TX, NY, American Society of Civil Engineers, pp. 536–545.

Grimstad, G., Andresen, L. & Jostad, H.P. (2012) NGI-ADP—Anisotropic shear strength model for clay. *International Journal of Numerical and Analytical Methods in Geomechanics* 2012(36), 483–497.

Hong, Y., He, B., Wang, L.Z., Ng, C.W.W. & Masin, D. (2017) Cyclic lateral response and failure mechanisms of semi-rigid pile in soft clay—Centrifuge tests and numerical modelling. *Canadian Geotechnical Journal* 54, 806–824.

Hoogeveen, M., Hofstede, H. & Kaynia, A.M. (2018) Enhanced kinematic hardening model for load-dependent stiffness and damping of jack-up foundations. *Proc. ASME 2018 37th Int. Conf. on Ocean, Offshore and Arctic Engineering OMAE2018*, Paper No. OMAE2018-77285, pp. V009T10A016; 9 pages, June 17–22, 2018, Madrid.

ISO (2016) ISO 19901-4. Petroleum and natural gas industries—Specific requirements for offshore structures—Part 4—Geotechnical and foundation design considerations. Edition 2016.

Jeanjean, P. (2009). Re-assessment of p-y curves for soft clays from centrifuge testing and finite element modeling. *Offshore Technology Conference,* Houston, Texas, OTC 20158.

Jeanjean, P., Zhang, Y., Zakeri, A., Andersen, K.H., Gilbert, R. & Senanayake, A. (2017) A framework for monotonic p-y curves in clays. Keynote lecture. *Proc. OSIG SUT Conference*, London.

Jostad, H.P., Grimstad, G., Andersen, K.H., Saue, M., Shin, Y. & You, D. (2014) A FE procedure for foundation design of offshore structures—Applied to study a potential OWT monopile foundation in the Korean Western Sea. *Geotechnical Engineering Journal of the SEAGS & AGSSEA* 45(4), 63–72.

Kaynia, A.M. (2019) Seismic considerations in design of offshore wind turbines. *Soil Dynamics and Earthquake Engineering* 124, 399–407.

Kaynia, A.M. & Andersen, K.H. (2015) Development of nonlinear foundation springs for dynamic analysis of platforms. In Meyer (ed.) *Proc. Frontiers in Offshore Geotechnics III.* IS-FOG, Taylor & Francis Group, London, pp. 1067–1072.

Kaynia, A.M., Norén-Cosgriff, K., Andersen, K.H. & Tuen, K.A. (2015) Nonlinear foundation spring and calibration using measured dynamic response of structure. *Proc. ASME 2015 34th Int. Conf. on Ocean, Offshore and Arctic Engineering.* OMAE2015, paper OMAE2015-41236, May 31–June 5, St. John's, Newfoundland, Canada.

Klar, A. (2008) Upper bound for cylinder movement using "elastic" fields and its possible application to pile deformation analysis. *International Journal of Geomechanics* 8, 162–167.

Klar, A. & Osman, A.S. (2008) Load-displacement solutions for piles and shallow foundations based on deformation fields and energy conservation. *Géotechnique* 58(7), 581–589.

Lai, Y.Q., Wang, L.Z., Zhang, Y. & Hong, Y. (2020) Site-specific soil reaction model for monopiles in soft clay based on laboratory element stress-strain curves. *Ocean Engineering.* Doi: 10.1016/j.oceaneng.2020.108437.

Matlock, H. (1970) Correlations for design of laterally loaded piles in soft clays. *Proc. Offshore Tech. Conf.* OTC 1204, Houston, TX.

Matlock, H. & Tucker, R. (1961) Lateral-load tests of an instrumented pile as Sabine. Texas. Report to Shell Development Company.

Murff, J.D. & Hamilton, J.M. (1993) P-ultimate for undrained analysis of laterally loaded piles. *Journal of Geotechnical Engineering* 119(1), 91–107.

O'Neil, M.W. & Murchison, J.M. (1983) An Evaluation of p-y Relationships in Sands, prepared for the American Petroleum Institute. Report PRAC 82-41-1. Houston, University of Houston, March 1983.

Osman, A.S. & Bolton, M.D. (2005) Simple plasticity-based prediction of the undrained settlement of shallow circular foundations on clay. *Géotechnique* 55(6), 435–47.

Osman, A.S., White, D.J., Britto, A.M. & Bolton, M.D. (2007) Simple prediction of the undrained displacement of a circular surface foundation on non-linear soil. *Géotechnique* 57(9), 729–737.

Page, A.M., Nass, V., De Vaal, J.B., Eiksund, G.R. & Nygaad, T.A. (2019) Impact of foundation modelling in offshore wind turbines—Comparison between simulations and field data. *Ocean Engineering* 64, 379–400.

Poulos, H.G. & Davis, E.H. (1974). *Elastic Solutions for Soil and Rock Mechanics*, Wiley, New York.

Randolph, M.F. & Houlsby, G.T. (1984) The limiting pressure on a circular pile loaded laterally in cohesive soil. *Géotechnique* 34(4), 613–623.

Randolph, M.F. & Wroth, C.P. (1978) Analysis of deformation of vertically loaded piles. *Journal of the Geotechnical Engineering Division*, 1465–1488.

Reese, L., Cox, W. & Koop, F. (1975) Field testing and analysis of laterally loaded piles in stiff clays. *Proc. Offshore Technology Conference.* OTC 2312, Houston, TX.

Tempel, J. van der & Molenaar, D.P. (2002) Wind turbine structural dynamics—A review of the principles for modern power generation, onshore and offshore. *Wind Engineering* 26(4), 211–220.

Vardanega, P.J. & Bolton, M.D. (2011) Strength mobilization in clays and silts. *Canadian Geotechnical Journal* 48, 1485–1503.

Vardanega, P.J., Lau, B.H., Lam, S.Y., Haigh, S.K., Madabhushi, S.P.G. & Bolton, M.D. (2012) Laboratory measurement of strength mobilisation in kaolin—Link to stress history. *Géotechnique Letters* 2(1), 9–15.

Wang, L.Z., Lai, Y.Q., Hong, Y. & Masin, D. (2020 A unified lateral soil reaction model for monopiles in soft clay considering various length-to-diameter (L/D) ratios. *Ocean Engineering* 212, 107492.

Yu, J., Huang, M., Li, S. & Leung, C.F. (2017) Load-displacement and upper-bound solutions of a loaded laterally pile in clay based on a total-displacement-loading EMSD method. *Computers and Geotechnics* 83, 64–76.

Yu, J., Huang, M. & Zhang, C. (2015) Three-dimensional upper-bound analysis for ultimate bearing capacity of laterally loaded rigid pile in undrained clay. *Canadian Geotechnical Journal* 52(11), 1775–1790.

Zhang, Y., Aamodt, K.K. & Kaynia, A.M. (2021) Hysteretic damping model for laterally loaded piles. *Marine Structures* 76, 102896.

Zhang, Y. & Andersen, K.H. (2017) Scaling of lateral pile p-y response in clay from laboratory stress-strain curves. *Marine Structures* 53, 124–135.

Zhang, Y. & Andersen, K.H. (2019) Soil reaction curves for monopiles in clay. *Marine Structures* 65, 94–113.

Zhang, Y., Andersen, K.H. & Jeanjean, P. (2019) Cyclic p-y curves in clays for offshore structures. *Offshore Technology Conference*, Texas. OTC-29346-MS.

Zhang, Y., Andersen, K.H. & Jeanjean, P. (2020b) Verification of a framework for cyclic p-y curves in clay by hindcast of Sabine River, SOLCYP and centrifuge laterally loaded pile tests. *Applied Ocean Research* 97, 102085.

Zhang, Y., Andersen, K.H., Jeanjean, P., Karlsrud, K. & Haugen, T. (2020a) Validation of a monotonic and cyclic p-y framework by lateral pile load tests in a stiff over-consolidated clay at the Haga site. *ASCE Journal of Geotechnical and Geoenvironmental Engineering* 146(9), 04020080.

Zhang, Y., Andersen, K.H., Jeanjean, P., Mirdamadi, A., Gundersen, A.S. & Jostad, H.P. (2017) A framework for cyclic p-y curves in clay and application to pile design in GoM. *OSIG SUT Conference*, London.

Zhang, Y., Andersen, K.H., Klinkvort, R.T., Jostad, H.P., Sivasithamparam, N., Boylan, N.P. & Langford, T. (2016b) Monotonic and cyclic p-y curves for clay based on soil performance observed in laboratory element tests. *Offshore Technology Conference*. OTC 26942, Houston, TX.

Zhang, Y., Andersen, K.H. & Tedesco, G. (2016a) Ultimate bearing capacity of laterally loaded piles in clay—Some practical considerations. *Marine Structures* 50, 260–275.

Zhu, B., Zhu, Z.J., Li, T., Liu, J.C., Liu, Y.F. (2017). Field tests of offshore driven piles subjected to lateral monotonic and cyclic loads in soft clay. *Journal of Waterway, Port, Coastal, and Ocean Engineering* 143(5), 05017003.

Chapter 3

Design of monopiles supporting offshore wind turbines using finite element analysis

Rasmus T. Klinkvort and Nallathamby Sivasithamparam
Norwegian Geotechnical Institute (NGI)

CONTENTS

3.1 INTRODUCTION

The focus on alternative sustainable energy has increased in the past decades, and renewable energy from offshore wind turbines (OWT) is one of the main contributors to this development. There is a continuous search for design optimization in order to bring the levelized cost of energy from OWT down. OWTs are placed in a harsh environment with loads from wind and waves acting on the structure, see Figure 3.1. The wind turbines are tall and slender constructions and are dynamically sensitive. The primary forces on an OWT are lateral loads from wind and waves. The two forces will both result in cyclic lateral reactions in the foundation. A typical power spectrum of the forces acting on an OWT is given in Figure 3.1. The loading frequencies from wind and waves are given here, with a peak wave frequency of about ~0.1 Hz and a peak wind frequency of about ~0.01 Hz. The rotor frequency range, often called 1P, and the blade passing frequency range from a three-blade wind turbine, called 3P, are also shown in Figure 3.1. It is important in the design to achieve a first natural frequency of the structure which lies outside these frequencies. This is normally achieved through a design where the entire wind turbine structure has an eigenfrequency in between 1P and 3P, normally called a soft-stiff structure. Considering typical turbines this range is rather narrow and can be difficult to obtain, especially for wind turbines installed at large water depths. Many developments in the recent years have been devoted to determining this initial foundation stiffness to a higher degree. This chapter describes a robust methodology to estimate the initial foundation stiffness and also the foundation stiffness during cyclic loading for monopile foundations.

Monopiles are the most popular foundation concept for OWTs. The continuous improvement of design methodology allows the piles to grow alongside the turbines. In recent years, the PISA project (Byrne et al. 2017) has especially contributed to industry acceptance of the use and benefits of adopting the finite element method

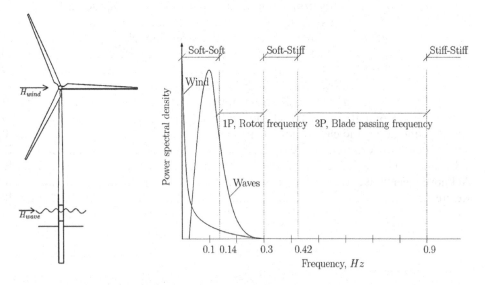

Figure 3.1 Illustration of the forces acting on a wind turbine and the typically forcing frequencies.

(FEM) for monopile design. In contrast to traditional pile design approaches, the use of FEM, together with a proper material model for the soil, allows for a much higher degree of accuracy and optimization of monopile geometry.

Most offshore wind farms comprise layered soil profiles which vary from location to location. Furthermore, OWTs are subjected to continuous cyclic loading from wind and wave actions. A consistent and rigorous design approach for monopiles needs to be therefore able to handle both cyclic loading as well as soil layering and base its input on the results measured from advanced soil element testing. Andersen (2015) presented a cyclic accumulation methodology to design offshore foundations subjected to cyclic loading, often referred to as the NGI cyclic accumulation procedure. The approach uses cyclic element test data to establish cyclic contour diagrams as inputs to the foundation design. The methodology was originally developed for offshore gravity-based structures but has since been developed for other foundation types. Jostad et al. (2014) implemented this approach into a material model and performed 3D FEM analyses of the monopile response for serviceability limit state (SLS) and ultimate limit state (ULS) cyclic load histories. The material model describes the cyclic stress-strain behavior of the soil at each integration point based on cyclic contour diagrams and provides a model for monopile design that properly accounts cyclic loading and soil layering. This approach, however, comes with a cost of calculation speed and complexity. Pisanò (2019) provides a good overview of other constitutive cyclic accumulation methods, for example, the high cyclic accumulation (HCA) method (Niemunis et al. 2015) or the SANISAND model by Dafalias & Manzari (2004).

This chapter presents an alternative, simpler design approach that is based on the NGI cyclic accumulation procedure and used together with 3D finite element analysis (FEA). The FEA results are then used to calibrate springs that can then be used in the structural design.

3.2 DESIGN PHILOSOPHY

During its lifetime, a monopile will experience cyclic loading from current, wind, waves, and operational actions, often from a predominant direction. This loading may cause a temporary or permanent change in stiffness and potentially leads to an accumulation of rotation. Consequently, the effects from cyclic loading must be accounted in the design. The methodology presented here is outlined in Klinkvort et al. (2020) and is based on the well-established NGI method for cyclic loading, Andersen (2015).

One assumption is that the loads and corresponding soil stresses can be decomposed into an average (av) and a cyclic (cyc) component (Eq. 3.1) and that these load components can be treated separately using representative average and cyclic soil (stiffness and strength) parameters.

$$F_{tot} = F_{av} + F_{cyc} \qquad \tau_{tot} = \tau_{av} + \tau_{cyc} \qquad (3.1)$$

Where F is load and τ is soil shear stress.

The separation of loads and stresses and the corresponding displacements and strains allows for a detailed analysis of different effects during the cyclic loading. This is used in the design in the following way. The monopile sizing from a geotechnical perspective is based on three aspects:

Serviceability Limit State (SLS): Sizing of the monopile based on a permanent rotational criterion which shall not be exceeded in the lifetime of the monopile.

Ultimate Limit State (ULS): Sizing of the monopile based on a displacement criterion which shall not be exceeded at peak load. The ULS analysis comprises further calibration of springs for the structural analysis.

Fatique Limit State (FLS): Sizing of the monopile based on a criterion of the system eigenfrequency and calibration of springs for the structural fatigue analysis.

3.2.1 SLS

The sizing of the monopile using a SLS criterion is based on a maximum allowable rotation of the foundation at the seafloor, which shall not be exceeded during the lifetime of the OWT. An allowable rotation for the foundation of 0.5° is often used in the design, where a value between 0.10° and 0.25° is typically used as installation tolerance, leaving 0.25° to 0.4° as the limit state criteria for the accumulated rotation of the foundation during the lifetime. This value is the permanent rotation when the foundation is unloaded. The operational load cases with many load cycles are most critical to the accumulation of rotation. The reason for this is the high average load that leads to a higher degree of accumulation. The analysis in SLS is therefore performed using average loads and average stress-strain curves that account for the effect of cyclic loading. Figure 3.2 shows a simplified example of the idea behind the analysis whereby a single push over analysis is performed using loads and stress-strain curves that take the cyclic loading into account. Material and load factors are taken equal to 1. This will result in a load-deflection curve as illustrated, where the final point is the accumulated rotation at the end of the considered load history. It is then unloaded with the elastic stiffness of the system to find the permanent rotation.

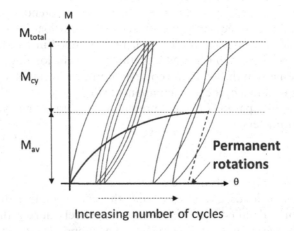

Figure 3.2 Illustration of the SLS approach, here with $M_{av}/M_{cy} = 1$.

3.2.2 ULS

The ULS of a monopile is often not well defined from a geotechnical perspective. Thus, a capacity criterion based on a displacement or rotational criterion is commonly applied, which is often a pile head displacement at the seafloor of D/10. In the ULS condition, the displacement at the peak load is evaluated. Maximum loads are therefore used together with total (sum of cyclic and average) stress-strain curves.

This is illustrated in Figure 3.3. The displacement found at the maximum load corresponds to the maximum displacement taking the previous cyclic loading into account. The analysis is again performed as a single push-over analysis but here with material and load factors appropriate for the ULS situation.

3.2.3 FLS

The calculation of the monopile stiffness used for FLS analyses is important for the evaluation of the eigenfrequency of the entire OWT structure and for the fatigue analysis of the structural components. The foundation stiffness that governs the eigenfrequency is related to the cyclic part of the load and soil response. These loads and stress-strain curves are therefore used in the evaluation of the foundation stiffness. The stiffnesses will often be used as input to aero-elastic models where it is not possible to de-couple average and cyclic loads. This is accounted for by scaling the cyclic repose to an equivalent total response making sure that the secant stiffness values remain the same, as illustrated in Figure 3.4.

Figures 3.2–3.4 illustrate the design philosophy for the three limit states and how the average-cyclic-total load and average-cyclic-total soil parameters are used in the different analyses. In these examples, it is assumed that the ratio between average and cyclic load is 1 for all three situations. This is simply chosen for illustrative purposes, and the actual ratio needs to be evaluated in the design and may be different for the different limit states. It is also noted that the cyclic load history illustrated in the figures are simplified since only cycles with constant average and cyclic moment amplitude are shown. This is also for illustrative purposes only.

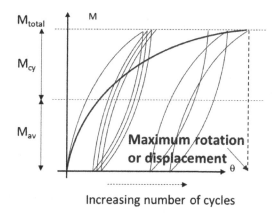

Figure 3.3 Illustration of the ULS approach, here with $M_{av}/M_{cy} = 1$.

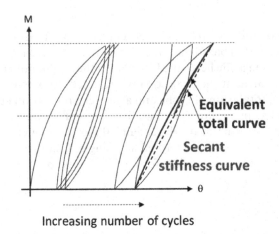

Figure 3.4 Illustration of the FLS approach, here with $M_{av}/M_{cy} = 1$.

3.3   SIMPLIFIED CYCLIC ACCUMULATION

The design philosophy outlines the ideas used in the different design limit states. This section describes how to establish the corresponding stress-strain curves used in the FEA. Stress-strain curves can be established based on a simplified geotechnical load history and a cyclic contour diagram established from cyclic element tests. The approach is given in the following steps:

1   Extract cyclic and average foundation loads from time-domain analysis and establish a simplified geotechnical load history. This is often done using the rain-flow counting procedure, but other procedures may be more appropriate for soils. A detailed discussion on this is presented in Norén-Cosgriff et al. (2015).
2   Carry out a cyclic soil element test program to establish a cyclic contour diagram that relates average and cyclic shear stresses and strains to the number of cycles for each soil unit. A detailed description of how to establish these diagrams are given in (Andersen 2015).
3   Perform cyclic accumulation to find an equivalent number of cycles (N_{eq}) of the peak shear stress that gives the same failure strain as the idealized load. The method is described in (Andersen 2015). For simplicity, this can be evaluated in the cross-section where $\tau_{av} = 0$.
4   The ratio between cyclic and average load, at peak load, is assumed to be the same for the cyclic and average shear stress, and this is used to find representative stress curves in the given N_{eq} cross section. Representative average and cyclic stress-strain curves are established along this stress ratio, and these are then used in the 3D FEA to describe the soil response at each integration point.

This outlines the methodology for a single soil element. For monopiles, the load history applied to the pile top is not transformed identical to all soil elements. There is a redistribution of the cyclic history, and that needs to be considered. The cyclic loading of the monopile will have a much larger effect on the topsoil because it is mobilized to a higher degree than lower layers. One way to take this into account in a simplified way is to use a spring model to establish N_{eq} and then use this to determine the stress-strain curves used in the 3D FEA. Zhang et al. (2016) describes a spring model that performs the cyclic accumulation along the monopile. The spring model uses soil layering and cyclic contours for the soil profile together with the monopile geometry and pile head load history to calculate N_{eq} along the pile. A simple method as described in Klinkvort et al. (2020) that can also be used here is the pile head load applied for each load level in the load history, and the soil resistance along the pile depth is stored. The soil resistance is then normalized, and local soil stress histories in selected points along the pile can then be established. In each of these points, a cyclic accumulation is performed, and a normalized stress-strain curve for the given N_{eq} and average to cyclic load ratio is then determined and used in further analyses.

The procedure described above to establish appropriate stress-strain curves is used together with a 3D FEM program for single pile push-over analysis. Here is the cyclic loading accounted for by the established stress-strain curves and is used to establish the response under the final load in the three design limit states.

3.4 3D FINITE ELEMENT TOOL FOR MONOPILE DESIGN

3D finite element can be performed in different ways, and we will here use an NGI-developed program called INFIDEP. The methodologies are though general and can be used using other software packages also.

INFIDEP is a nonlinear three-dimensional FEM tool developed especially for the efficient analysis of cylindrical foundations. This section presents the basic theory that the program is based on. The program uses an axis-symmetric geometry with asymmetric loading and displacements. This makes it very suitable for analyses of monopile foundations. This is used together with nonlinear elastic soil models and an explicit solver. All these features make the calculation time short, approximately ~1 minute for a monopile with a fine mesh and a sufficient number of load steps.

This chapter outlines the general theory used in program, and more details can be found in NGI (1999) and in Sivasithamparam & Jostad (2020).

3.4.1 Axis-symmetric geometry with asymmetric loading

The formulation of axis-symmetric geometry with asymmetric loading reduces the required degrees of freedom to describe the 3D displacement field and thus the computer time significantly. Displacements in the two-dimensional vertical planes are described by standard element interpolation functions, while the variations of displacements in the hoop direction are approximated by Fourier terms. In addition to this, the far-field is modelled by linear elastic submodels.

While the shell and skirt element formulations are conventional displacement formulations based on the principle of minimum potential energy, the soil elements (near field and far field) are based on a mixed variational principle with independent assumptions on isotropic stress as well as displacements. The global system of equations expresses the volumetric compressibility constraints in the soil as well as the equilibrium conditions.

In the vertical and radial directions, simple iso-parametric interpolation is used for the displacements, while the pressure is assumed constant within the near-field elements. In the ring direction, an expansion in Fourier series is used.

These principles are not new, and more information can be, for example, found in Wilson (1965) and Potts & Zdravkovic (1999).

3.4.1.1   Near- and far-field soil elements

The soil volume is represented by near-field soil volume and far-field soil volume as illustrated in Figure 3.5. The structural elements used to model the monopile are represented by skirt and shell elements.

The near-field soil volume is represented by "pineapple-slice" elements. The upper and lower faces of the element coincide with horizontal planes, while cylindrical surfaces constitute the inner and outer surfaces. The far-field soil volume is represented by layer elements, which have a constant thickness in the horizontal direction and extend from the cylindrical boundary between near field and far field to infinity in the radial direction.

In order to improve the stress predictions and allow for incompressible materials, the near-field soil elements have internal pressure degrees of freedom as well as displacement degrees of freedom. In the two-dimensional vertical planes, standard four-noded element interpolation functions are used for the three Cartesian displacement components, while the pressure is assumed to be constant within an element. Fourier terms are used to model the variations of the displacements and pore pressure in the hoop direction (Winnicki & Zienkiewicz 1979).

A pressure field that is discontinuous from element to element is allowed, and the solution of the global system of equations with respect to the pressure degrees of freedom implies an "overall best" satisfaction of the pressure/volume change properties

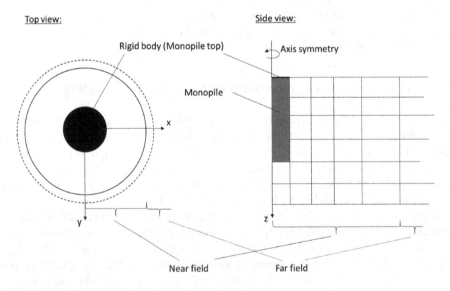

Figure 3.5 General computational topology model.

of the system. Any amount of the pressure change will maintain equilibrium at the center of the element. Macro elements are employed to overcome problems with a system of ill-conditioned systems of equations. Each macro element consists of four pineapple-slice elements: two in the vertical direction and two in the radial direction. Each macro element has three pressure degrees of freedom for each Fourier term.

The far-field is treated as a substructure, where the internal degrees of freedom are eliminated before adding the reduced stiffness matrix into the global system of equations. Thus, no external loads can be applied to the far field. The material in the far-field elements is assumed to be linear elastic. When nonlinear material types are specified, the initial modulus values are used in the far field. A global cylindrical coordinate system is used for the far-field elements. Consistent with the near-field assumptions, the displacements and pressures are expanded in Fourier series in the hoop direction.

3.4.1.2 Skirt-cell and shell element

The skirt elements represent circular rings of cylindrical skirt cells. The intention is to include the flexibility of the skirts and their principal manner of transmitting the loads in the computational model in a realistic manner. The bending stiffness of the skirt walls is assumed to be negligible compared to membrane effects. The flexibility aimed at in the model reflects the effects of membrane forces in the skirt walls.

3.4.1.3 Solution procedure

The nonlinear global set of equations is solved by a "corrected" Euler process, equivalent to the second-order Runge-Kutta method by Zienkiewicz (1977). This explicit algorithm is a robust method for efficient convergence. However, one should be careful to use sufficiently small load increments.

3.4.2 Material models

A linear elastic model is used to model the monopile and the far field elements, a nonlinear total stress-based model, for undrained soil response and an effective stress-based model, for drained soil response. A summary of the material models is outlined here.

3.4.2.1 Isotropic linear elastic model

This material model is used to model the steel behavior and the far field elements where only small strains are seen.

The tangential stiffness is constant and does not change with mobilization.

The input parameters to the isotropic linearly elastic material model are the following:

- Initial shear modulus, G_{max}
- Poisson's ratio, v
- Submerged unit weight, γ

3.4.2.2 Anisotropic nonlinear elastic total stress model

The material model is described in terms of a tangential stiffness as a function of the degree of mobilization. The model is total stress-based and takes anisotropy and effect of stress path into account in a simple way, by linking directly to the results from element testing.

The tangential shear modulus for this material model is given by

$$\log\left[\frac{G_t}{\tau_f(\alpha)}\right] = \log\left[\frac{G_{max}}{\tau_f(45°)}\right] - C_1\xi - C_2\xi^2 - C_3\xi^2 \tag{3.2}$$

where

$$\xi = \frac{\tau}{\tau_f(\alpha)} \text{ if } \xi \leq \text{the limit value } \tau/s_u \text{ (input)}$$

$$C_1 = 12\log(G_{max}/G_{25}) - 6\log(G_{max}/G_{50}) + 4/3\log(G_{max}/G_{75})$$

$$C_2 = -40\log(G_{max}/G_{25}) + 32\log(G_{max}/G_{50}) - 8\log(G_{max}/G_{75})$$

$$C_3 = 32\log(G_{max}/G_{25}) - 32\log(G_{max}/G_{50}) + 32/3\log(G_{max}/G_{75})$$

For the model, the shear stress at failure τ_f is dependent upon a failure plane angle α which is determined from an "equivalent two-dimensional" stress space. The "equivalent two-dimensional" stresses are computed from the full three-dimensional stresses as follows:

$$\sigma_h = \frac{1}{2}(\sigma_x + \sigma_y) \quad \sigma_v = \sigma_z \quad \tau_{vh} = \sqrt{\tau_{yz}^2 + \tau_{zx}^2 + \tau_{xy}^2 + \frac{(\sigma_x - \sigma_y)^2}{4}} \tag{3.3}$$

The angle α is the angle between the maximum compressive principal stress in the equivalent stress space and the vertical axis. The anisotropic shear strength is defined by input of the shear strengths for $\alpha = 0°$, $45°$, and $90°$ (triaxial compression, direct simple shear, and triaxial extension). The program determines constants S_i such that the interpolation fits for the three input values:

$$\tau_f(\alpha) = S_0 + S_1\cos(2\alpha) + S_2\cos(4\alpha) \tag{3.4}$$

The model takes the following input parameters:

- Initial shear modulus, G_{max}
- Tangential stiffness ratio at 25%, 50%, and 75% mobilization, G_{max}/G_{25}, G_{max}/G_{50}, and G_{max}/G_{75}
- Undrained DSS shear strength, s_u^{DSS}
- Anisotropy strength ratio in compression and extension, s_u^c/s_u^{DSS}, s_u^E/s_u^{DSS}
- Poisson's ratio, v
- Submerged unit weight, γ
- Earth pressure coefficient at rest, K_0.

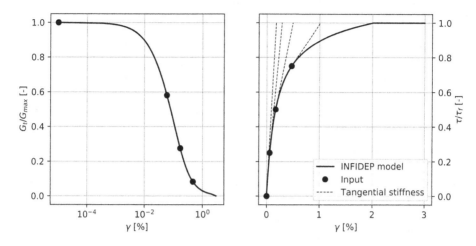

Figure 3.6 Response of total–stress-based model.

The response of the model is shown in Figure 3.6; these two plots should also be used for the calibration of the model to the observed stress-strain response from the laboratory results.

3.4.2.3   Tension cut-off model

The tension cut-off interface model limits the tensile stresses with a tensile strength value in the interface. An isotropic nonlinear elastic material model is used to model tension cut-off. The model is in principle like the model presented above. The tangential shear modulus (G_t) of this material model is given by

$$\log\left[\frac{G_t}{\tau_f(45°)}\right] = \log\left[\frac{G_{max}}{\tau_f(45°)}\right] - C_1 \frac{\tau}{\tau_f(45°)} - C_2 \sin\left[\frac{2\pi}{C_3} \cdot \frac{\tau}{\tau_f(45°)}\right] \tag{3.5}$$

where G_o is the initial shear modulus, $\tau_f(45°)$ is the undrained DSS shear strength at the interface, τ is the maximum shear stress, and C_1, C_2, and C_3 are constants used to fit a nonlinear shear stress-strain curve. The model uses automatically the input from the soil described above and only takes the strength ratio between soil and interface as input.

3.4.2.4   Nonlinear elastic effective stress model

The effective stress-based nonlinear model is based on a model originally proposed by Duncan et al. (1980). The model is based on the principle of hyperbolic stress-strain relation after the work of Kondner (1963) and was developed for triaxial soil tests. The hyperbolic stress-strain relation is shown by the equation below:

$$\sigma_3 - \sigma_1 = \frac{\varepsilon}{\dfrac{1}{E_i} + \dfrac{\varepsilon}{(\sigma_1 - \sigma_3)_u}} \tag{3.6}$$

where σ_1 and σ_3 are the major and minor principal stresses, ε is the major principle strain (axial strain), E_i is the initial tangent modulus, and $(\sigma_1-\sigma_3)_u$ is the ultimate deviator stress. Klinkvort & Hededal (2014) showed a good match between centrifuge monopile lateral load tests in dry sand and a Winkler type model using p–y curves with a hyperbolic shape. Knowing that there is a link between stress-strain soil behavior and soil resistance displacements, it therefore seems appropriate to use a hyperbolic model for monopiles in sand.

A commonly used relationship between the initial tangent modulus, E_i, and the confining stress, σ_3, is used:

$$E_i = K_i \cdot p_a \cdot \left[\frac{\sigma_3}{p_a}\right]^{n_i} \tag{3.7}$$

where p_a is the atmosphere pressure, K_1 is the bulk modulus number, and n is the bulk modulus exponent. K_1 and n_1 are nondimensional parameters. The relationship between ultimate deviator stress, $(\sigma_1-\sigma_3)_u$, and the deviator stress at failure, $(\sigma_1-\sigma_3)_f$, is shown below:

$$\left(\sigma_1 - \sigma_3\right)_f = R_f\left(\sigma_1 - \sigma_3\right)_u \tag{3.8}$$

where R_f is the failure ratio and always takes a value less than or equal to 1.0. This can be used to fit the hyperbolic shape soil response seen in element test and is typically in the range of 0.5–1.0 for most soils. The relationship between classical Mohr-Coulomb shear deviator stress at failure and confining stress is given below:

$$\left(\sigma_1 - \sigma_3\right)_f = \frac{2 \cdot c \cdot \cos\varphi + 2\sigma_3 \sin\varphi}{1 - \sin\varphi} \tag{3.9}$$

where c and φ are the effective stress Mohr-Coulomb cohesion intercept and friction angle, respectively. This equation is used, along with others, to determine the slope at any point on the strain hyperbola. The resulting equation for the tangent Young's modulus in a loading situation is therefore

$$E_t = \left[1 - \frac{R_f\left(1 - \sin\varphi\right)\left(\sigma_1 - \sigma_3\right)}{2 \cdot c \cdot \cos\varphi + 2 \cdot \sigma_3 \cdot \sin\varphi}\right]^2 \cdot K_1 \cdot p_a \cdot \left[\frac{\sigma_3}{p_a}\right]^{m_1} \tag{3.10}$$

Here, φ = friction angle of material,
c = cohesion of material,
σ_1, σ_3 = highest, lowest principal (compression) stresses,
p_a = atmospheric pressure,
K_i, n_i, R_f = dimensionless material parameters

In an unloading/reloading situation, the tangential stiffness is calculated as

$$E_{t,ur} = K_2 \cdot p_a \cdot \left[\frac{\sigma_3}{p_a}\right]^{n_2} \tag{3.11}$$

where K_2 and n_2 are additional and dimensionless material parameters.
Regardless of loading or unloading, the bulk modulus is given by

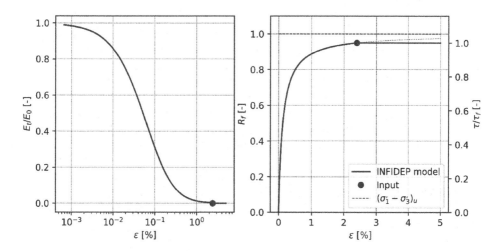

Figure 3.7 Response of effective–stress-based model.

$$B = \frac{p_a}{K_b} \left(\frac{\sigma_3}{p_a} \right)^{m_b} \tag{3.12}$$

where K_b and m are dimensionless material parameters.

The minimum value of the confining pressure of 1% of p_a is used when evaluating the simulations.

The model therefore takes the following input parameters:

- Nondimensional coefficient to determine the initial tangent modulus, K_1 and n_1
- Nondimensional coefficient to determine the unloading re-loading tangent modulus, K_2 and n_2
- Nondimensional coefficient to determine the bulk modulus, K_b and n_b
- Friction angle of material, φ'
- Cohesion of material, c
- The failure ratio, R_f
- Poisson's ratio, v
- Submerged unit weight, γ
- Earth pressure coefficient at rest, K_0

The response of the model is shown in Figure 3.7; these two plots should also be used for the calibration of the model to the observed stress-strain response from the laboratory results.

3.5 VALIDATION OF THE 3D FEA TOOL

To validate the FEM formulation and the presented soil models, the 3D FEA tool is here used to predict the response form the PISA tests.

3.5.1 PISA field tests

The PISA project was a joint industry research project aimed at improving the design methods for laterally loaded monopiles. The PISA project was coordinated by Ørsted and run through the Carbon Trust's Offshore Wind Accelerator programme.

The extent of the PISA project can be divided into three parts:

- Small- to medium-scale field tests of a laterally loaded monopile. A test series in a clay site (Cowden) and a test series in a sand site (Dunkirk);
- Verification of 3D FEM using the field test results. A subsequent numerical extension of the test domain;
- Calibration of a set of empirical equations to establish updated springs for a Winkler spring model. These comprise rotational and horizontal springs, both along the length of the pile and at the toe, based on the field test results and extended test domain.

The field test campaign primarily focused on the monotonic response of the monopiles; however, a small number of cyclic lateral load tests were also performed. A selection of the results has been made publicly available in a series of scientific papers:

- Ground conditions for the two test sites are described in Zdravkovic et al. (2020);
- Monotonic pile test results for the Cowden test can be found in Byrne et al. (2020);
- Monotonic pile test results for the Dunkirk test can be found in McAdam et al. (2020).

The specific details on soil conditions and test setup can therefore be found in these papers. The experimental setup and methodology of the field tests at both Cowden and Dunkirk followed the same philosophy. Piles with different outer diameters (D), penetration depths (L), and load eccentricities (h) were tested. The test pile in question was laterally loaded (H) by being pulled against a larger diameter reaction pile. For the largest diameters ($D = 2$m), both the reaction and test piles served as the test pile since they were pulled against each other. The load was for these tests applied 10m above the ground surface.

The medium- and large-diameter ($D = 0.762$ and 2.0m respectively) piles were installed in a two-stage process, with an initial vibration stage used to embed the piles to a stable depth (1.0–1.5m), followed by pile driving with a hydraulic hammer until reaching the target embedment depth.

3.5.1.1 Cowden site conditions

The test site at Cowden comprises 40m of overconsolidated clay till with a varying degree of weathering, fissuring and stone inclusions. A layer of sand of approximately 1m thick is located at a depth of around 12m. The Cowden site was chosen due to the overconsolidated nature of the clay which behaves like the clays found offshore. The groundwater table was at 1m below the ground surface. Based on the soil data from the test site, soil input to the total stress-based model was established. The soil input profiles for the Cowden site are given in Table 3.1.

Table 3.1 Summary of soil parameters from the Cowden site and input to anisotropic nonlinear elastic total stress model

z_0 [m]	γ [kN/m³]	G_{max} [MPa]	ΔG_{max} [MPa/m]	G_{max}/G_{25} [-]	G_{max}/G_{50} [-]	G_{max}/G_{75} [-]	s_u^D [kPa]	Δs_u^D [kPa/m]	s_u^C/s_u^D [-]	s_u^E/s_u^D [-]	K_0 [-]
0.0	21.6	12	17	81	103	118	35	48.1	1.143	0.857	1.50
1.0	21.6	32	17	65	83	96	83.1	48.1	1.143	0.857	1.50
2.0	21.6	49	17	59	76	87	140	0	1.143	0.857	1.50
2.55	21.6	59	16	71	91	104	140	-35.0	1.143	0.857	1.50
3.9	21.6	80	10	64	162	281	91.9	6.1	1.143	0.857	1.45
5.0	21.6	91	10	25	119	317	98.9	4.4	1.143	0.857	1.20
8.0	21.6	121	20	29	141	372	112	6.1	1.143	0.857	0.97
12.2	21.6	144	20	29	141	372	137.4	2.6	1.143	0.857	0.85

3.5.1.2 Dunkirk site conditions

The Dunkirk site is a site in northern France with a normally consolidated marine sand deposit present in the topsoils. The upper 3 m consists of a hydraulically backfilled layer of sand, which was placed in the without compaction or surcharge to raise the ground level. The sand has high quartz content and contains shell fragments. From several CPTs, it was seen that the ground water table is located at a depth between 4 and 5.4 m, with hydrostatic conditions beneath. Above the ground water table, the conditions are uncertain, and the CPTs clearly indicate the presence of suction above the ground water table, though Taborda et al. (2020) conclude that this is not expected to develop full hydrostatic suction above the ground water table. The presence of suction is here modelled using cohesion in the soil model. The soil input profiles for the Dunkirk site are given in Table 3.2.

The drained soil model cannot model the effect of dilation, and instead, it is assumed here that the dense sand has an incompressible bulk modulus.

Table 3.2 Summary of soil and model input parameters from the Dunkirk site and input to the nonlinear elastic effective stress model

z_0 [m]	γ' [kN/m³]	ϕ' [°]	c [kPa]	$K_1 = K_2$ [-]	$n_1 = n_2$ [-]	R_f [-]	K_b
0.0	17.1	51	5	1500	0.45	0.99	0
1.0	17.1	51	5	1500	0.45	0.99	0
1.6	17.1	48	5	1500	0.45	0.99	0
3.4	17.1	47	13	3500	0.3	0.99	0
3.9	17.1	48	13	3500	0.3	0.99	0
4.5	17.1	48	13	3600	0.3	0.99	0
5.8	9.9	46	8	3600	0.47	0.99	0
7.0	9.9	46	0	3600	0.47	0.99	0
8.0	9.9	46	0	3600	0.47	0.99	0
9.0	9.9	46	0	3600	0.47	0.99	0
12.0	9.9	46	0	3600	0.47	0.99	0

3.5.2   Comparison of model and field tests

Results from the pile test on the monopiles with a dimeter of 2 m and a penetration depth of $L/D = 5.25$ used in the PISA project are presented here. The pile thickness of these piles was $t = 0.025$ m. Because of the size of the diameter, these are the most reliable tests. The discretization used to model the two tests are shown in Figures 3.8 and 3.10. Comparison against the smaller test piles can be found in Klinkvort et al. (2020). The same input parameters have been used for all analyses both for the small and larger pile diameters.

Figure 3.9 shows a comparison of the model prediction using the 3D FEA tool and the results from the two $D = 2.0$ m test piles. The model shows a good agreement with the results; the initial stiffness is captured perfectly followed by slight underestimation of the stiffness until around a horizontal load of 1700 kN. At the end of the test,

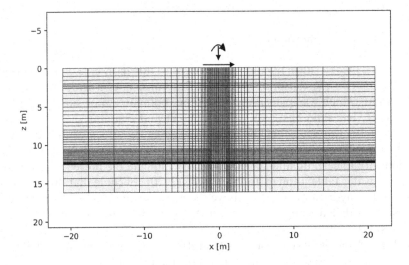

Figure 3.8 Topology of near-field elements used for the modelling of the $D = 2.0$ m pile tests at the Cowden site.

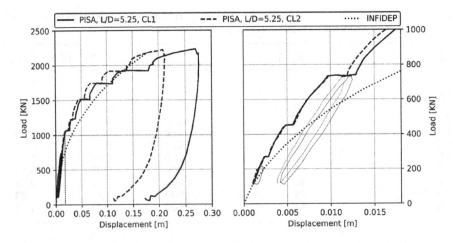

Figure 3.9 Modelling of pile test at the Cowden site.

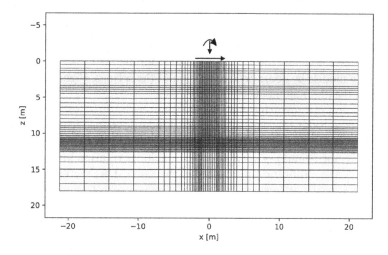

Figure 3.10 Topology of near-field elements used for the modelling of the *D* = 2.0 m pile tests at the Dunkirk site.

significant creep is seen which is not captured in the FE modelling. This sustained load is not representative for the load situation offshore and is therefore not considered to be a problem. The difference seen in the test results highlights the uncertainties associated with the test results (Figure 3.10).

Figure 3.11 shows the comparison of the model prediction using 3D FEA and the results from the two *D* = 2.0 m test piles. The model shows a good agreement with the results; the stiffness is captured perfectly until ~3000 kN where the model starts to under-predict the capacity. It is believed that this is related to that the soil model cannot model dilation. This is a downside of the simple drained model but is acceptable keeping in mind that this is only for large load close to failure.

The 3D FE analysis of the monopile test from the two PISA test sites has demonstrated the capability of 3D FEA with the different material models.

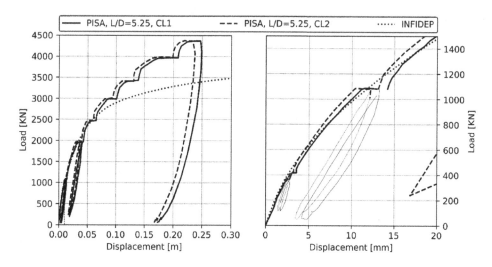

Figure 3.11 Modelling of pile test at the Dunkirk site.

3.6   SPRING MODEL AND CALIBRATION OF SPRINGS

Other engineering disciplines may want to model the response of the monopile. For these analyses, simpler spring models as, for example, described in Zhang (2020) are often used. This section described how the response found from the 3D FEA can be transferred to simpler spring models by dedicated calibration of the springs. Figure 3.12 illustrates the differences in the different models that can be used.

The calibration is done using the results from the 3D FEA together with a routine that tries to match the response seen in FEA with response from the simple spring model. In principle, it does not matter what spring model is used and if the spring model consist of only horizontal p-y springs or also includes rotation m-r springs, as shown in Figure 3.12. The calibration procedure is the same. This section will therefore not describe theory for the different spring models but just refer to some commonly known models as, for example, the one described in API (2014) or Zhang & Andersen (2019).

The principle is that monopile and soil are modelled in a simplified manner as a beam and a series of uncoupled springs. The springs are then manipulated to match the pile response seen in the FE analysis. Each soil spring is described by a mathematical equation and takes, for example, a form as illustrated in Figure 3.13. For each spring, two calibration constants are used to change the shape of the spring: one factor changing the spring stiffness (f_{stiff}) and one factor changing the capacity of the spring (f_{max}). By changing the two factors for each spring along the pile, the pile response can then be changing until a match is seen between 3D FEA and spring model.

The calibration can, for example, be done using a residual vector that is minimized using a least-squares optimization algorithm.

The residual vector is here chosen to be constructed of the pile head response in the form of load-displacement and moment-rotation together with the pile response along the pile during the final load. The pile response along the pile at the final load can, for example, be displacement, rotation, bending moment, shear force, and pile resistance.

The residual is then calculated based on the two vectors from the two models:

$$\text{residual} = \frac{Y_{Spring\,model} - Y_{3D\,FEA}}{\varepsilon} \tag{3.13}$$

where $Y_{3D\,FEA}$ is the normalized pile head response and pile response during the final load, $Y_{Spring\,model}$ is a vector with the normalized response from the spring model, and ε is

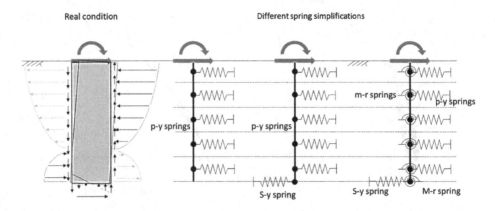

Figure 3.12 Schematic illustration of linking 3D finite element to 1D spring models.

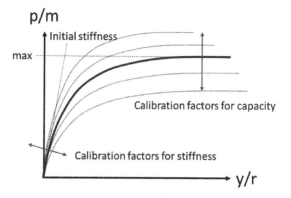

Figure 3.13 Illustration of spring manipulation.

a weighting vector that is used to put more weight on some part of the response than other. The weighting vector is chosen so more importance is put on the pile head response.

The iterative process of this optimization and the result are illustrated in Figure 3.14. The calibrated springs can then be used for structural or load calculations.

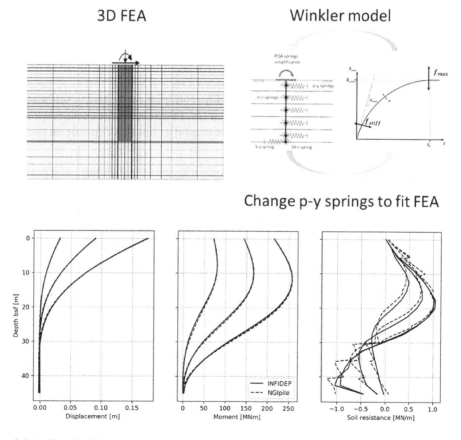

Figure 3.14   Sketch of the optimization process.

3.7   CALCULATION EXAMPLE

In this section, the monopile example used in Klinkvort et al. (2020) is used to illustrate how 3D FEA modelling with cyclic accumulation and spring calibration can be applied.

The soil profile for this example consists of a soft clay with undrained monotonic DSS strength of $s_u^{DSS}/\sigma_{vo}' = 0.38$, together with undrained anisotropic shear strength ratios of $s_u^{DSS}/s_u^C = 0.90$ and $s_u^E/s_u^C = 0.45$. This is representative for a slightly overconsolidated silty clay often seen in Asia. The submerged unit weight was taken as $\gamma' = 8$ kN/m³. In the simplified accumulation procedure presented here, only the DSS contour diagram is used to evaluate the effect of cyclic loading. The DSS contour is shown in Figure 3.15.

The ULS situation is analyzed here, and the following loads were applied to the pile head at seafloor; $(H, V, M) = (4.2\,\text{MN}, 9.1\,\text{MN}, 159.2\,\text{MNm})$ together with the normalized load history given in Table 3.3. The pile geometry is given by $D = 5.2$ m, $L = 45$ m, and $t = 0.054$ m.

The results of the analysis are shown in Figure 3.16 for three different load situations to illustrate the outcome of using the accumulation procedure in different ways;

1 INFIDEP mono.; which shows the response if monotonic soil parameters are used ($\tau_{cy} = 0$).
2 INFIDEP $N = 1$; which shows the cyclic response with no soil degradation, and
3 INFIDEP cyc.; which shows the response after cyclic loading.
From these analyses, some general conclusions can be made.

1   The cyclic response with no soil degradation is stiffer than the monotonic response. This is because of rate effects. For cyclic loading, the load is assumed to be applied with a period like that performed in the lab (often $T = 10$ s)
2   The cyclic response for pile surrounded by soil with degradation is always softer than the cyclic response without degradation. In this case, the cyclic response with degradation is softer than the monotonic response, but this depends on the specific soil, pile geometry, and load history.

The final part is to relate the cyclic pile response to the simplified spring model. The spring calibration is typically performed to provide input to aero-elastic load calculations and structural design. For this example, only horizontal springs are calibrated. Figure 3.17 shows the outcome of the calibration and demonstrates that a very good match between pile head response and pile forces is obtained. The zigzagging of the calibrated soil resistance response is a consequence of the layering, given that each layer has its own set of calibration values.

3.8   SUMMARY

For optimal design of monopiles, detailed analyses of the response are needed. We have in this chapter demonstrated how this can be done using a 3D FE program together with a simple spring calibration. If appropriate input parameters are determined, 3D FEA provides a good tool to estimate the response of monopiles. This was

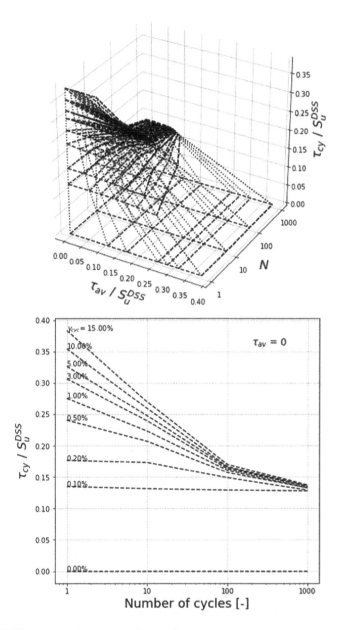

Figure 3.15 3D DSS contour diagram and $\tau_{av} = 0$ cross section.

Table 3.3 Normalized load history in ULS

Number of cycles	500	200	90	50	30	15	8	4	2
Normalized moment M/M_{max}	0.37	0.49	0.58	0.64	0.70	0.77	0.83	0.89	0.96

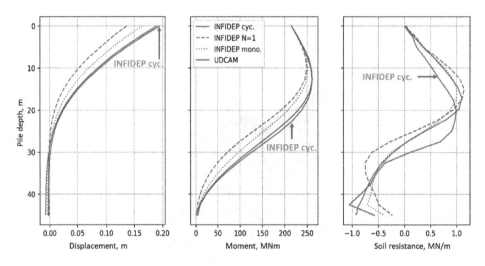

Figure 3.16 Demonstration of the monopile response assuming different loading situations.

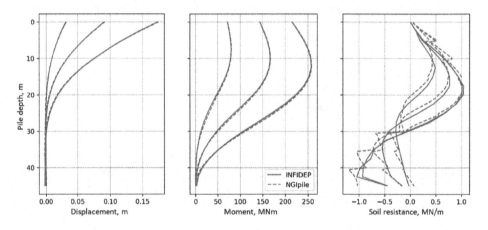

Figure 3.17 Comparison of 3D FEA (INFIDEP) response and calibrated simplified spring model (NGIpile).

demonstrated by a comparison of the response from pile tests and 3D FE predictions. For a detailed design, three limit states need to be considered: the permanent rotation of the monopile at the end of the lifetime (SLS), the capacity under maximum loads (ULS), and the pile response during operation (FLS). By appropriate treatment of the cyclic loads acting on the monopile and advanced element test results, the effect of cyclic loading can be included in the 3D FEA in a relatively simple way. This can be used for a pure geotechnical design, and together with spring calibration, the response seen in the advanced analysis can also be transferred to other calculation tools where the monopile response is important, for example, in load calculations.

ACKNOWLEDGEMENTS

We would like to acknowledge the contribution by colleagues at the Norwegian Geotechnical Institute and collaborators on the various works summarized in this chapter. This includes Dr Hans Petter Jostad, Dr Hendrik Sturm, Dr Ana M. Page, and Dr. Youhu Zhang.

REFERENCES

American Petroleum Institute (2014) API recommended practice 2GEO/ISO 19901-4. Geotechnical and foundation design considerations, 1st Edition, April 2011, Addendum 1, October 2014.

Andersen, K.H. (2015) Cyclic soil parameters for offshore foundation design. In Meyer (ed.) *Frontiers in Offshore Geotechnics III*, Taylor & Francis Group, London, pp. 5–82.

Byrne, B.W., McAdam, R.A., Burd, H.J., Houlsby, G.T., Martin, C.M., Beuckelaers, W.J.A.P., Zdravkovic, L., Taborda, D.M.G., Potts, D.M., Jardine, R.J., Ushev, E., Liu, T., Abadias, D., Gavin, K., Igoe, D., Doherty, P., Gretlund, J.S., Andrade, M.P., Wood, A.M., Schroeder, F.C., Turner, S. & Plummer, M.A.L. (2017) PISA—New design methods for offshore wind turbine monopiles. *Proc. OSIG SUT Conference*, 2017, London.

Byrne, B.W., McAdam, R.A., Burd, H.J., Beuckelaers, W.J.A., Gavin, K.G., Houlsby, G.T., Igoe, D.J.P., Jardine, R.J., Martin, C.M., Muir Woods, A., Potts, D.M., Gretlund, J.S., Taborda, D.M.G. & Zdrawkoviz, L. (2020). Monotonic laterally loaded pile testing in a stiff glacial clay. *Géotechnique*, 70(11), 970–985.

Dafalias, Y.F. & Manzari, M.T. (2004). Simple plasticity sand model accounting for fabric change effects. *Journal of Engineering Mechanics*, 130(6), 622–634.

Duncan, J., Byrne, O., Wong, K. & Mabry, P. (1980). Strength, stress/strain and bulk modulus parameters for finite element analyses of stresses and movements in soil masses. Report No. UCB/GT/80 01, Dept. of Civil Engineering, University of California Berkeley.

Jostad, H.P., Grimstad, G., Andersen, K.H., Saue, M., Shin, Y. & You, D. (2014). FE procedure for foundation design of offshore structures—Applied to study a potential OWT monopile foundation in the Korean Western Sea. *Geotechnical Engineering Journal of the SEAGS & AGSSEA*, 45(4), 66–72 in December 2014.

Kondner, R.L. (1963). Hyperbolic stress-strain response—Cohesive soils. *Journal of the Soil Mechanics and Foundations Division*, 89(1), 115–143.

Klinkvort, R.T. & Hededal, O. (2014). Effect of load eccentricity and stress level on monopile support for offshore wind turbines. *Canadian Geotechnical Journal*, 51, 966–974.

Klinkvort, R.T., Sturm, H., Page, A.M., Zhang, Y. & Jostad, H.P. (2020). A consistent, rigorous and super-fast monopile design approach. *Proceedings from the 4th International Symposium on Frontiers in Offshore Geotechnics.*

McAdam, R.A., Byrne, B.W., Houlsby, G.T., Beuckelaers, W.J.A.P., Burd, H.J., Gavin, K.G., Igoe, D.J.P, Jardine, R.J., Martin, C.M., Muir Woods, A., Potts, D.M., Gretlund, J.S., Taborda, D.M.G. & Zdrawkovic, L. (2020). Monotonic laterally loaded pile testing in a dense marine sand at Dunkirk. *Géotechnique*, 70(11), 986–998.

NGI (1999). Description of INFIDEL – A non-linear, 3D Finite Element Program. NGI report 514090-4, 27 August 1999.

Niemunis, A., Wichtmann, T. & Triantafyllidis, Th. (2015). A high-cycle accumulation model for sand. *Computers and Geotechnics*, 32(4), June 2005, 245–263.

Norén-Cosgriff, K., Jostad, H.P. & Madshus, C. (2015). Idealized load composition for determination of cyclic undrained degradation of soils. *ISFOG*, 2015, 1097–1102.

Pisanò, F. (2019). Input of advanced geotechnical modelling to the design of offshore wind turbine foundations. *Conference—European Conference on Soil Mechanics and Geotechnical Engineering* (ECSMGE2019).

Potts, D.M. & Zdravkovic, L. (1999). Finite element analysis in geotechnical engineering—Theory. Thomas Telford, 1999- Technology & Engineering.

Sivasithamparam, N. & Jostad, H.P. (2020). An efficient FEA tool for offshore wind turbine foundations. *Proceedings from the 4th International Symposium on Frontiers in Offshore Geotechnics.*

Taborda, D.M.G., Zdravkovic, L., Potts, D.M., Burd, H.J., Byrne, B.W., Gavin, K., Houlsby, G.T., Jardine, R.J., Liu, T., Martin, C.M. & McAdam, R. (2020). Finite-element modelling of laterally loaded piles in a dense marine sand at Dunkirk. *Géotechnique*, 70(11), 1014–1029.

Wilson, E.L. (1965). Structural analysis of axisymmetric solids. *AIAA Journal*, 3(12), 2269–2274.

Winnicki, L.A. & Zienkiewicz, O.C. (1979). Plastic (or visco-plastic) behaviour of axisymmetric bodies subjected to non-symmetric loading; Semi-analytical finite element solution. *International Journal of Numerical Methods in Engineering*, 14(9), 1399–1412.

Zdravkovic, L., Jardine, R.J., Taborda, D.M.G., Abadias, D., Burd, H.J., Byrne, B.W., Gavin, K.G., Houlsby, G.T., Igoe, D.J.P., Liu, T., Martin, C.M., McAdam, R.A., Muir Wood, A., Potts, D.M., Gretlund, J.S. & Ushev, E. (2020) Ground characterisation for PISA pile testing and analysis. *Géotechnique*, 1–16.

Zhang, Y. (2020) Analysis of laterally loaded pile foundations using subgrade reaction method. Pile Engineering Book.

Zhang, Y. & Andersen, K.H. (2019) Soil reaction curves for monopiles in clay. *Marine Structures*, 65(2019), 94–113.

Zhang, Y., Andersen, K.H., Klinkvort, R.T., Jostad, H.P., Sivasithamparam, N., Boylan, N.P. & Langford, T. (2016). Monotonic and cyclic p-y curves for clay based on soil performance observed in laboratory element tests. *Offshore Technology Conference*, Houston, TX. OTC 26942.

Zienkiewicz O. C. (1977). *The finite element method*. Third edition. McGraw Hill Book Company, UK.

Chapter 4

Design of suction piles for offshore applications

Hans Petter Jostad and Knut H. Andersen
Norwegian Geotechnical Institute (NGI)

CONTENTS

4.1   INTRODUCTION

Suction piles may be a competitive alternative to more traditional anchor solutions as drag anchors. Reasons for using suction piles are that they may give potential for significant cost saving due to cheaper fabrication and are less expensive to install and more accurate positioning and thus shorter anchor lines compared to drag anchors.

Suction anchors, also called suction piles, bucket anchors, or caissons, are normally cylindrical and made by steel. The cylinders are open at the bottom and closed by a cap or lid at the top as shown in Figure 4.1 (Sparrevik, 1998).

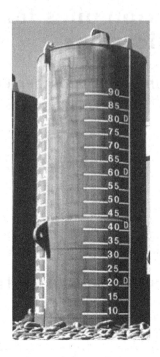

Figure 4.1 Suction pile.

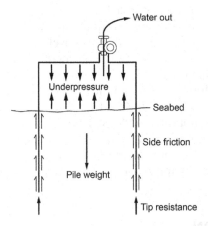

Figure 4.2 Suction pile with forces acting during penetration by under-pressure.

Suction piles are installed by first lowering the pile to the seabed, and then they penetrate somewhat into the soil due to the weight of the pile. Further penetration is achieved by pumping water out of the cylinder as shown in Figure 4.2 (Sparrevik, 1998). This creates an under-pressure (for simplicity often called suction) in the trapped water between the soil plug and the top cap. The difference between the hydrostatic water pressure at the top of the cap and the under-pressure gives a driving force in addition to the weight that is used to overcome the penetration resistance.

After installation, the vent is closed which results in an undrained pull-out capacity that is significantly larger than the penetration resistance. This is because the soil plug will then be pulled-out together with the suction pile.

4.2 PENETRATION ANALYSES

4.2.1 Introduction

The installation of a suction pile to the required depth below seabed is achieved first by some self-weight penetration, then followed by pumping water out of the top of the pile as shown in Figure 4.2. The driving forces due to the submerged weight of the pile and the under-pressure (also often called suction) within the water filled space in the pile created by a pump is used to overcome the soil reaction forces given by shear forces along the outer and inner walls (including any vertical sections of inner stiffeners or bulkhead) and wall tip resistance (including tip resistance of inner stiffeners or bulkhead).

4.2.2 Penetration resistance and necessary suction

The necessary under-pressure u_n within the suction pile to be generated by the pump in order to penetrate the pile is calculated using the following expression:

$$u_n = \left(Q_{tot} - W'\right)/A_{in} \tag{4.1}$$

where
Q_{tot} = total penetration resistance
W' = submerged weight of suction pile
A_{in} = plan view inside area where under-pressure is acting

The total soil resistance without any internal stiffeners is given by the shear stresses along the inner and outer walls and the wall tip resistance:

$$Q_{tot} = Q_{side} + Q_{tip} = A_{wall} \cdot r_{int} \cdot s_u^{\overline{DSS}} + \left(N_c \cdot s_{u,tip}^{AV} + \gamma' \cdot z\right) \cdot A_{tip} \tag{4.2}$$

where
Q_{side} = resistance along inner and outer walls
Q_{tip} = resistance at wall tip
A_{wall} = total wall area (inside and outside)
A_{tip} = skirt tip area
r_{int} = adhesion factor (see discussion below)
z = penetration depth
$s_u^{\overline{DSS}}$ = average undrained direct simple shear (DSS) strength over the penetration depth
$s_{u,tip}^{AV}$ = average undrained shear strength at wall tip (see discussion below)
γ' = average submerged unit weight of soil over the penetration depth
N_c = bearing capacity factor (7.5 for a strip loading some distance below seabed)

The s_u^{AV} is the average of the undrained compression, extension, and DSS strengths at depth z. However, in many cases, s_u^{AV} can in these calculations be approximated by s_u^{DSS}.

A high estimated value of $s_u^{\overline{DSS}}$ should be used to calculate the required suction u_n and corresponding necessary capacity of the pump.

The adhesion factor (or interface roughness) r_{int} is often taken as the inverse of the sensitivity S_t of the clay, meaning that a totally remoulded undrained shear strength is used at the interface between the wall and the soil. Due to uncertainties in the adhesion factor, a conservatively high r_{int}-value should be used in the calculation of the penetration resistance. This high value should also account for the effect of strain rate close to the wall interface if the strain rate is very different from the one used in standard laboratory tests.

Furthermore, the shear resistance given by the remolded undrained shear strength $s_{ur} = s_u^{DSS}/S_t$ should not be larger than the interface shear strength, $\tau_{int} = \tan(\delta) \cdot \sigma_n'$, given by the interface friction angle δ and the effective normal stress σ_n'. This interface shear strength can for instance be determined from ring shear tests accounting for the actual roughness of the pile surface. This interface strength may for instance be governing if the wall is painted (Andersen et al., 2005). When estimating the effective normal stress σ_n', the effect of shear-induced pore pressure before the interface strength is mobilized should be taken into account.

4.2.3 Maximum allowable suction

To reduce the potential of large soil heave inside the suction pile, and to include some safety margin against penetration refusal, the maximum allowable under-pressure u_a that may be applied within the suction pile is given as

$$u_a = N_c \cdot \beta \cdot s_{u,plug}^{AV} + A_{inside} \cdot r_{int} \cdot s_u^{\overline{DSS}} \Big/ A_{in} \tag{4.3}$$

where the utilization factor β controls the degree of mobilized average undrained shear strength along a critical slip surface involving soil flowing into the pile (reverse end-bearing of the soil plug). In general, $\beta = 0.67$ is a suitable utilization. The bearing capacity factor N_c may be taken as 9 at penetration depths larger than 2.5 times the suction pile diameter. For suction piles with inner stiffeners, the shear and tip resistance along these elements (similar as in the calculation of total penetration resistance) should be included.

Note, the shear strength $s_{ur} = r_{int} s_u^{\overline{DSS}}$ along the inside of the wall is the same whether the skirt wall penetrates, or the soil is flowing into the suction pile. A requirement that the necessary under-pressure u_n should be less than the maximum allowable under-pressure u_a (calculated with a utilization factor $\beta = 1.0$) divided by for instance a factor of 1.5 is therefore not suitable at large penetration depths, since the ratio u_n/u_a approaches the ratio between the total skirt wall area and the inside skirt wall area A_{wall}/A_{inside} at very large depths, independent of other parameters.

In addition, in shallow water depths, the generated under-pressure cannot be lower than the cavitation pressure given by the water pressure at seabed plus atmospheric pressure.

4.2.4   Inner stiffeners

In cases with structural elements as for instance inner stiffeners, bulkheads, etc. (see Figure 4.3), the increased penetration resistance of these elements should be added to the wall and tip resistance.

For suction piles with ring stiffeners, it may be interaction between ring stiffeners at different depths and the inner wall friction, resulting in a resultant resistance that is lower than the sum of all elements (Andersen et al., 2005; Jostad et al., 2015). In addition, the clay needs to be soft enough to deform back to the wall after passing a ring stiffener. If not, water may be trapped between two ring stiffeners at different depths. Furthermore, soft soil from one layer may after falling back also be trapped between two ring stiffeners. Therefore, for suction piles with several ring stiffeners at different heights, the minimum inside penetration resistance may therefore be close to the penetration resistance of the first ring stiffener. When calculating the required under-pressure, the effect of these rather complex mechanisms should be accounted for in a conservative way.

The effect of any changes in the suction pile diameter over the height needs to be accounted for in the same way as other structural elements.

4.2.5   Soil plug heave

The soil plug heave inside the suction pile during installation may be estimated by assuming that the clay displaced by skirts and stiffeners goes into the pile. It is then assumed that the applied under-pressure is below the utilization factor $\beta = 1/1.5$ against inverse bearing capacity at pile tip level. If not, the soil heave may be larger.

For suction piles with ring stiffeners, and where the soil not fully deforms back to the inner wall, the contribution of the volume of trapped water between the ring stiffeners should be added to the soil heave volume.

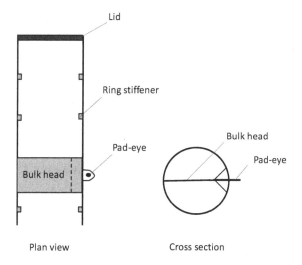

Figure 4.3 Suction pile with bulkhead at pad-eye depth and ring stiffeners.

The suction pile length must be increased by the soil heave height to achieve the target penetration depth. The optimum pad-eye depth along the wall should also consider this soil heave height.

4.2.6 Example

To demonstrate the use of the above equations, the penetration of a suction pile with a diameter $D = 5$ m, wall and lid thickness $t = 0.05$ m, and varying length z in a clay with an isotropic constant undrained shear strength of $s_u = 100$ kPa (i.e. a highly over-consolidated clay), a submerged unit soil weight $\gamma' = 10$ kN/m^3, and a clay sensitivity $S_t = 2$ which gives an adhesion factor $r_{int} = 0.5$ is considered. No internal stiffener is assumed for simplicity. More example calculations with and without stiffeners are given in Andersen et al. (2005).

The submerged unit steel weight γ_{steel}' is 68.5 kN/m^3, which gives a submerged pile weight versus pile length z of

$$W' = \left(A \cdot t + A_{tip} \cdot z\right) \cdot \gamma_{steel}' = \left(\pi/4 \cdot 5 \cdot 5 \cdot 0.05 + \pi \cdot 5 \cdot 0.05 \cdot z\right) \cdot 68.5$$

$$= 67.2 \,\text{kN} + 53.8 \,\text{kN/m} \cdot z$$

The inside and outside wall resistance:

$$Q_{side} = A_{wall} \cdot r_{int} \cdot s_u^{\overline{DSS}} = 2 \cdot \pi \cdot 5 \cdot z \cdot 0.5 \cdot 100 = 1570 \,\text{kN/m} \cdot z$$

The skirt tip resistance:

$$Q_{tip} = \left(N_c \cdot s_{u,tip}^{AV} + \gamma' \cdot z\right) \cdot A_{tip} = \left(7.5 \cdot 100 + 10 \cdot z\right) \cdot \pi \cdot 5\text{m} \cdot 0.05$$

$$= 589 \,\text{kN} + 7.85 \,\text{kN/m} \cdot z$$

Necessary under-pressure:

$$u_n = \left(Q_{tot} - W'\right)/A_{in} = \left(1570 \cdot z + 589 + 7.85 \cdot z - 67.2 - 53.8 \cdot z\right)/\left((\pi/4) \cdot 4.9 \cdot 4.9\right)$$
$$= \left(522 + 1524 \cdot z\right)/18.85 = 28 \,\text{kPa} + 81 \,\text{kN/m}^3 \cdot z$$

Allowable under-pressure:

$$u_a = N_c \cdot \beta \cdot s_{u,plug}^{AV} + A_{inside} \cdot r_{int} \cdot s_u^{\overline{DSS}}/A_{in} = 9 \cdot 1/1.5 \cdot 100 + 0.5 \cdot 1570 \cdot z/18.85$$

$$= 600 \,\text{kPa} + 42 \,\text{kN/m}^3 \cdot z$$

Based on this, the maximum calculated penetration depth is

$$u_a - u_n = 600 + 42 \cdot z - 28 - 81 \cdot z = 0, \text{ which gives } z = 572/39 = 15\,\text{m}$$

The results from these equations are plotted in Figure 4.4. From Figure 4.4a, it is seen that the contribution of the tip resistance in clay is very small compared to the contribution from shear stresses along the walls.

The self-weight penetration for a pile with a length of 15 m and submerged weight $W' = 874$ kN is

$$W' - Q_{tot} = 874 - 1570 \cdot z - 589 - 7.85 \cdot z = 0, \text{ which gives } z = 285/1578 = 0.2 \text{ m}$$

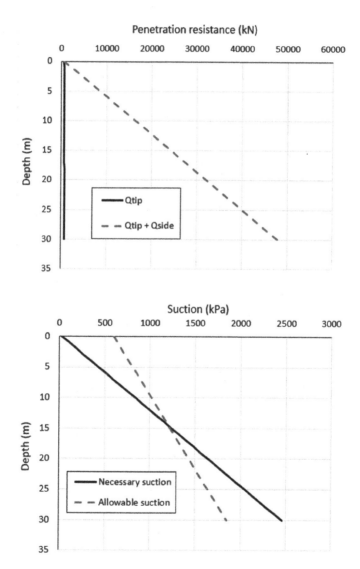

Figure 4.4 Calculated penetration resistance (a) and necessary and allowable suction (b) versus penetration depth for $s_u = 100$ kPa constant with depth, $D = 5$ m, $t = 0.05$ m, and $r_{int} = 0.5$.

The self-weight penetration for this special case is very small due to the high undrained shear strength of 100 kPa at seabed.

Figure 4.5 shows an example for a normally consolidated clay where the isotropic undrained shear strength increases linearly with depth by 2.0 kPa/m and where the submerged pile weight $W' = 1000$ kN. All the other parameters are the same as in the example above. The calculated self-weight penetration depth is then 7.4 m, and the calculated maximum penetration depth is 31 m.

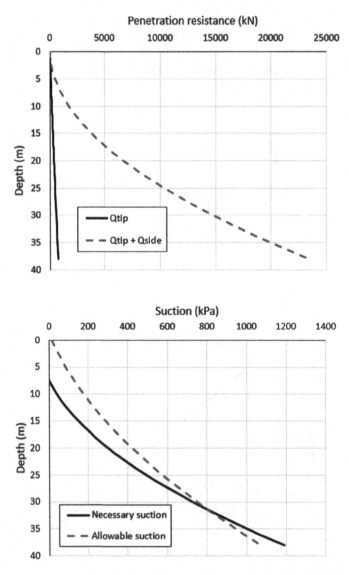

Figure 4.5 Calculated penetration resistance (a) and necessary and allowable suction (b) versus penetration depth for $s_u = 2.0$ kPa/m times depth, $W' = 1000$ kN, $D = 5$ m, $t = 0.05$ m, and $r_{int} = 0.5$.

4.3 HOLDING CAPACITY

4.3.1 Introduction

The holding capacity of suction piles may be calculated using equations developed for piles, plastic solutions, and finite element analyses (e.g., Andersen & Jostad, 1999; Randolph & House, 2002; Bang et al., 2011; Zdravkovic & Potts, 2005; Aubeny et al., 2003; Jostad & Andersen, 2015; Li & Wang, 2013; Edgers et al., 2009; Huang et al., 2003). The equations presented in Section 4.3.2 are based on equations initially developed for embedded and skirted foundations in clay. These equations are extended with formulations valid for deep foundations and piles (Andersen & Jostad, 1999). The advantage of these equations is that they account for the coupling between horizontal and vertical components of the soil resistance. The 3D effect is calibrated by 3D finite element analyses (Jostad & Andersen, 2015).

4.3.2 Limit equilibrium approach

The undrained holding capacity of a suction pile depends on the loading direction given by the anchor line and position of the load attachment point (pad-eye), the time-dependent shear strength (set-up) along the walls, the anisotropic or stress path-dependent cyclic undrained shear strength variation with depth, a potential gap developed at the back of the pile, tilt and mis-orientation of the suction pile after installation, and the drainage condition through the top of the suction pile. The top cap is generally closed after installation, which means that under-pressure or suction can be generated during short-term tension loading.

The soil resistance is given by active (acting at the back of the suction pile) and passive (acting at the front of the suction pile) earth pressure accounting for 3D effects, vertical and horizontal shear stresses along the walls, and vertical normal stress and horizontal shear stress at the bottom of the soil plug or pile tip level. For a deep suction pile, the active and passive earth pressure may beneath a given depth be merged into a resultant component resulting from that the clay flows around the pile. These reaction forces are illustrated in Figure 4.6. In this figure, the suction pile is idealized by a rectangular cross section $A = L \cdot B$ to use limit equilibrium equations for plane strain condition. The most representative idealization of the cross section depends on the actual failure mode. However, for situations where the horizontal resistance is fully mobilized, the width into the plane or normal to the loading direction B is taken equal to the suction pile diameter D, and the width parallel to the loading direction L is calculated from the cross-section area $A = \pi/4 \cdot D^2$, i.e., $L = A/D$.

These soil reactions need to be in equilibrium with the applied external forces (anchor line force at pad-eye and submerged weight of the suction pile).

Vertical and horizontal equilibriums are given as

$$F \cdot \cos(\alpha_l) \cdot \gamma_l = H_f \tag{4.4}$$

$$F \cdot \sin(\alpha_l) \cdot \gamma_l - W' = V_f \tag{4.5}$$

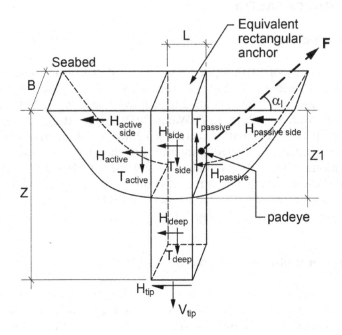

Figure 4.6 Soil reaction components on a suction pile loaded at the optimum loading point.

where F is the resultant chain load, γ_l is a load scaling factor, α_l is the loading angle at the load attachment point, W' is the submerged weight of the suction pile, and H_f and V_f are the resultant horizontal and vertical soil reaction forces.

For an idealized condition where the load attachment point or pad-eye is located at a point that gives no rotations of the suction pile at failure, the resultant reaction forces shown in Figure 4.6 may be estimated by the equations shown below.

The horizontal capacity H_f is obtained from seven contributions (see Figure 4.6):

$$H_f = H_{active} + H_{passive} + H_{tip} + H_{active,\ side} + H_{passive,\ side} + H_{side} + H_{deep} \tag{4.6}$$

where the contribution from the active earth pressure H_{active} is

$$H_{active} = \int_0^{Z1} K(r_{active}) \cdot s_u(\alpha) \cdot B \cdot dz \tag{4.7}$$

and the contribution from the passive earth pressure $H_{passive}$ is

$$H_{passive} = \int_0^{Z1} K(r_{passive}) \cdot s_u(\alpha) \cdot B \cdot dz \tag{4.8}$$

and the contribution from the shear at the pile tip level H_{tip} is

$$H_{tip} = r_{tip} \cdot s_{u,tip}^{DSS} \cdot B \cdot L \tag{4.9}$$

and the contribution from the horizontal shear along the upper part of the pile H_{side} (both sides) is

$$H_{side} = 2 \int_{0}^{Z1} r_{h,anchor,side} \cdot s_{u}^{DSS} \cdot L \cdot dz \tag{4.10}$$

and the contribution from the deep part of the pile H_{deep} is

$$H_{deep} = \int_{Z1}^{Z} N_p \left(r_{anchor,\,deep} \right) \cdot s_{u}^{DSS} \cdot B \cdot dz \tag{4.11}$$

In the same way, the vertical capacity V_f is obtained from five contributions, see Figure 4.6:

$$V_f = T_{active} - T_{passive} + T_{side} + T_{deep} + V_{tip} \tag{4.12}$$

where the contribution from the active side T_{active} is

$$T_{active} = \int_{0}^{Z1} r_{active} \cdot s_{u}^{DSS} \cdot B \cdot dz \tag{4.13}$$

and the contribution from the passive side $T_{passive}$ is

$$T_{passive} = \int_{0}^{Z1} r_{passive} \cdot s_{u}^{DSS} \cdot B \cdot dz \tag{4.14}$$

and the contribution from the vertical side shear along the upper part of the pile T_{side} (both sides) is

$$T_{side} = 2 \cdot \int_{0}^{Z1} r_{v,\,side} \cdot s_{u}^{DSS} \cdot L \cdot dz \tag{4.15}$$

and the contribution from the vertical shear along the deep part of the pile T_{deep} is

$$T_{deep} = \int_{Z1}^{Z} r_{deep} \cdot r_{int} \cdot s_{u}^{DSS} \cdot \pi \cdot D \cdot dz \tag{4.16}$$

and the contribution from the vertical tension capacity at pile tip level V_{tip} is

$$V_{tip} = N_c \left(r_{tip} \right) \cdot s_u^{AV} \cdot B \cdot L \tag{4.17}$$

The parameters in the above equations will be discussed in the next section.

For a linearly increasing shear strength with depth, s_u^{AV} may be taken at depth $z = Z + B/4$.

For a suction pile without a gap at the active side, the resulting soil reaction forces due to the submerged weight of the soil are equal on the active and the passive side of the suction pile and therefore give no contribution to the holding capacity. For a suction pile with an open gap to a given depth, the corresponding net effect of the submerged weight of soil must be added.

Since the horizontal and vertical holding capacities are coupled, an iterative procedure is required to find the combination of reaction forces that gives the largest resultant chain force F or load scaling factor γ_l that can be applied to the suction pile. The above equations are described in more detail in the following section.

4.3.2.1   Limit equilibrium equations

For an idealized case where the load attachment point is at a point that gives no rotation of the suction pile at failure, the holding capacity can be estimated by equations from classical plasticity theory. The 3D effect is taken into account by side shear on the active and passive failure wedges and the side of the suction pile as shown in Figure 4.6.

The earth pressure coefficient N is then given as

$$K(r) = 1 + \text{Arcsin}(r) + \sqrt{1 - r^2} \quad \text{for } r > 0 \tag{4.18a}$$

$$K(r) = 2\sqrt{1 + r} \quad \text{for } r < 0 \tag{4.18b}$$

where $r = r_{active}$ on the active side and $r = r_{passive}$ on the passive side. The factor r is defined as the ratio between the mobilized vertical interface shear stress τ_v and the DSS strength s_u^{DSS} along the wall. The vertical shear stress τ_v is positive downwards at the active side and upwards at the passive side. The vertical interface shear stress τ_v cannot be larger than given by the time-dependent undrained shear strength in the re-consolidated remoulded zone after installation, i.e., $\tau_v < r_{int}(t) \cdot s_u^{DSS}$, or limited by the interface friction angle δ and the normal effective stress σ_n', i.e., $\tau_v < \tan(\delta) \cdot \sigma_n'$. The time-dependent set-up factor $r_{int}(t)$ will be discussed in more detail in Section 4.3.4.

A simplified approach to account for the 3D effect (often used for shallow foundations) is to add contributions from side shear on the idealized plane strain active and passive earth pressure wedges. The contribution from the side shear of the active zone $H_{active, side}$, (both sides) is

$$H_{active, side} = 2 \cdot \int_{A_{active}} r_{active, side} \cdot s_u^{DSS} \cdot dA \tag{4.19}$$

and the contribution from the side shear of the passive zone $H_{passive, side}$ (both sides) is

$$H_{passive,\,side} = 2 \cdot \int\limits_{A_{passive}} r_{passive,\,side} \cdot s_u^{DSS} \cdot dA \tag{4.20}$$

The empirical side shear factors $r_{active,\,side}$ and $r_{passive,\,side}$ need to be calibrated based on full 3D plasticity analyses. This type of calibration is for instance presented in Jostad & Andersen (2015). However, as a conservative assumption, $r_{active,\,side} = r_{passive,\,side} = 0.5$ may be used.

A_{active} and $A_{passive}$ are the areas of the plane vertical sides of the active and passive zones (which depend on the roughness factors $r = r_{active}$ and $r = r_{passive}$), see Figure 4.7. For simplicity, linear interpolation may be used between the side area for a perfectly smooth wall ($r = 0$) where $A = \frac{1}{2} \cdot Z1^2$ and the side areas for a fully rough wall ($r = 1$) where $A = (\pi + 4)/8 \cdot Z1^2$. For the condition of a fully mobilized negative roughness ($r = -1$), the side area is taken equal to zero ($A = 0$).

To calculate the position of the resultant horizontal reaction forces due to this side shear components, it is important to account for that the area of the wedge increases with increasing depth $Z1$ following the critical slip line as shown in Figure 4.7.

To account for that the undrained shear strength, $s_u(\alpha)$ is anisotropic or stress path-dependent, i.e., a function of the angle α between the direction of the maximum compressive principal stress and the vertical axis (e.g., compression $s_u(\alpha = 0°) = s_u^C$, DSS $s_u(\alpha = 45°) = s_u^{DSS}$, and extension $s_u(\alpha = 90°) = s_u^E$, a weighted average undrained shear strength on the active and passive side may be estimated from

$$s_{ua} = 0.3 \cdot r_{active} \cdot s_u^{DSS} + \left(1 - 0.3 \cdot r_{active}\right) \cdot s_u^C \tag{4.21a}$$

$$s_{up} = 0.3 \cdot r_{passive} \cdot s_u^{DSS} + \left(1 - 0.3 \cdot r_{passive}\right) \cdot s_u^E \tag{4.21b}$$

The above weighting factor of 0.3 for the DSS strength s_u^{DSS} is found by a plane strain analysis of passive earth pressure with the finite element program Bifurc described in

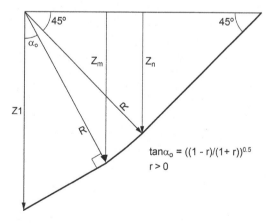

Figure 4.7 Geometry of the passive (and active) earth pressure wedges as a function of the roughness factor r.

Jostad & Andersen (2015) using a ratio of $s_u^E/s_u^{DSS} = 0.7$ and a fully rough wall ($r = 1$). The same weighting function is assumed to also be valid on the active side. The effect of the anisotropic undrained shear strength is important for inclined loading with different contributions from the active and passive side or the extreme case with a gap at the active side.

The resultant horizontal reaction forces from the active and passive wedges and corresponding side shear forces should not be larger than the contribution from soil flowing around the suction pile, i.e., the deep part of the suction pile. The resulting horizontal earth pressure factor for the deep part of the suction pile is approximated by the following equation:

$$N_p = 11.9 - 0.8 \cdot (1 - r_{int}) - 2.0 \cdot (1 - r_{int})^4 - 0.8 \cdot r_{int} \cdot (r_{deep}) - 2.8 \cdot r_{int} \cdot (r_{deep})^4 \qquad (4.22)$$

where the roughness factor r_{deep} is the ratio between the vertical shear stress τ_v (positive downwards) and the shear strength s_u^{DSS} at the deep part of the suction pile ($r_{deep} = \tau_v/s_u^{DSS}$). The roughness factor $r_{int} = s_u(t)/s_u^{DSS}$ takes into account the effect that the shear strength $s_u(t)$ at the wall-soil interface may be smaller than the intact DSS strength s_u^{DSS}. The above equation is found by 3D finite element analyses of a thin horizontal disc of the pile and the surrounding soil.

The (inverse) bearing capacity factor N_c at skirt tip may be taken from Brinch Hansen (1970):

$$N_c(r_{tip}) = (2 + \pi) \cdot (1 + s_c + d_c - i_c) \qquad (4.23)$$

where

$$s_c = 0.2 \cdot (1 - 2 \cdot i_c) \cdot L/B$$

$$d_c = 0.4 \cdot \text{Arctan}(Z/L)$$

$$i_c = 0.5 - 0.5\sqrt{1 - r_{tip}}$$

The roughness factor r_{tip} depends on the loading angle at the load attachment point.

The undrained shear strength profiles should account for any requirements to material coefficients or safety factor, cyclic degradation, and strain rate effects. How to account for cyclic loading and strain rate effect is for instance explained in Andersen (2015).

The holding capacity obtained from the equations above is found by varying the roughnesses r_{active}, $r_{passive}$, $r_{v, side}$, $r_{h, side}$, r_{deep}, and r_{tip} until the maximum holding capacity that is in equilibrium with the scaled anchor force (Eq. 4.4) is found.

The calculated holding capacities (H_f and V_f) given by Eq. (4.4) are only valid when the resulting moment at center skirt tip level M_{tip} is zero, i.e.,

$$M_{tip} = F \cdot \gamma_l \cdot \cos(\alpha_l)(Z - z_l) - F \cdot \gamma_l \cdot \sin(\alpha_l)x_l - M_{soil} \qquad (4.24)$$

where F is the resulting chain force, γ_l is the load scale factor, z_l is the depth below the seabed to the load attachment point, x_l is the horizontal distance from the vertical centerline of the suction pile to the load attachment point, and α_l is the loading angle from the horizontal. M_{soil} is the resultant moment at center pile tip from the contribution of all soil reaction forces on the pile.

The above equations may for instance be included in an Excel sheet. We have included the equations into a Fortran code. The integration of the reaction forces over the pile depth is solved numerically by dividing the suction pile into a suitable number of sublayers. In the next section, the use of the above equations is demonstrated for an idealized case.

4.3.3   Example calculation of capacity envelope

To demonstrate the use of the above plasticity equations, the capacity envelope of a suction pile with a dimeter $D = 5\,\text{m}$ and penetration depth $Z = 10\,\text{m}$ in a clay with an isotropic constant undrained shear strength of $s_u = 100\,\text{kPa}$ over the actual depth is calculated. An interface set-up factor $r_{int} = 0.5$ is used. The 3D effect is accounted for by $r_{active, side} = r_{passive, side} = r_{side} = 0.5$. The suction pile is assumed to be perfectly sealed at the top cap and loaded at the optimum load attachment point.

First, the maximum vertical pull-out capacity is calculated. This is given by the contribution from the outer side shear along the wall and the inverse bearing capacity at the pile tip level.

$$T_{side} = r_{int} \cdot s_u \cdot \pi \cdot D \cdot L = 0.5 \cdot 100 \cdot \pi \cdot 5 \cdot 10 = 7850\,\text{kN}$$

$$V_{tip} = N_c\left(r_{tip} = 0\right) \cdot s_u \cdot A = 8.44 \cdot 100 \cdot \pi/4 \cdot 5 \cdot 5 = 16{,}580\,\text{kN}$$

where

$$N_c\left(r_{anchor,\ tip}\right) = \left(2 + \pi\right) \cdot \left(1 + s_c + d_c - i_c\right) = 5.14 \cdot \left(1 + 0.2 + 0.44 - 0\right) = 8.44$$

$$s_c = 0.2 \cdot \left(1 - 2 \cdot i_c\right) \cdot L/B = 0.2$$

$$d_c = 0.4 \cdot \text{Arctan}\left(L/D\right) = 0.44$$

$$i_c = 0.5 - 0.5\sqrt{1 - r_{tip}} = 0$$

The total vertical pull-out capacity (excluding the submerged weight of the suction pile) becomes

$$V_f = T_{side} + V_{tip} = 7850\,\text{kN} + 16{,}580\,\text{kN} = 24{,}430\,\text{kN}$$

The corresponding maximum horizontal holding capacity assuming pure horizontal translation of the suction pile is given by six components. In this case, an idealized geometry is used where $B = D$ and $L = A/D = \pi/4 \cdot 5 = 3.93\,\text{m}$.

$$H_{active} = H_{passive} = N\left(r_{active}\right) \cdot s_u \cdot Z = 2.39 \cdot 100\,\text{kPa} \cdot 5\,\text{m} \cdot 10\,\text{m} = 11{,}948\,\text{kN}$$

$$H_{tip} = s_u \cdot A = 100\,\text{kPa} \cdot \pi/4 \cdot 5\,\text{m} \cdot 5\,\text{m} = 1963\,\text{kN}$$

$$H_{side} = 2 \cdot r_{h,\,side} \cdot s_u^{DSS} \cdot L \cdot Z = 2 \cdot 0.5 \cdot 100\,\text{kPa} \cdot 3.93\,\text{m} \cdot 10\,\text{m} = 1965\,\text{kN}$$

$$H_{active,\,side} = H_{passive,\,side} = 2 \cdot r_{active,\,side} \cdot s_u^{DSS} \cdot$$
$$A_{active,\,side} = 2 \cdot 0.5 \cdot 0.5 \cdot 100\,\text{kPa} \cdot 69.6\,\text{m}^2 = 1963\,\text{kN}$$

where

$$N(r) = 1 + \text{Arcsin}(r) + \sqrt{1 - r^2} = 1 + \text{Arcsin}(0.5) + (1 - 0.25)^{0.5} = 2.39$$

$$A_{active,\,side,\,min} = \tfrac{1}{2} \cdot H^2 = \tfrac{1}{2} \cdot 10\,\text{m} \cdot 10\,\text{m} = 50\,\text{m}^2$$

$$A_{active,\,side,\,max} = \pi/8 \cdot H^2 + \tfrac{1}{2} \cdot H^2 = \pi/8 \cdot 10\,\text{m} \cdot 10\,\text{m} + \tfrac{1}{2} \cdot 10\,\text{m} \cdot 10\,\text{m} = 89.3\,\text{m}^2$$

$$A_{active,\,side} = 0.5 \left(A_{active,\,side,\,min} + A_{active,\,side,\,max} \right) = 69.6\,\text{m}^2$$

The total horizontal holding capacity becomes

$$H_f = H_{active} + H_{passive} + H_{active,\,side} + H_{passive,\,side} + H_{side} + H_{tip}$$
$$= 11,948\,\text{kN} + 11,948\,\text{kN} + 6963\,\text{kN} + 6963\,\text{kN} + 1965\,\text{kN} + 1963\,\text{kN} = 41,750\,\text{kN}$$

The resultant average horizontal resistance factor against the pile walls is $N = (H_f - H_{tip})/(D \cdot Z \cdot s_u) = 8.0$. The optimum load attachment point for pure horizontal loading is calculated to be $z_l = 5.55\,\text{m}$ below seabed.

Similar calculations are then performed for different loading directions in order to establish the full capacity envelope as shown in Figure 4.8.

The effect that a potential tension gap may develop at the active side to a certain depth needs to be considered in this case. This effect is considered in the next section.

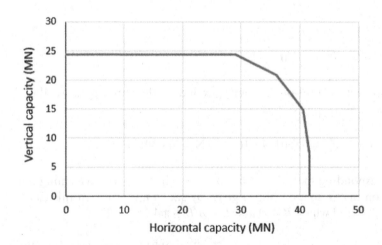

Figure 4.8 Calculated holding capacity diagram (H_f, V_f) for a suction pile with $D = 5\,\text{m}$, $Z = 10\,\text{m}$, $s_u = 100\,\text{kPa}$ (isotropic and constant with depth), and $r_{int} = 0.5$ loaded at the optimum load attachment point.

4.3.4 Other effects that need to be considered

The holding capacity may also be affected by other aspects as the development of an open gap at the active side, drainage through the top cap, mis-orientation and tilt during installation, nonoptimal pad-eye depth, and set-up effects. In addition, the loading angle will change from seabed down to the load attachment point due to the soil resistance against the mooring line. These effects are briefly discussed below. In addition, it may be effects of inclined seabed, the development of trenching due to mooring line movements (Hernandez-Martinez et al., 2015; Sassi et al., 2018), long term loading, etc. that also need to be considered.

4.3.4.1 Open gap at the active side

If the undrained shear strength on the active side is too large compared to the submerged weight of the soil, a tensile gap may develop to a certain depth at the active side. To simulate or calculate the development of a tensile crack starting from seabed is not straightforward; therefore, it is often assumed that a gap may develop if the resultant tensile stress due to the undrained shear strength of the soil is larger than the contribution from the submerged weight of the soil. This means that if a calculation assuming a gap at the active side gives lower capacity than without, a gap is conservatively assumed. A more detailed study of the effect of a tensile gap at the active side performed by 3D finite element analyses may for instance be found in (Fu et al., 2020).

4.3.4.2 Drainage through the top cap

If the top cap cannot be considered as sealed, it should be checked that the resultant tensile capacity of the soil plug V_{tip} is not larger than the vertical shear resistance along the inner walls. This shear force can be calculated by the same equation as used for the penetration resistance, however, with a time dependent set-up factor $r_{set-up, inside}$ that may be taken from the study in Andersen & Jostad (2004).

4.3.4.3 Misorientation and tilt

The effect of mis-orientation after installation, which means that a vertical plane that goes through the vertical centerline and the load attachment point (pad-eye) has an inclination β with the vertical plane defined by the loading direction, is accounted for by reducing the set-up factor r_{int} along the outer walls based on the following equation:

$$r_{int, mis} = r_{int} \left(1.0 - \left(F \cos(\alpha_l) \cdot \sin\beta \cdot x_p / M_{max} \right)^2 \right)^{0.5} > 1/S_t$$

where M_{max} is the calculated maximum torsional capacity of the suction pile.

The effect that the suction pile may be installed with an inclination α_{tilt} from the vertical axis is often accounted for in a simplified way by modifying the loading angle accordingly, i.e., the effective loading angle is $\alpha_l \pm \alpha_{tilt}$. However, the effect that the optimum load attachment point at the same time is changing is then neglected. This effect is considered in the next section.

4.3.4.4 Nonoptimal pad-eye depth

The effect that the load attachment point is not at the optimum load attachment point is difficult to account for by the simplified equations given in Section 4.3.2. In these situations, some of the soil resistance needs to be used to stabilize the overturning moment given by the eccentricity between the actual loading point and the optimum load attachment point. The effect is illustrated in Figure 4.9 for the suction pile considered in Section 4.3.3 subjected to pure horizontal loading ($\alpha_l = 0$). The effect is calculated by a finite element formulation where the 3D effect is accounted for by side shear in a similar way as used in the limit equilibrium analyses presented in Sections 4.3.2 and 4.3.3. This formulation is described in more detail in Jostad & Andersen (2015). From Figure 4.9, it is seen that the holding capacity may be rather sensitive to the actual load attachment point. A load attachment point that is located 1.0 m above or below the optimum depth $z_{l,opt} = 5.55$ m may reduce the capacity by more than 10%. Figure 4.10 shows the calculated characteristic failure mode with a rotational slip surface around skirt tip level for a top loaded suction pile.

For a suction pile that potentially may develop a gap at the active side when loaded at the optimum load attachment point, it may be more efficient to lower the load attachment point and create a rotational failure mode that prevent a gap to develop. This is discussed in more detail in (Fu et al., 2020).

4.3.4.5 Set-up factor

The holding capacity of suction piles is generally affected by the time after installation. During penetration, the shear strength in the interface between the suction pile and the soil is remoulded. In a normally consolidated clay, the effective stress in this

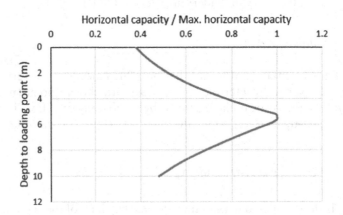

Figure 4.9 Calculated normalized holding capacity $H_f/H_{f,max}$ for a suction pile with $D = 5$ m, $Z = 10$ m, and $s_u = 100$ kPa (isotropic and constant with depth) loaded horizontally at different depths below seabed.

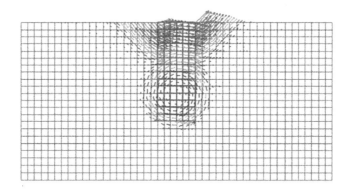

Figure 4.10 Calculated failure mode (resultant displacement vectors) for a suction pile with $D = 5$ m, $Z = 10$ m, and $s_u = 100$ kPa (isotropic and constant with depth) loaded horizontally at seabed level ($z_l = 0$).

remoulded zone is close to zero if the shear stress applied to the soil is close to zero. This means that there is large excess pore pressure in the remoulded zone after installation. This excess pore pressure will gradually reduce with time due to pore pressure dissipation into the surrounding soil. However, the remoulded zone will not regain its full strength after 100% pore pressure dissipation. This time-dependent process is studied in more detail in Andersen & Jostad (2002). For some considered offshore normally consolidated clays with a plasticity index $Ip > 30\%$, the set-up factor r_{int} after 90% pore pressure dissipation is found to be about 0.65. For a low plastic silty Norwegian clay ($Ip =12\%-15\%$), the corresponding set-up factor was found to be 0.58. The time for 90% consolidation is typically between 2 and 3 months for a suction pile with a diameter of around 4.5 m. For shorter times after installation, a smaller set-up factor is recommended to be used. For over-consolidated clays, the full set-up factor will generally be smaller than for normally consolidated clays. For a clay with OCR of 4, the set-up factor may be around 30% lower than for a normally consolidated clay. The set-up factor for the part of the soil interface where the suction pile penetrates by the self-weight was not considered in Andersen & Jostad (2002). If this contribution becomes important for the capacity, recommendations for piles may for instance be used, e.g., API (2014).

4.3.4.6 Chain configuration

The holding capacity of a suction pile depends on the loading angle at the load attachment point as shown in Figure 4.8. The chain angle and slightly also the load will change from the touch down point at seabed down to the load attachment point at the suction pile depending on the strength of the soil and the effective diameter of the mooring line (Figure 4.11).

The changes in load angle and load are calculated based on equilibrium of a number of linearized sections of the chain, starting from seabed. The forces acting on a section with length ΔL is the chain force F_{i-1} at the connection to the previous chain section,

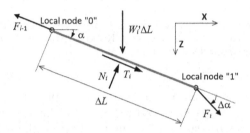

Figure 4.11 Illustration of chain configuration from seabed to the load attachment point at the suction pile.

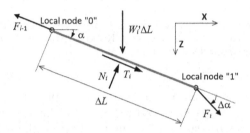

Figure 4.12 Forces acting on a chain section with length ΔL.

the resultant submerged weight of the section W_i', the resultant normal reaction force from the soil N_i and corresponding shear force T_i, and the chain force F_i at the other end of the section, as shown in Figure 4.12. The chain load F_i and change in loading angle $\Delta\alpha_i$ are found by considering axial and lateral equilibrium of the considered chain section:

$$F_i \cdot \cos(\Delta\alpha_i) = F_{i-1} - T_i - W_i' \cdot \sin(\Delta\alpha_i)$$

$$F_i \cdot \sin(\Delta\alpha_i) = N_i - W_i' \cdot \cos(\Delta\alpha_i)$$

The reaction forces from the soil (N_i and T_i) are calculated assuming that the chain is dragged into the soil and thus given by the strength of the soil. For a clay with undrained shear strength s_u, the soil reaction forces may be calculated from

$$T_i = d_e \cdot \Delta L \cdot r \cdot s_u$$

$$N_i = d_e \cdot N_c \cdot \Delta L \cdot s_u$$

where d_e is the equivalent chain diameter depending on the chain link geometry and diameter d; r is an effective roughness factor depending on the shape of the chain links, diameter d, relative displacement between the chain and the soil and the sensitivity of the clay; and N_c is an undrained bearing capacity factor varying between 5 and 12 depending on the relative depth of the chain section compared to seabed, the chain geometry, and the mobilized shear force T_i along the chain. Some theoretical studies of this may for instance be found in Steenfelt (1975) and back calculations in Degenkamp & Dutta (1989). Often an equivalent diameter of 2.5 times the chain link diameter d and a maximum $N_c = 9$ is often used for simplicity together with an average peak anisotropic undrained shear strength, $s_u = \left(s_u^C + s_u^{DSS} + s_u^E\right)/3$.

The change of chain tension F and orientation given by α are calculated by considering several sections with length ΔL required to define the configuration of the chain from seabed to the load attachment point at a defined depth z_l beneath the seabed. The chain force F_0 and angle α_0 at seabed are assumed to be known. The length ΔL needs to be adjusted to satisfy the depth of the load attachment point z_l.

4.3.5 Finite element approaches

The holding capacity of suction piles may be calculated by a 3D finite element program (e.g., Jostad & Andersen, 2015; Zdravkovic & Potts, 2005; Edgers et al., 2009; Fu et al., 2020). However, it is then important that the effect of discretization error is properly accounted for. Figure 4.13 shows the calculated holding capacity for the top loaded suction pile considered in Section 4.3.7 as a function of the average element length within the calculated failure zone. The capacity is herein defined at a horizontal displacement of the suction pile at seabed of 1.0 m. The soil is modeled by 10-noded tetrahedral elements, and 12-noded zero thickness interface elements are used between the suction pile and the surrounding soil with a set-up factor $r_{int} = 0.65$. Based on simulations using a model with 20-noded isoparametric quadrilateral elements with reduced integration and consistent 16-noded zero thickness interface elements, Figure 4.14 indicates that the capacity tends to converge to a value of $H_f = 18.5$ MN upon mesh refinements.

The advantages of using the finite element method for calculation of the holding capacity of suction piles are that the coupling between horizontal and vertical loading, effect of tilt and mis-orientation, nonoptimal loading point, interface set-up factor, anisotropic or stress path-dependent undrained shear strength, trenching, etc. may automatically be considered. However, it is more challenging to properly account for a

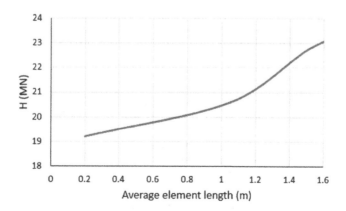

Figure 4.13 Calculated horizontal capacity H_f for a suction pile with $D = 5$ m, $Z = 10$ m, $s_u = 100$ kPa (isotropic and constant with depth), and $r_{int} = 0.65$ loaded horizontally at seabed level as a function of the average element length of the 15-noded tetrahedral element used to model the surrounding soil.

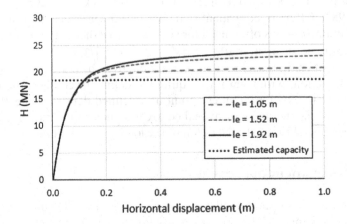

Figure 4.14 Calculated horizontal load *H* versus horizontal displacement at seabed for a suction pile with $D = 5\,m$, $Z = 10\,m$, $s_u = 100\,kPa$ (isotropic and constant with depth), and $r_{int} = 0.65$ loaded horizontally at seabed level for element meshes with different average element lengths of the 15-noded tetrahedral element used to model the surrounding soil.

potential tensile cap at the active side of the suction pile (Fu et al., 2020). Finally, there exist several benchmark examples in the literature (e.g. Andersen et al., 2005; Jostad & Andersen, 2015) that may be used to check the finite element calculation approach.

REFERENCES

American Petroleum Institute (2014) API recommended practice 2GEO/ISO 19901-4. Geotechnical and Foundation Design Considerations, 1st Edition, April 2011, Addendum 1, October 2014.

Andersen, K. H. (2015) Cyclic soil parameters for offshore foundation design. In: Meyer (ed) *Frontiers in Offshore Geotechnics III*, Taylor & Francis Group, London, pp. 5–82.

Andersen, K. H. & Jostad, H. P. (1999) Foundation design of skirted foundations and anchors in clay. In *Proceedings Offshore Technology Conference*: OTC Paper No. 10824, Houston, USA.

Andersen, K. H. & Jostad, H. P. (2002) Shear strength along outside wall of suction anchors in clay after installation. In *Twelfth International Offshore and Polar Engineering Conference*. International Society of Offshore and Polar Engineers.

Andersen, K. H. & Jostad, H. P. (2004) Shear strength along inside of suction anchor skirt wall in clay. In *Proceedings Paper 16844, Offshore Technology Conference*, OTC, Houston, May 2004.

Andersen, K. H., Murff, J. D., Randolph, M. F., Clukey, E. C., Erbrich, C. T., Jostad, H. P., ... & Supachawarote, C. (2005). Suction anchors for deepwater applications. In *Proceedings of the 1st International Symposium on Frontiers in Offshore Geotechnics, ISFOG, Perth* (pp. 3–30).

Aubeny, C. P., Han, S. W. & Murff, J. D. (2003) Inclined load capacity of suction caissons. *International Journal for Numerical and Analytical Methods in Geomechanics*, 27(14), 1235–1254.

Bang, S., Jones, K., Kim, Y. S. & Cho, Y. (2007) Horizontal capacity of embedded suction anchors in clay. In *Proc. Int. Conf. Offshore Mechanics and Arctic Eng, ASME*. Paper OMAE2007-29115 (pp. 353–360). https://doi.org/10.1115/OMAE2007-29115.

Brinch Hansen, J. (1970) A Revised and extended formula for bearing capacity. *Geoteknisk Institutt, Bulletin* No. 28, 5–11. Copenhagen.

Degenkamp, G. & Dutta, A. (1989, October) Soil resistance to embedded anchor chain in soft clay. *Journal of Geotechnical Engineering*, 115(10), 1420–1438.

Edgers, L., Andresen, L. & Jostad, H. P. (2009) Capacity analysis of suction anchors in clay by 3D finite element analysis. In *Proceedings, 1st International Symposium on Computational Geomechanics (ComGeo I), Juan-les-Pins*, France.

Fu, D., Zhang, Y., Yan, Y. & Jostad, H. P. (2020) Effects of tension gap on the holding capacity of suction anchors. *Marine Structures*, 69, 102679.

Hernandez-Martinez, F. G., Saue, M., Schroder, K. & Jostad, H. P. (2015, January) Trenching effects on holding capacity for in-service suction anchors in high plasticity clays. In *SNAME 20th Offshore Symposium*. The Society of Naval Architects and Marine Engineers.

Huang, J., Cao, J. & Audibert, J. M. (2003) Geotechnical design of suction caisson in clay. In *Thirteenth International Offshore and Polar Engineering Conference*. International Society of Offshore and Polar Engineers.

Jostad, H. P. & Andersen, K. H. (2015) Calculation of undrained holding capacity of suction anchors in clays. *Frontiers in Offshore Geotechnics*, III, 263–268.

Jostad, H. P., Andersen, K. H., Khoa, H. D. V. & Colliat, J. L. (2015, May) Interpretation of centrifuge tests of suction anchors in reconstituted soft clay. In *Proceedings of the 3rd International Symposium on Frontiers in Offshore Geotechnics,* Oslo, Norway (pp. 269–276).

Li, S. & Wang, J. (2013, June) Analysis of suction anchors bearing capacity in soft clay. In *The Twenty-Third International Offshore and Polar Engineering Conference*. International Society of Offshore and Polar Engineers.

Randolph, M. F. & House, A. R. (2002) Analysis of suction caisson capacity in clay. *OTC Paper 14236, Proceedings Offshore Technology Conference*, Houston, TX, May 6–9.

Sassi, K., Zehzouh, S., Blanc, M., Thorel, L., Cathie, D., Puech, A. & Colliat-Dangus, J. L. (2018, April) Effect of seabed trenching on the holding capacity of suction anchors in soft deepwater Gulf of Guinea clays. In *Offshore Technology Conference*.

Sparrevik, P. (1998) Suction anchors - a versatile foundation concept finding its place in the offshore market. In *American Society of Mechanical Engineers, 17th International Conference on Offshore Mechanics and Arctic Engineering (USA)*, (p. 13).

Steenfelt, J. S. (1975) Some plane strain failure problems in undrained clay. PhD thesis, Technical University of Denmark, Copenhagen.

Zdravkovic, L. & Potts, D. M. (2005) Parametric finite element analyses of suction anchors. *Frontiers in Offshore Geotechnics*. Perth: Taylor & Francis, 297–302.

Chapter 5

Simplified models for axial static and dynamic analysis of pile foundations

Jamie J. Crispin
University of Bristol

George E. Mylonakis
University of Bristol
Khalifa University
University of California at Los Angeles

CONTENTS

5.1   INTRODUCTION

Pile response to axial loads is critical for the safety and performance of critical infrastructure and has been thoroughly explored for a long time. Relevant research approaches fall into three main groups:

A   Rigorous numerical methods, mainly using the finite element method (FEM) and boundary element method (BEM) formulations:
  - FEMs are general-purpose continuum formulations capable of handling a wide range of pile-related problems including elastic analyses (Ottaviani 1975; Blaney et al. 1975; Randolph 1977; Kuhlemeyer 1979; Roesset & Angelides 1980; Kaynia 1980; Sanchez-Salinero 1982, 1983; El-Sharnouby & Novak 1990; Comodromos et al. 2003; Maheshwari et al. 2004; Syngros 2004; Kanellopoulos & Gazetas 2020)
  - On the other hand, BEMs and other semi-analytical formulations are best suited for linear elastostatic/elastodynamic analyses and can better handle infinite domains (especially in three dimensions) and dynamic loads (Poulos 1968; Butterfield & Banerjee 1971; Banerjee & Davies 1977, 1978; Kaynia 1982; Kaynia & Kausel 1982, 1991; Davies et al. 1985; Rajapakse 1990; Mamoon et al. 1990; Gazetas et al. 1991,1993; El-Marsafawi et al. 1992; Padrón et al. 2007, 2009, 2012);
B   Experimental methods, mainly field testing including reduced-scale models, monitoring of piles supporting real structures, or analysis of case histories (Seed & Reese 1957; Coyle & Reese 1966; O'Neill et al. 1982; Blaney et al. 1987; Mizuno 1987; El-Marsafawi et al. 1992; Mandolini & Viggiani 1997; Syngros 2004; Rollins et al. 2005; Voyagaki et al. 2019, 2021);
C   Approximate analytical models, mainly Winkler formulations. These can be further classified into the following:
  - Linear elastic Winkler models developed for static (Cooke 1974; Randolph & Wroth 1978,1979; Baguelin & Frank 1980; Mylonakis 1995; Guo & Randolph 1997; Mylonakis & Gazetas 1998a; Crispin et al. 2018, 2019; Crispin & Leahy 2019) or dynamic (Baranov 1967; Novak 1974; Novak et al. 1978; O'Rourke & Dobry 1978; Nogami 1983; Gazetas & Makris 1991; Makris & Gazetas 1993; Mylonakis & Gazetas 1998b; Dezi et al. 2009, 2016; Di Laora et al. 2017) conditions;
  - Non-linear Winkler models encompassing classic "*t-z*" formulations for static loads (Kraft et al. 1981; Motta 1994; Guo & Randolph 1998; Vardanega et al. 2012; Crispin et al. 2018, 2019; Bateman et al. 2021) or special "*t-z*" formulations for dynamic loads (Akiyoshi 1982; Nogami & Konagai 1987; El Naggar & Novak 1994a, b; Michaelides et al. 1998);
  - More rigorous analytical solutions involving the spatial condensation model originally proposed for surface footings by Vlasov & Leontiev (1966) (Mylonakis 2000, 2001b; Gupta & Basu 2018) or the constrained elasticity models originally suggested by Tajimi (1969) for laterally loaded piles (Mylonakis 2001a; Anoyatis & Mylonakis 2012, 2020; Anoyatis 2013; Anoyatis et al. 2019);
  - Other approximate solutions based on theory of elasticity and wave propagation including cone models (Dobry & Gazetas 1988; Wolf et al. 1992; Wolf 1994; Cairo et al. 1999).

Review papers and reports on the subject have been published (Roesset 1980; Poulos 1989; Novak 1991; Pender 1993; Gazetas & Mylonakis 1998; O'Neill 2001; Randolph 2003; NEHRP 2012). In addition, several books on the subject are available (Poulos & Davis 1980; Scott 1981; Prakash & Sharma 1990; Wolf 1994; Reese et al. 2005; Salgado 2008; Fleming et al. 2009; Viggiani et al. 2011; Madabhushi et al. 2010; Guo 2012; Poulos 2017; Jia 2018).

The methods in Group C are deemed particularly advantageous given their ability to reduce the dimensionality of the problem from two to a single dimension (or a superposition of a series of one-dimensional problems, that of a "horizontal soil slice" and an elastic rod), which requires significantly smaller computational effort than rigorous continuum formulations. Despite this simplification, these methods are still able to (Mylonakis 2001a)

- Yield predictions that are in satisfactory agreement with more rigorous solutions (both in the elastic and inelastic regime) after calibration of the governing parameters
- Incorporate variation of soil properties with load amplitude (non-linearity) and depth (in-homogeneity)
- Be extended to dynamic loads by adding pertinent distributed dampers to the spring bed, accounting for wave radiation and material energy dissipation
- Incorporate group effects through pertinent pile-to-pile interaction models
- Require small computational effort than rigorous numerical formulations

With reference to the last item in the above list, this work recognises that rigorous numerical methods such as FEM are not ideally suited for routine design as they are time-consuming and limited by the lack of general-purpose constitutive models as well as theoretical and computational complexities associated with the underlying solution procedures, which can make them unappealing to practicing engineers.

In this chapter, simplified models based on the Winkler approach for axially loaded piles are reviewed. Static and dynamic harmonic loading at the pile head is considered for both single piles and pile groups. The derivation of these methods is explained and modifications for layered and inhomogeneous soil provided. A suite of novel approximate energy solutions, capable of handling general soil inhomogeneity, is also provided. Various calibrations of the Winkler spring and dashpot values are implemented and compared to existing rigorous numerical solutions (Group A above). Application examples are presented.

5.2 SINGLE PILE RESPONSE

The problem, modelled using Winkler springs, is shown in Figure 5.1. The governing differential equation can be obtained by taking equilibrium of an infinitesimal element of the pile, modelled as a rod, in the vertical direction (Scott 1981):

$$E_p A_p \frac{d^2 w(z)}{dz^2} - k(z) w(z) = 0 \qquad (5.1)$$

where $w(z)$ is the displacement of the pile at depth z, E_p and A_p are the axial stiffness and cross-sectional area of the pile, respectively, and $k(z)$ describes the stiffness of

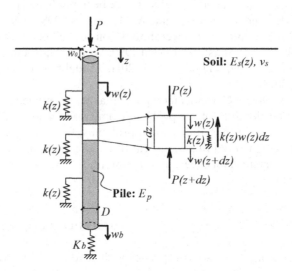

Figure 5.1 The Winkler model for a pile under axial load. (Adapted from Crispin et al. (2018).)

the Winkler springs as a function of depth. The Winkler spring stiffness is assumed proportional to the soil stiffness, and various approaches are available to calibrate it, which are discussed later in this chapter.

Of most interest is normally the response at the pile head, which can be described using a spring of stiffness $K_0 = P/w_0$, where P is the applied load, a common boundary condition, and w_0 is the resulting head settlement. The other boundary condition is the response at the pile base, represented here with a lumped spring of stiffness K_b. This boundary condition can model both an end-bearing pile $\left(K_b \to \infty\right)$ and a fully floating one carrying no load at the tip $\left(K_b = 0\right)$. Following the arguments of Randolph and Wroth (1978) and Scott (1981), the pile base response can be considered separately of the shaft response and acts as a rigid disc on the surface of an elastic medium. For a homogeneous half-space, this results in Eq. 5.2:

$$K_b = \frac{E_{sb}D}{1-v_s^2} \tag{5.2}$$

where D is the pile diameter, E_{sb} is the soil stiffness as the pile base, and v_s is the soil Poisson's ratio.

5.2.1   Homogeneous soil

For the simplest case of homogeneous soil [$k(z) = k$], solving Eq. 5.1 and applying the boundary conditions yield Eq. 5.3 (Mylonakis 1995; Mylonakis & Gazetas 1998a):

$$K_0 = E_p A_p \lambda \frac{\Omega + \tanh\left(\lambda L\right)}{1 + \Omega \tanh\left(\lambda L\right)} \tag{5.3}$$

where L is the pile length and λ and Ω are a load transfer parameter (units of length^{-1}) and normalised base stiffness constant, respectively:

$$\lambda = \sqrt{\frac{k}{E_p A_p}}, \quad \Omega = \frac{K_b}{E_p A_p \lambda} \tag{5.4}$$

The emergence of the parameter λ (which can be interpreted either as a characteristic length or a wavenumber controlling the attenuation of pile settlement with depth) is a fundamental property of Winkler models which allows direct normalisation of the independent variable z. The product λL can be viewed as a dimensionless "mechanical pile length" encompassing both geometric slenderness, L/D, and pile-soil relative stiffness, E_p/k, thus reducing the number of dimensionless parameters governing pile response by one (Anoyatis 2013).

Figure 5.2 shows how the normalised pile head stiffness varies with dimensionless pile length λL for different base conditions. It is intuitive that as the pile length increases, the pile head stiffness becomes less dependent on the base stiffness. This indicates an "active length", beyond which the pile head stiffness is independent of both pile length and base conditions (note that this is different from the corresponding length for lateral loading). If this is assumed to occur when the pile stiffness becomes an asymptotic function of length (e.g., $\lambda L = 2$, Figure 5.2), this gives an active length of approximately 70 pile diameters for $E_p/E_s = 1000$. This is well above most pile lengths used in practice (excluding more unconventional pile types such as micro-piles). Therefore, contrary to laterally loaded piles, for which actual length and base conditions are usually irrelevant as far as head response is concerned, both total pile length and base conditions are key to determining pile head response due to axial loading. The stiffness of an infinitely long pile can be obtained by substituting $\lambda L \rightarrow \infty$ into Eq. 5.3 (O'Rourke & Dobry 1978; Mylonakis 1995; Mylonakis & Gazetas 1998a):

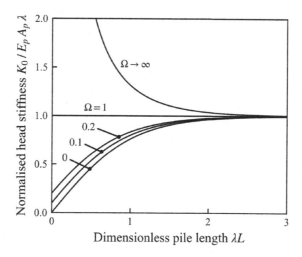

Figure 5.2 Normalised head stiffness of a pile embedded in a homogeneous soil, Eq. 5.3. (Adapted from Mylonakis (1995), and Mylonakis & Gazetas (1998a).)

$$K_0 = E_p A_p \lambda \tag{5.5}$$

For end-bearing piles, $\Omega \to \infty$, substituting this into Eq. 5.3 yields (O'Rourke & Dobry 1978; Mylonakis 1995)

$$K_0 = E_p A_p \lambda / \tanh(\lambda L) \tag{5.6}$$

For fully floating piles with no base resistance, $\Omega = 0$, therefore (O'Rourke & Dobry 1978; Mylonakis 1995)

$$K_0 = E_p A_p \lambda \tanh(\lambda L) \tag{5.7}$$

Finally, as the pile length approaches zero $(\lambda L \to 0)$, the pile head stiffness approaches that of the pile base spring.

5.2.2   Selection of k(z)

Various methods are available to establish the Winkler spring stiffness, most of which involve calibration against a continuum solution. In principle, Winkler stiffness can be established by dividing pile shaft resistance ("side friction") by the corresponding pile displacement using experimental, numerical, or analytical data (e.g., Syngros 2004; Mylonakis 2001a, b; Salgado 2008; Anoyatis 2013). Such an approach, however, will invariably lead to depth-dependent results which are difficult to implement in practice. As an alternative, depth-independent "average" values can be derived by matching (calibrating) a key response parameter (usually settlement at pile head) between a Winkler and a more rigorous solution.

Table 5.1 includes a number of formulations available in the literature. O'Rourke & Dobry (1978) and Roesset & Angelides (1980) provide expressions calibrated against boundary element and finite element numerical results, respectively. Randolph and Wroth (1978) and Guo (2012) employ the concentric cylinder model (Cooke 1974; Baguelin & Frank 1980), in which the variation with depth of the vertical normal stress is neglected to give a simple plane-strain model. This yields a simple analytical expression for the Winkler spring stiffness. However, in order to obtain a finite stiffness, an empirical mobilised radius, r_m, must be selected beyond which the settlement due to the pile head load is assumed to be zero. Randolph & Wroth (1978), Guo & Randolph (1998), and Guo (2012) provide expressions for r_m based on comparison with finite element and boundary element results, respectively. Mylonakis (2001a) and Anoyatis & Mylonakis (2012) obtained more rigorous analytical solutions for end-bearing piles without the simplifying assumption of the concentric cycled model, negating the need for r_m. Both authors matched these solutions to the Winkler model, to obtain simplified expressions for the Winkler spring stiffness.

The aforementioned expressions for Winkler spring stiffness are compared in Figure 5.3 for the cases of a pile embedded in a homogeneous half-space and a homogeneous layer with rigid bedrock at the pile base.

Recasting Eq. 5.3 into Eq. 5.8 allows for convenient comparison between Winkler solutions and numerical continuum solutions:

$$\frac{K_0}{E_s D} = \frac{4}{\pi} \frac{L}{D} \left(\frac{E_p}{E_s}\right)^{-1} \frac{\Omega + \tanh(\lambda L)}{\lambda L + \lambda L \Omega \tanh(\lambda L)} \tag{5.8}$$

where (Mylonakis 1995):

$$\lambda L = \frac{2}{\sqrt{\pi}} \frac{L}{D} \left(\frac{k}{E_s}\right)^{1/2} \left(\frac{E_p}{E_s}\right)^{-1/2} \tag{5.9}$$

$$\Omega = \frac{2}{\left(1 - v_s^2\right)\sqrt{\pi}} \frac{E_{sb}}{E_s} \left(\frac{k}{E_s} \frac{E_p}{E_s}\right)^{-1/2} \tag{5.10}$$

Table 5.1 Static Winkler spring stiffness equations

Source	Winkler spring stiffness, k
Roesset & Angelides (1980)	$k \approx 0.6 E_s$
Guo (2012), Guo & Randolph (1998)	$k \approx 2\pi G_s / \ln(2r_m/D)$ where $r_m = A(1-v_s)\dfrac{1}{1+n}L + D/2$, $A \approx 1 + 1.1e^{-\sqrt{n}}\left(1 - e^{1-\frac{H}{L}}\right)$, n is an exponent describing soil stiffness inhomogeneity with depth, and H is the depth to an underlying hard layer
Floating piles only Randolph & Wroth (1978)	$k \approx 2\pi G_s / \ln(2r_m/D)$ where $r_m = 2.5\rho L(1-v_s)$ and $\rho = G_s(L/2)/G_s(L)$
O'Rourke & Dobry (1978)	$\dfrac{k}{G_s} = 1 + 5.5\left(\dfrac{L}{D}\right)^{-0.74} + 0.6\left(\dfrac{L}{D}\right)^{1.3}\left(\dfrac{E_p}{E_s}\right)^{-1.04}$
End-bearing piles only O'Rourke & Dobry (1978)	$\dfrac{k}{G_s} = 1.48 + 17\left(\dfrac{L}{D}\right)^{-1}$
Mylonakis (2001a)	$\dfrac{k}{G_s} = 1.3\left(\dfrac{E_p}{E_s}\right)^{-1/40}\left[1 + 7\left(\dfrac{L}{D}\right)^{-3/5}\right]$
Anoyatis & Mylonakis (2012), Mylonakis (2001b)	$\dfrac{k}{G_s} = \dfrac{2\pi}{\ln(4/s_{st}) - \gamma} - s_{st}^2$ where $\gamma\ (\approx 0.577)$ is Euler's number, $s_{st} = \eta_s \dfrac{\pi}{2}\left(\dfrac{L}{D}\right)^{-1}$ and $\eta_s = \sqrt{\dfrac{2}{1-v_s}}$

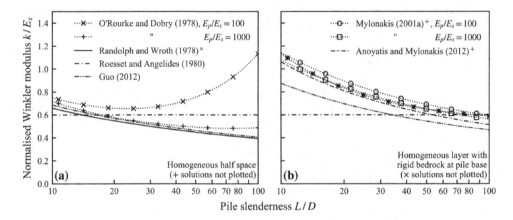

Figure 5.3 Static Winkler modulus (Table 5.1) for piles embedded (a) in a homogeneous half-space and (b) in a homogeneous soil layer with rigid bedrock at the pile base $\left(v_s = 0.4\right)$.

Figure 5.4 shows a comparison of the performance of the Winkler solutions described in Eq. 5.8 and Table 5.1 against some existing continuum solutions. In general, the continuum solutions all provide results in good agreement with each other. For the low pile stiffness ($E_p/E_s = 100$), the results are naturally independent of pile slenderness and base conditions, indicating that in this case, the pile is acting as infinitely long. However, an increase in stiffness with pile slenderness is observed in the results by Poulos & Davis (1980). This is due to the sensitivity of the boundary element method to the discretisation of the pile (Mylonakis 2001a). For end-bearing piles, all of the Winkler spring stiffness calibrations perform well. Nevertheless, for the low pile stiffness case in the homogeneous half-space, the results using the O'Rourke & Dobry (1978) spring stiffness deviate in a similar way to the results presented in Poulos & Davis (1980), against which they were calibrated. For the homogeneous half-space, there is good agreement for the remaining Winkler results, with the possible exception of the simple expression provided by Roesset & Angelides (1980), which does not perform as well in this case.

5.2.3   Layered soil

Piles are often embedded in soil that can be discretised into a number of piece-wise layers based on geological history and engineering properties. An example of a pile installed in M soil layers is shown in Figure 5.5.

Considering the boundary conditions applied to obtain Eq. 5.3, it can be seen that K_b can be substituted with the head stiffness of a pile that terminates immediately below the pile section being considered. In this way, Eq. 5.3 can be rewritten for pile section m of thickness h_m:

$$K_m = E_p A_p \lambda_m \frac{\Omega_m + \tanh\left(\lambda_m h_m\right)}{1 + \Omega_m \tanh\left(\lambda_m h_m\right)} \tag{5.11}$$

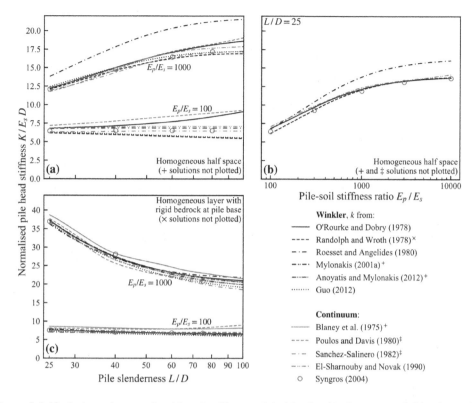

Figure 5.4 Variation of normalised head stiffness with (a) pile slenderness and (b) pile-soil stiffness ratio for piles embedded in a homogeneous half-space; (c) variation of normalised head stiffness with pile slenderness for piles embedded in a homogeneous soil layer with rigid bedrock at the pile base $(v_s = 0.5)$. (Numerical data from El-Sharnouby & Novak (1990) and Syngros (2004).)

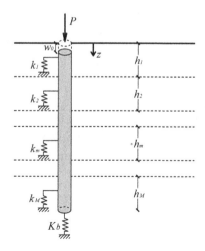

Figure 5.5 Winkler model for a pile embedded in M soil layers.

where

$$\lambda_m = \sqrt{\frac{k_m}{E_p A_p}}, \quad \Omega_m = \frac{K_{m+1}}{E_p A_p \lambda_m} \tag{5.12}$$

The above equations can then be applied iteratively, from the base of the pile (layer M), with $K_{M+1} = K_b$ (calculated from Eq. 5.2) to get the overall head stiffness, $K_0 = K_1$. For example, the head stiffness of a pile embedded in a two-layer soil is given by (Mylonakis & Gazetas 1998a):

$$K_0 = E_p A_p \lambda_1 \frac{\lambda_1 \tanh(\lambda_1 h_1) + \lambda_2 \tanh(\lambda_2 h_2) + \lambda_1 \Omega \, \tanh(\lambda_1 h_1) \tanh(\lambda_2 h_2) + \lambda_2 \Omega}{\lambda_1 + \lambda_1 \Omega \, \tanh(\lambda_2 h_2) + \lambda_2 \Omega \, \tanh(\lambda_1 h_1) + \lambda_2 \tanh(\lambda_1 h_1) \tanh(\lambda_2 h_2)} \tag{5.13a}$$

where λ_1 and λ_2 can be calculated by substituting 1 and 2, respectively, for m in Eq. 5.12, and Ω is given by Eq. 5.4. Eq. 5.13a can generate the solution for a three-layer soil by replacing Ω with Eq. 5.13b:

$$\Omega = \frac{\lambda_3}{\lambda_2} \left[\frac{K_b + E_p A_p \lambda_3 \tanh(\lambda_3 h_3)}{E_p A_p \lambda_3 + K_b \tanh(\lambda_3 h_3)} \right] \tag{5.13b}$$

where λ_3 can be calculated from Eq. 5.12 as before. In this way, complex inhomogeneous soil profiles can be modelled.

5.2.4   Inhomogeneous soil layers

In many cases, soil properties vary in ways that would be inconvenient to describe using homogeneous layers, and a continuous function may be preferred. Whilst the method described above for layered soil could be employed using small homogeneous layers to describe any continuous function, a closed form solution usually gives more insight into pile behaviour and, in certain cases, can be easier to employ. However, for each family of inhomogeneity functions, a solution to Eq. 5.1 must be found and the boundary conditions applied. This has only been carried out for a limited set of inhomogeneity functions.

5.2.4.1   Linear and power-law stiffness profiles

Scott (1981), Guo (2012), and Crispin et al. (2018) have developed solutions for $k(z)$ following linear and power-law functions of depth. These can be written in a general form similar to Eq. 5.14:

$$k(z) = k_L \left[a + (1-a) \frac{z}{L} \right]^n, \quad a^n = \frac{k_0}{k_L} \tag{5.14}$$

where k_0 and k_L are the Winkler spring stiffness at the pile head and tip, respectively, a is a dimensionless surface stiffness, and n is a dimensionless exponent. Any power-law or linear function that has been fitted through soil properties can be recast in this form to allow the following solutions to be applied. Setting $n = 1$ gives a linear variation of $k(z)$ with depth, and setting $n = 0$ or $a = 1$ simplifies the problem to that of a homogeneous soil.

Inputting Eq. 5.14 into the governing differential equation (Eq. 5.1) and applying the boundary conditions yield Eq. 5.15 for pile head stiffness (Crispin et al. 2018). This can be modified to apply to a soil layer in a sequence in a similar way that Eq. 5.3 was modified into Eq. 5.11. Alternative forms of this solution are available in Scott (1981), Guo & Randolph (1997), and Crispin et al. (2018).

$$K_0 = E_p A_p \lambda_0$$
$$\frac{\left[I_{v-1}(\chi_0)I_{1-v}(\chi_L) - I_{1-v}(\chi_0)I_{v-1}(\chi_L)\right] + \Omega_L\left[I_{v-1}(\chi_0)I_{-v}(\chi_L) - I_{1-v}(\chi_0)I_{+v}(\chi_L)\right]}{\left[I_{-v}(\chi_0)I_{v-1}(\chi_L) - I_{+v}(\chi_0)I_{1-v}(\chi_L)\right] + \Omega_L\left[I_{-v}(\chi_0)I_{+v}(\chi_L) - I_{+v}(\chi_0)I_{-v}(\chi_L)\right]}$$

$$(5.15)$$

where λ_0 and λ_L are analogous to λ, Ω_L is analogous to Ω, $I_\mu(\cdot)$ is the modified Bessel function of the first kind of order v, and χ_L and χ_0 are the arguments of the Bessel functions (Crispin et al. 2018):

$$\lambda_0 = \sqrt{\frac{k_0}{E_p A_p}}, \quad \Omega_L = \frac{K_b}{E_p A_p \lambda_L}, \quad \lambda_L = \sqrt{\frac{k_L}{E_p A_p}}, \quad v = \frac{1}{n+2}, \quad \chi_0 = \frac{2v\lambda_0 La}{1-a}, \quad \chi_L = \frac{2v\lambda_L L}{1-a}$$

$$(5.16)$$

The variation of normalised pile head stiffness with dimensionless pile length $\lambda_L L$ is shown in Figure 5.6 for different base conditions and two soil profiles: linear ($n = 1$) and parabolic ($n = 1/2$), both with vanishing surface stiffness ($a = 0$). This shows that the behaviour of a pile embedded in inhomogeneous profile described by Eq. 5.14 is

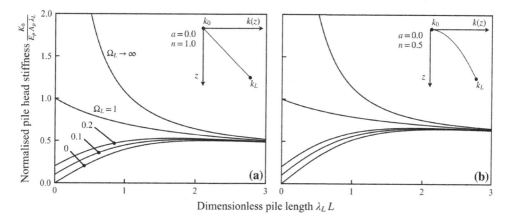

Figure 5.6 Normalised head stiffness of a pile embedded in inhomogeneous soil (Eq. 5.15) for (a) a linear soil profile with vanishing surface stiffness ($n = 1$, $a = 0$) and (b) a parabolic soil profile with vanishing surface stiffness ($n = 1/2$, $a = 0$). (Adapted from Crispin et al. (2018)).

similar to that of one embedded in a homogeneous soil. In general, an equal length pile with the same base conditions appears to have a lower stiffness in inhomogeneous soils. However, this is due to the choice k_L to normalise the solution, which means the average Winkler spring stiffness over the pile length can be as low as half of that for the homogeneous equivalent, depending on the profile. Table 5.2 shows different forms of Eq. 5.15 attained at the limits of the parameters discussed for Eqs. 5.5–5.8. Note that to prevent the trivial solution of $K_0 = 0$ as $L \rightarrow \infty$ that occurs due to the dependence of Eq. 5.14 on L, this solution is defined based on the Winkler spring stiffness at a depth of one diameter, k_D.

$\Gamma(\cdot)$ is the gamma function and $K_\mu(\cdot)$ is the modified Bessel function of the second kind of order μ.

As a approaches zero, the asymptotic behaviour of the Bessel functions dominates, and Eq. 5.15 may prove difficult to compute. Table 5.2 includes alternative forms for the above equations when $a = 0$ in terms of Gamma functions to enable easier computation. Alternatively, a suitable lower bound for a can be chosen, such as $a = 10^{-3}$, which still allows computation at a reasonable accuracy. A similar issue occurs as a approaches 1 (a homogeneous soil). In this case, accurate results are produced by employing the homogeneous solutions when $a > 0.9$.

Figure 5.7 shows a comparison of the performance of the Winkler solutions for inhomogeneous soil described in Eq. 5.15 against some existing continuum solutions. Note that some of the Winkler spring stiffness equations in Table 5.1 are calibrated using

Table 5.2 Summary of closed-form solutions for the head stiffness, K_0, of a pile embedded in an inhomogeneous soil (Adapted from Crispin et al. (2018))

	General case $(a \neq 0)$		Vanishing surface stiffness $(a = 0)$
General case	Eq. 5.15		$E_p A_p \lambda_L \dfrac{\Gamma(1-v)}{\Gamma(v)(v\lambda_L L)^{nv}}$ $\times \dfrac{I_{1-v}(\chi_L) + \Omega_L I_{-v}(\chi_L)}{I_{v-1}(\chi_L) + \Omega_L I_{+v}(\chi_L)}$
Fully floating pile, $\Omega_L = 0$	$E_p A_p \lambda_0 \dfrac{I_{v-1}(\chi_0)I_{1-v}(\chi_L) - I_{1-v}(\chi_0)I_{v-1}(\chi_L)}{I_{-v}(\chi_0)I_{v-1}(\chi_L) - I_{+v}(\chi_0)I_{1-v}(\chi_L)}$		$E_p A_p \lambda_L \dfrac{\Gamma(1-v)}{\Gamma(v)(v\lambda_L L)^{nv}}$ $\times \dfrac{I_{1-v}(\chi_L)}{I_{v-1}(\chi_L)}$
End-bearing pile, $\Omega_L \rightarrow \infty$	$E_p A_p \lambda_0 \dfrac{I_{v-1}(\chi_0)I_{-v}(\chi_L) - I_{1-v}(\chi_0)I_{+v}(\chi_L)}{I_{-v}(\chi_0)I_{+v}(\chi_L) - I_{+v}(\chi_0)I_{-v}(\chi_L)}$		$E_p A_p \lambda_L \dfrac{\Gamma(1-v)}{\Gamma(v)(v\lambda_L L)^{nv}}$ $\times \dfrac{I_{-v}(\chi_L)}{I_{+v}(\chi_L)}$
Infinitely long pile, $L \rightarrow \infty$	$E_p A_p \lambda_0 \dfrac{K_{v-1}(\chi_D)}{K_v(\chi_D)}$, $\chi_D = \dfrac{2v\lambda_0 D a_D}{1-a_D}$, $a_D^n = \dfrac{k_0}{k_D}$		$E_p A_p \lambda_D \dfrac{\Gamma(1-v)}{\Gamma(v)(v\lambda_D D)^{nv}}$, $\lambda_D = \sqrt{\dfrac{k_D}{E_p A_p}}$

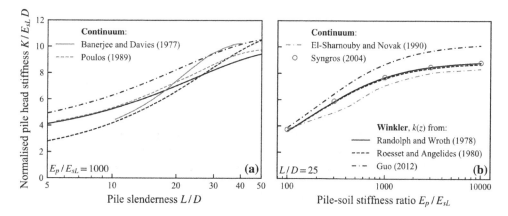

Figure 5.7 Variation of normalised head stiffness with (a) pile slenderness and (b) pile-soil stiffness ratio for piles embedded an inhomogeneous half-space with a linear soil stiffness profile (Eq. 5.14, $a = 0$, $n = 1$; $v_s = 0.5$). (Numerical data from Poulos (1989) and Syngros (2004).)

the pile soil stiffness ratio E_p/E_s for homogeneous soil and as such are omitted. Care should be taken when employing these expressions in inhomogeneous soil; selecting an appropriate equivalent homogeneous pile soil stiffness ratio is discussed in Syngros (2004). In general, the Winkler solution performs well. However, the results using Guo's (2012) springs tend to overestimate stiffness compared to the other solutions except for long or low stiffness piles (i.e., closer to acting like an infinitely long pile). This is likely due to differences between the finite difference solutions (FLAC) this method was calibrated against, and the boundary-element and finite-element continuum solutions presented here. The largest differences between solutions are for short piles, where the boundary element methods shown are more sensitive to pile discretisation.

5.2.4.2 Other inhomogeneous profiles

Due to the presence of Bessel functions, the solution presented in the previous section might not be appealing to geotechnical engineers and can also be difficult to employ. In addition, solutions may be desired for more general soil stiffness profiles that do not require solving Eq. 5.1 in closed form. An approximate solution that allows for general soil inhomogeneity can be obtained using an energy method similar to that employed by Karatzia & Mylonakis (2017) for laterally loaded piles. By considering the work done by a small virtual displacement, $w_v(z)$, an integral form of Eq. 5.1 can be obtained:

$$\int_0^L E_p A_p w''(z) w_v(z) dz - \int_0^L k(z) w(z) w_v(z) dz = 0 \tag{5.17}$$

A dimensionless shape function, $\chi(z)$, can then be selected to describe the variation of both the real and virtual displacement with depth, relative to the values at the surface.

This allows the pile head stiffness to be expressed as a virtual work equation (Crispin et al. 2019):

$$K_0 = E_p A_p \int_0^L \left[\chi'(z) \right]^2 dz + \int_0^L k(z) \left[\chi(z) \right]^2 dz + \left[\chi(L) \right]^2 K_b \tag{5.18}$$

which was derived from the weak form of Eq. 5.17. Note that each of the three terms on the right-hand side of Eq. 5.18 expresses a different component of the energy stored in the pile. These are the energy stored in the pile body, Winkler springs and base spring, respectively. Once a shape function is chosen, Eq. 5.18 can be resolved numerically for any soil stiffness profile. Inspiration for a suitable shape function can be taken from the closed-form solution for a pile embedded in homogeneous soil:

$$\chi(z) = \frac{w(z)}{w(0)} = \frac{\Omega_\mu \sinh(\mu L - \mu z) + \cosh(\mu L - \mu z)}{\Omega_\mu \sinh(\mu L) + \cosh(\mu L)} = \frac{w_v(z)}{w_v(0)} \tag{5.19}$$

where the parameter μ is analogous to the various λ parameters in Eqs. 5.4, 5.12, and 5.16, selected based on k_{avg}, the average of the Winkler spring stiffness over the pile length, yielding

$$\mu = \sqrt{\frac{k_{avg}}{E_p A_p}}, \Omega_\mu = \frac{K_b}{E_p A_p \mu} \tag{5.20}$$

For homogeneous soil, Eqs. 5.18 and 5.19 both yield the corresponding expressions derived by solving Eq. 5.1 in closed form. For a linear stiffness profile (Eq. 5.14, $n = 1$), this method gives the following approximate expression:

$$\frac{K_0}{E_p A_p \mu} \approx \frac{1}{2} \frac{k_L + 3k_0}{k_L + k_0} \frac{\Omega_\mu + \tanh(\mu L)}{1 + \Omega_\mu \tanh(\mu L)} + \frac{1}{2\mu L} \frac{k_L - k_0}{k_L + k_0} \left[1 - \frac{\left[1 + \mu L \Omega_\mu \right] / \cosh^2(\mu L)}{\left[1 + \Omega_\mu \tanh(\mu L) \right]^2} \right] \tag{5.21}$$

In this case, μ is given by

$$\mu = \sqrt{\frac{k_L + k_0}{2 E_p A_p}} \tag{5.22}$$

which corresponds to the value of $k(z)$ at the mid-length of the pile. The advantages of this expression compared to Eq. 5.15 are evident. Figure 5.8 shows the error in Eq. 5.21 compared to Eq. 5.15 for different base conditions as a function of both dimensionless length and dimensionless surface stiffness. In addition, the error in Eq. 5.18 (resolved numerically) is shown for a parabolic soil profile (Eq. 5.14, $n = 1/2$). As would be expected, the error approaches zero as the profile gets closer to homogeneous (i.e., as

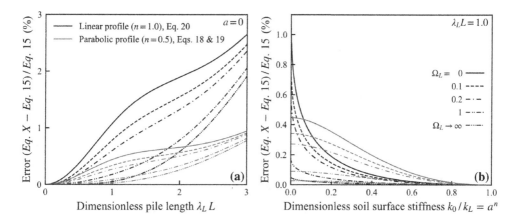

Figure 5.8 Error in energy method solutions relative to the closed form expression from Eq. 5.15 with (a) dimensionless pile length and (b) dimensionless surface stiffness.

a approaches 1). In general, the error is very small (<3% in the worst case), and considering the variance shown in Figures 5.4 and 5.7, the solution is suitable for most design problems.

5.2.5 Dynamic impedance

The solutions presented so far can be modified to account for dynamic loading. Considering an applied head load, $P(t)$, as a series of harmonics, the steady-state response for a head load applied at a particular angular frequency, ω, expressed by the phasor $Pe^{i\omega t}$, is illustrated in Figure 5.9. The pile is supported by distributed Winkler springs and dashpots of stiffness $k(z)$ and viscosity $c(z)$, respectively, and a base spring and dashpot of stiffness K_b and viscosity C_b, respectively. These impedances are dependent on the frequency of the load applied, soil stiffness, $E_s(z)$, shear wave velocity, $V_s(z)$, Poisson's ratio, v_s, mass density, ρ_s and hysteretic (material) damping factor, β_s, defined through the complex soil stiffness $E_s^* = E_s(1 + 2i\beta_s)$. The pile itself is modelled as an elastic rod with cross sectional area A_p, mass density ρ_p, and complex stiffness $E_p^* = E_p(1 + 2i\beta_p)$, where β_p is the hysteretic material damping factor for the pile. In this representation, the springs account for soil stiffness and the dashpots for energy dissipation due to friction and radiation damping (NEHRP 2012).

Considering steady state harmonic oscillations, the governing differential equation is given by

$$E_p^* A_p \frac{d^2 w^*(z)}{dz^2} - \left[k(z) + i\omega c(z) - \rho_p A_p \omega^2 \right] w^*(z) = 0 \tag{5.23}$$

where $w^*(z) = w^*(z, t)e^{-i\omega t}$, and $w^*(z, t)$ is the time-dependent displacement profile along the pile.

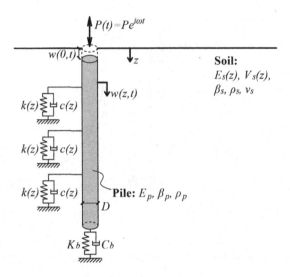

Figure 5.9 Winkler model for a pile under a harmonic load of angular frequency ω.

As before, of primary interest is the response at the pile head. In this case, it can be modelled using a lumped spring and dashpot of stiffness K_0 and viscosity C_0, respectively. It is convenient to express these as a complex valued stiffness (impedance), K^*, which can be expressed in various forms as shown in Eq. 5.24:

$$K^* = \frac{P}{w^*(0)} = K_0 + i\omega C_0 = |K^*|e^{i\theta} = \Re[K^*](1 + 2i\beta_0) \tag{5.24a}$$

$$|K^*| = \sqrt{\left(\Re[K^*]\right)^2 + \left(\Im[K^*]\right)^2}, \theta = \tan^{-1}\left(\frac{\Im[K^*]}{\Re[K^*]}\right), \beta_0 = \Im[K^*]/2\Re[K^*] \tag{5.24b}$$

where $\Re[K^*], \Im[K^*], |K^*|$, and θ are the real component, imaginary component, amplitude (modulus), and phase (argument) of the complex stiffness, respectively, and β_0 is the pile head damping coefficient.

Various approaches that are available to calculate the distributed Winkler impedances (stiffness intensities), $k(z)$ and $c(z)$, are shown in Table 5.3 and plotted in Figure 5.10. As with the overall pile impedance, it is convenient to express these as a complex valued stiffness, $k^* = k + i\omega c$.

Based on early work by Baranov (1967), Novak (1974) presented an analytical solution where the soil is modelled as an infinite number of uncoupled horizontal layers. The soil reaction due to unit vertical pile movement is calculated assuming in a way similar to the concentric cylinder model for the static case. For very low frequencies, this solution yields zero static stiffness (as is obtained in the static case). Novak et al. (1993) applies an empirical correction to the solution by assuming a constant spring value below a certain cut-off frequency. Alternatively, O'Rourke & Dobry (1978),

Table 5.3 Dynamic Winkler impedance equations

Source	Real component	Imaginary component
Baranov (1967), Novak (1974)[x]	$\dfrac{k}{G_s} = \pi a_0$ $\times \dfrac{J_1\left(a_0/2\right)J_0\left(a_0/2\right)+Y_1\left(a_0/2\right)Y_0\left(a_0/2\right)}{J_0^2\left(a_0/2\right)+Y_0^2\left(a_0/2\right)}$	$\dfrac{c_r \omega}{G_s} = \dfrac{4}{J_0^2\left(a_0/2\right)+Y_0^2\left(a_0/2\right)}$
Novak et al. (1978), Mylonakis (2001b)	$k^* = 2\pi G_s^* s^* \dfrac{K_1\left(s^*\right)}{K_0\left(s^*\right)}$, where $s^* = \dfrac{ia_0}{2\sqrt{1+2i\beta_s}}$	
O'Rourke & Dobry (1978)[x]	$\dfrac{k}{G_s} \approx 1 + 5.5\left(\dfrac{L}{D}\right)^{-0.74} + 0.6\left(\dfrac{L}{D}\right)^{1.3}\left(\dfrac{E_p}{E_s}\right)^{-1.04}$	$c_r \approx \pi \rho_s V_s D$
Roesset & Angelides (1980), Roesset (1980)[x]	$k \approx 0.75 E_s$	$c_r \approx \dfrac{C_1}{2}\rho_s V_s D$ where $6.3 \le C_1 \le 7.5$
Gazetas & Makris (1991), Makris & Gazetas (1993)	$k \approx 0.6 E_s\left(1 + \dfrac{1}{2}\sqrt{a_0}\right)$	$c \approx c_r + 2\beta_s \dfrac{k}{\omega}$ $c_r \approx 1.2 a_0^{-\frac{1}{4}} \pi \rho_s V_s D$

End-bearing piles only $\left(\omega > \omega_c\right)$

Source	Real component	Imaginary component
O'Rourke & Dobry (1978)[x]	$\dfrac{k}{G_s} \approx 1.48 + 17\left(\dfrac{L}{D}\right)^{-1}$	$c_r \approx \pi \rho_s V_s D$
Anoyatis & Mylonakis (2012)	$\dfrac{k^*}{G_s} = \dfrac{2\pi\left(1+2i\beta_s\right)}{\ln\left(2/s\right)-\gamma} + \left(a_0^2 - a_{0,c}^2\right)\left(1+\dfrac{i}{2}\right)$ where $a_{0,c} = \dfrac{\omega_c D}{V_s}$, $\gamma\ (\approx 0.577)$ is Euler's number and $s = \dfrac{1}{2}\sqrt{a_{0,c}^2 - a_0^2/\left(1+2i\beta_s\right)}$	

Where $a_0 = \omega D/V_s$ is the well-known dimensionless frequency and $J_\mu(\cdot)$ and $Y_\mu(\cdot)$ are the Bessel functions of the first and second kind, respectively, of order μ.
For $\omega \le \omega_c$ for end-bearing piles, see Eq. 5.29.
[x] Solution neglects hysteretic damping or accounts for it separately (see discussion above).

Roesset & Angelides (1980), and Gazetas & Makris (1991) provide Winkler impedance functions fitted against numerical continuum results. Some of these functions employ the well-known dimensionless frequency, $a_0 = \omega D/V_s$.

Most of the aforementioned expressions (indicated with a superscript "x" in Table 5.3) are for an undamped medium ($\beta_s = 0$). In these cases, only a radiation damping term c_r is provided. A few different approaches are available to incorporate hysteretic (material) damping into these solutions. This can be done in closed form using the correspondence principle of viscoelasticity, (Pipkin 1972) which involves substituting the material stiffnesses in the elastic solution with a corresponding complex modulus as follows:

$$G_s \to G_s^* = G_s\left(1 + 2i\beta_s\right), \quad E_s \to E_s^* = E_s\left(1 + 2i\beta_s\right), \quad V_s \to V_s^* = V_s\sqrt{1 + 2i\beta_s} \qquad (5.25)$$

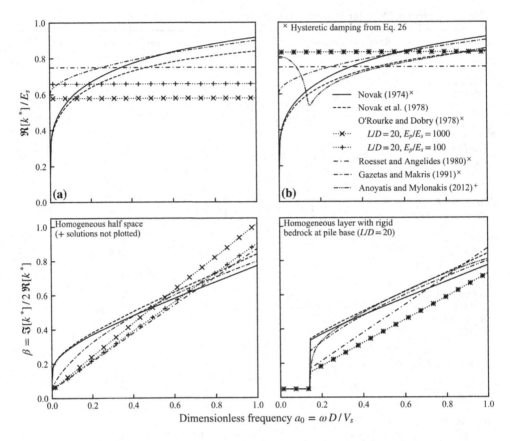

Figure 5.10 Dynamic Winkler modulus for a pile embedded (a) in a homogeneous half-space and (b) in a homogeneous soil layer with rigid bedrock at the pile base $\left(v_s = 0.4,\ \beta_s = 0.05\right)$.

This was applied to the Novak (1974) solution for an undamped soil medium by Novak et al. (1978). In most cases, it results in a complex stiffness that cannot be conveniently separated into real and imaginary parts in closed form (i.e., as the Bessel functions are not separable). Based on earlier work by Dobry et al. (1982) and Gazetas & Dobry (1984) for laterally loaded piles, Makris & Gazetas (1993) approximated this by simply adding a hysteretic damping term, c_h, to the radiation damping, as shown:

$$c \approx c_h + c_r, \quad c_h \approx 2\beta_s \frac{k}{\omega} \tag{5.26}$$

Alternatively, O'Rourke & Dobry (1978) suggest using a lumped dashpot at the top of the pile, C_h, given by

$$C_h \approx 2\beta_s \frac{K_0}{\omega} \tag{5.27}$$

The impedance functions discussed so far apply to piles embedded in an elastic half-space. However, a finite soil layer resting on a stiff deposit exhibits a cut-off frequency, ω_c, below which waves do not propagate and therefore, radiation damping does not exist. This cut-off frequency is nearly identical to the natural frequency of the soil layer in vertical compression-extension vibration, which is approximately given by (Gazetas et al. 1993)

$$\frac{\omega_c D}{V_s} \approx \frac{1.7}{1-v_s}\left(\frac{H}{D}\right)^{-1} \tag{5.28}$$

For frequencies below the cutoff $(\omega \le \omega_c)$, no radiation damping is generated, and therefore, the Winkler impedances can be obtained using the simplified expressions below (using the same assumptions as for Eq. 5.26):

$$k = \Re\left[k^*\right], \quad c = c_h \approx 2\beta_s \frac{k}{\omega} \tag{5.29}$$

Anoyatis & Mylonakis (2012) developed a more rigorous analytical solution for end-bearing piles, which does not require the independent horizontal soil layers assumption of Novak (1974) and Baranov (1967). In addition, they provided a simplified fitted function for the dynamic Winkler impedance by matching the resulting pile head response to the Winkler model, as well as an alternative expression for the cutoff frequency:

$$\frac{\omega_c D}{V_s} \approx \sqrt{\frac{2}{1-v_s}}\frac{\pi}{2}\left(\frac{H}{D}\right)^{-1} \tag{5.30}$$

This expression is in good agreement with Eq. 5.28.

Following the same arguments for the static case, the pile base can be modelled as a massless rigid disc on the surface of the underlying elastic half-space (Mylonakis 1995). For v_s close to 0.5, corresponding to the comment case of an incompressible medium, the complex base impedance is given by Eq. 5.31 (Veletsos & Verbic 1973; Roesset 1980). Hysteretic damping can be accounted for using the correspondence principle or approximated by simply adding an additional damping term as for the Winkler impedance.

$$K_b^* = K_b + i\omega C_b \approx \frac{E_{sb}D}{1-v_s^2}\left(\tilde{k}\left(1+2i\beta_s\right)+i\tilde{c}a_0\sqrt{1+2i\beta_s}\right)$$

$$\approx \frac{E_{sb}D}{1-v_s^2}\left(\tilde{k}+2i\tilde{k}\beta_s+i\tilde{c}a_0\right) \tag{5.31}$$

where $\tilde{k}$ and $\tilde{c}$ are dimensionless coefficients accounting for frequency dependence which can be taken as approximately 1 and 0.425, respectively (Veletsos & Verbic 1973).

5.2.5.1   Homogeneous soil

When $k(z) = k$ and $c(z) = c$, Eq. 5.23 is the same as Eq. 5.1 except for the value of the multiplier of the second term. As such, a solution can be obtained by simply replacing the static terms in Eq. 5.3 with their dynamic counterparts as shown in Eq. 5.32:

$$K^* = E_p^* A_p \lambda^* \frac{\Omega^* + \tanh\left(\lambda^* L\right)}{1 + \Omega^* \tanh\left(\lambda^* L\right)} \tag{5.32}$$

where

$$\lambda^* = \sqrt{\frac{k^* - \rho_p A_p \omega^2}{E_p^* A_p}}, \quad \Omega^* = \frac{K_b^*}{E_p^* A_p \lambda^*} \tag{5.33}$$

Figures 5.11 and 5.12 show the performance of this solution for different Winkler impedance equations (Table 5.3) against existing continuum solutions for piles embedded in a

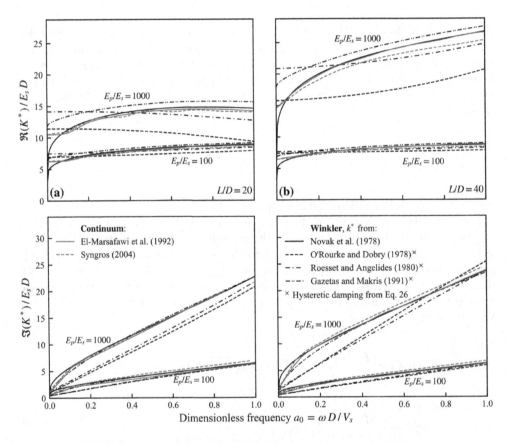

Figure 5.11 Normalised dynamic axial head stiffness of piles embedded in homogeneous half-space for (a) $L/D = 20$ and (b) $L/D = 40$ $\left(v_s = 0.4,\ \beta_s = 0.05,\ v_p = 0.25,\right.$ $\left.\beta_p = 0.01,\ \rho_p/\rho_s = 1.25\right)$. (Numerical data from Syngros (2004).)

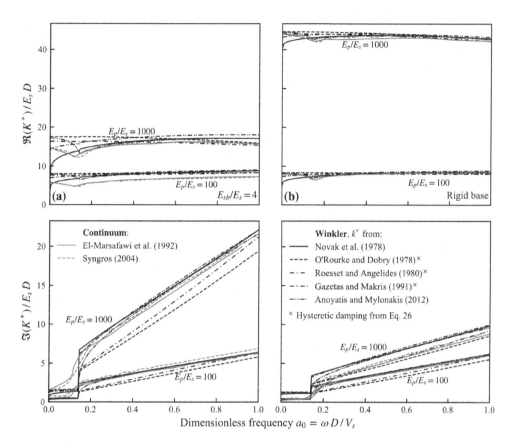

Figure 5.12 Normalised dynamic axial head stiffness of piles embedded in homogeneous soil bearing on (a) a stiff layer where $E_{sb}/E_s = 4$ and (b) rigid bedrock at the pile base $\left(L/D = 20, v_s = 0.4, \beta_s = 0.05, v_p = 0.25, \beta_p = 0.01, \rho_p/\rho_s = 1.25\right)$. (Numerical data from Syngros (2004).)

homogeneous half-space and homogeneous soil bearing on a stiff layer, respectively. The analytical expression from Novak et al. (1978) and the fitted functions that used a nonlinear relationship with a_0 (Gazetas & Makris 1991; Anoyatis & Mylonakis 2012) perform well and show good agreement with the numerical continuum results, with the exception of the real component of the stiffness at low frequencies. The simple fitted expressions from O'Rourke & Dobry (1978) and Roesset & Angelides (1980) vary significantly from the other solutions and do not perform well. The predicted cutoff frequency from Eq. 5.30 also matches well with the continuum results for the piles bearing on a stiff layer.

5.2.5.2   Layered soil

In a similar fashion, the head stiffness of a pile installed in M soil layers can be obtained from the iterative solution for the static case by substituting the static terms in Eqs. 5.11 and 5.12 with their dynamic counterparts:

$$K_m^* = E_p^* A_p \lambda_m^* \frac{\Omega_m^* + \tanh\left(\lambda_m^* h_m\right)}{1 + \Omega_m^* \tanh\left(\lambda_m^* h_m\right)} \tag{5.34}$$

$$\lambda_m^* = \sqrt{\frac{k_m^* - \rho_p A_p \omega^2}{E_p^* A_p}}, \quad \Omega_m^* = \frac{K_{m+1}^*}{E_p^* A_p \lambda_m^*} \tag{5.35}$$

The head stiffness can be calculated iteratively from the pile base (layer M) using Eq. 5.34 and $K_{M+1}^* = K_b^*$ to obtain $K^* = K_1^*$. In the realm of the Winkler analysis, this solution is exact.

5.2.5.3 Inhomogeneous soil layers

For an inhomogeneous soil, the closed form static solution cannot be employed to derive the dynamic solution. This is due to the different dependence on depth for each of the three terms multiplying the displacement in Eq. 5.23. In addition, no closed form solutions to this equation for commonly encountered inhomogeneous soil profiles are known to the authors. Instead, simplified approximate approaches are available. The inhomogeneous soil layer can be split into a number of layers and the approach described above can be applied with the properties of the mid-depth of each layer. With a sufficient number of layers, reasonable accuracy can be achieved, as shown by Mylonakis & Gazetas (1998b) for calculating dynamic interaction factors (discussed later).

Alternatively, the energy method employed in Section 5.2.4.2 can be modified to apply to dynamic problems. Following the same approach, Eq. 5.36 can be obtained from Eq. 5.23:

$$K^* = E_p^* A_p \int_0^L \left[\chi^{*\prime}(z)\right]^2 dz + \int_0^L \left[k^*(z) - \rho_p A_p \omega^2\right]\left[\chi^*(z)\right]^2 dz + \left[\chi^*(L)\right]^2 K_b^* \tag{5.36}$$

where $\chi^*(z)$ is generally a complex valued shape function. As for the static case, a suitable function can be based on the displaced shape of a pile in homogeneous soil:

$$\chi^*(z) = \frac{w^*(z)}{w^*(0)} = \frac{\Omega_\mu^* \sinh\left(\mu^* L - \mu^* z\right) + \cosh\left(\mu^* L - \mu^* z\right)}{\Omega_\mu^* \sinh\left(\mu^* L\right) + \cosh\left(\mu^* L\right)} \tag{5.37}$$

where μ^* is analogous to λ^* as μ is to λ, selected based on k_{avg}^*, the average of the dynamic Winkler impedance over the pile length, yielding

$$\mu^* = \sqrt{\frac{k_{avg}^* - \rho_p A_p \omega^2}{E_p^* A_p}}, \quad \Omega_\mu^* = \frac{K_b^*}{E_p^* A_p \mu} \tag{5.38}$$

This approach has been employed by the senior author to obtain the results presented in NEHRP (2012).

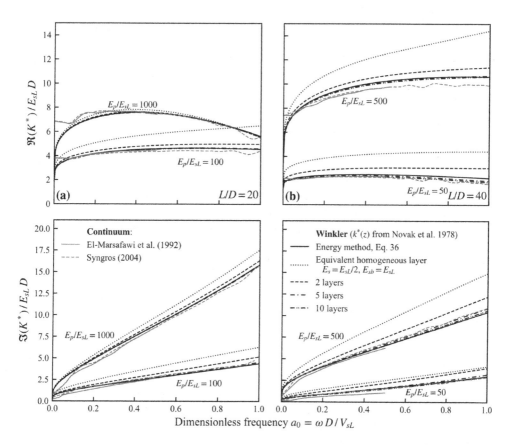

Figure 5.13 Normalised dynamic axial head stiffness of a pile embedded in an inhomogeneous half-space with a linear soil stiffness profile (Eq. 5.14, $a = 0, n = 1$); (a) $L/D = 20$ and (b) $L/D = 40$ $\left(v_s = 0.4,\ \beta_s = 0.05, v_p = 0.25, \beta_p = 0.01, \rho_p/\rho_s = 1.25\right)$. (Numerical data from Syngros (2004).)

These two approximate approaches are compared to continuum results in Figure 5.13. The Novak et al. (1978) impedance function is employed, resulting in a depth-dependent ratio between Winkler modulus and soil stiffness. The remaining Winkler impedance functions in Table 5.3 were calibrated specifically for homogeneous soil and so have not been considered. The two approximate approaches agreed well when a number of soil layers were employed. Only five soil layers were required to give good results, and in some cases, suitable results could be obtained using only two layers. The results also agree well with the numerical continuum methods presented.

5.2.6   Pile load and settlement distribution with depth

In addition to the pile head stiffness, other aspects of pile response can be predicted using the Winkler model. The variation of displacement with depth can be approximated by the shape function in Eq. 5.19 (repeated below). A similar approach can be employed to approximate the distribution of axial load down the pile:

$$\frac{w(z)}{w(0)} = \chi(z) = \frac{\Omega_\mu \sinh(\mu L - \mu z) + \cosh(\mu L - \mu z)}{\Omega_\mu \sinh(\mu L) + \cosh(\mu L)} \tag{5.19}$$

$$\frac{P(z)}{P} = \frac{\Omega_\mu \cosh(\mu L - \mu z) + \sinh(\mu L - \mu z)}{\Omega_\mu \cosh(\mu L) + \sinh(\mu L)} \tag{5.39}$$

For homogeneous soils, $\mu = \lambda$ and $\Omega_\mu = \Omega$, and in the realm of Winkler analyses, these expressions are exact. Equivalent exact expressions for the load and settlement of the pile base are tabulated in Crispin et al. (2018) for different Winkler spring stiffness profiles and pile base conditions. These expressions can also be modified in a similar way to Eq. 5.11 for use in layered soil profiles. In this case, the stiffnesses at each layer interface must be already calculated and the iterative procedure starts from the top of the pile, where the head load and displacement are known. If the loading is dynamic, these expressions can be used with $\mu = \mu^*$, $\Omega_\mu = \Omega_\mu^*$, which results in complex valued displacements and axial loads.

5.2.7   Applying these solutions

The solutions presented in this chapter require selection of an appropriate soil stiffness variation with depth, $E_s(z)$. Ideally, this would be based on in-situ measurements of soil stiffness supported by high-quality laboratory testing with small strain measurement. Small-strain shear modulus, $G_{s,0}$, can be evaluated from the corresponding shear wave velocity, $V_{s,0}$, using Eq. 5.40. Young's modulus for soil can be related to shear modulus using Eq. 5.41.

$$G_s = \rho_s V_s^2 \tag{5.40}$$

$$E_s(z) = 2\left(1 + v_s\right) G_s(z) \tag{5.41}$$

For very small strains, $G_{s,0}$ may be sufficient. However, soil shear modulus should be reduced relative to $G_{s,0}$ for larger strain effects. Soil strains around an axially loaded pile can be evaluated in an approximate manner as

$$\gamma(z) = c_1 \, w(z)/D \tag{5.42}$$

where $\gamma(z)$ denotes an average soil shear strain at depth z, $w(z)$ is the corresponding vertical pile displacement, and c_1 is a pertinent proportionality coefficient, which is analogous to the mobilisable strength design coefficient M_c according to Osman & Bolton (2005) and similar coefficients for laterally loaded piles (Matlock 1970; Kagawa & Kraft 1980). This might be taken at around 0.5–0.6 for clay soils (Bateman et al. 2021). On the basis of this equation, a strain-compatible soil shear modulus can be obtained through conventional modulus reduction curves (e.g., Darendeli 2001; Ishihara 1996; Vucetic & Dobry, 1991; Vardanega & Bolton 2011).

For large strains, various simple extensions to the static solutions presented here are available in the literature (Scott 1981; Motta 1994; Guo 2012; Crispin et al. 2018). These involve linear elastic perfectly plastic Winkler springs that have a capacity based on the stress at which slip occurs between the pile and soil. The pile can then be split into an "elastic region", where yielding has not yet occurred, below a "plastic region" where the capacity of the Winkler springs has been reached. For a chosen depth at which this change in behaviour is assumed to occur, each region can be considered independently, and the pile head response obtained. Varying the chosen depth from the ground surface to base of the pile and repeating the process yields a complete pile head load-settlement curve. Corresponding dynamic solutions are much harder to obtain, and hence, the available information is limited (Akiyoshi 1982; Nogami & Konagai 1987; El Naggar & Novak 1994a, b; Michaelides et al. 1998; Maheshwari et al. 2004; Kanellopoulos & Gazetas 2020).

5.3 PILE GROUP RESPONSE

Foundations normally consist of many piles, installed either supporting individual columns in close proximity, or as a group connected by a pile cap. Numerous studies have shown that when piles are installed in proximity to each other, their static stiffness is lower than if installed individually, and the overall efficiency of the foundation is reduced. Rigorous analysis of this interaction effect (even with only two piles) requires a numerical continuum analysis in three dimensions, such as that carried out by Ottaviani (1975). However, to simplify the problem, Poulos (1968) introduced the concept of interaction factors. These allow complex pile group effects involving an arbitrary number of piles to be assessed by superposition of the effects of two piles at a time. Poulos & Davis (1980) define the interaction factor between pile i and pile j, α_{ij}, as the additional settlement of pile i due to pile j, divided by the settlement of pile i under its own load. When the piles are identical, this is the same as the head settlement of an unloaded pile i due to pile j, $w_{ij}(0)$, divided by the settlement of pile j due to its own load, $w_{jj}(0)$, as shown in Eq. 5.43 for the loaded Pile 1 and unloaded Pile 2. Due to reciprocity, the interaction factors are symmetric, i.e., $\alpha_{ij} = \alpha_{ji}$.

$$\alpha_{21} = \frac{w_{21}(0)}{w_{11}(0)} = \alpha_{12} \qquad (5.43)$$

Once the interaction factors between each pile in the group and every other pile have been established, a set of simultaneous equations is obtained (one for each pile, although this can be reduced using symmetry) and solved using the boundary conditions describing how the pile heads are connected. This is detailed in Poulos & Davis (1980) and discussed later.

The interaction factor itself is a function of pile spacing, pile dimensions, pile stiffness, soil stiffness, and soil boundary conditions. Poulos (1968), Butterfield & Banerjee (1971), and Banerjee & Davies (1977) developed interaction factors numerically for different problem configurations using the integral equation method (a boundary element method). Randolph & Wroth (1979) introduced a simple iterative method for compressible piles based on a displacement attenuation function to give an approximate analytical method for calculating interaction factors. This was improved by

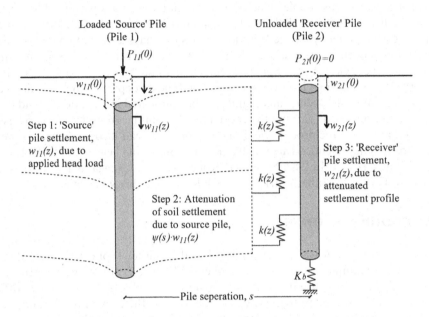

Figure 5.14 Three-step model for the calculation of interaction factors between two piles. (Adapted from Mylonakis & Gazetas (1998a).)

Mylonakis & Gazetas (1998a) who developed a closed form expression for interaction factors for pile groups in homogeneous and homogeneous layered soil. This solution is shown in Figure 5.14 and is split into three steps:

1   Calculate the settlement of a loaded "source" pile (pile 1) as a function of depth $w_{11}(z)$. This is discussed in the previous section on single piles.
2   Model the attenuation of this settlement with distance from the pile, s, absent other piles (including the "receiver" pile, pile 2). This is described using an attenuation function, $\psi(s)$, and is often referred to as the free-field response.
3   Calculate the head response of the unloaded "receiver" pile (pile 2) subject to the attenuated settlement profile, $w_{21}(0)$. This is carried out using the Winkler model.

The governing differential equation can be obtained using the same approach as for the single pile (Mylonakis & Gazetas 1998a):

$$E_p A_p \frac{d^2 w_{21}(z)}{dz^2} - k(z) w_{21}(z) = -\psi(s) k(z) w_{11}(z) \tag{5.44}$$

The concentric cylinder model developed by Cooke (1974) and Baguelin & Frank (1980) was employed by Randolph & Wroth (1978, 1979) and yields a logarithmic attenuation function:

$$\psi(s) \approx \frac{\ln(r_m/s)}{\ln(2r_m/D)} = 1 - \frac{\ln(2s/D)}{\ln(2r_m/D)}, \quad \frac{D}{2} \leq s \leq r_m \tag{5.45}$$

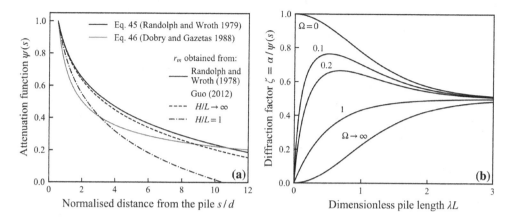

Figure 5.15 (a) Pile shaft displacement attenuation functions ($L/D = 20$, $n = 0$, $v_s = 0.5$) and (b) variation of diffraction factor, ζ, with dimensionless pile length for different pile base conditions for homogeneous soil. (Adapted from Mylonakis & Gazetas (1998a).)

where r_m is the empirical radius at which the displacement is assumed to reach zero, beyond which it is assumed that no interaction occurs. Randolph & Wroth (1978) and Guo (2012) provided the functions for r_m shown in Table 5.1 by matching two numerical continuum analyses. These expressions are compared in Figure 5.15a. They agree well for a floating pile in a half-space; however, the function provided by Guo (2012) explicitly accounts for a rigid layer at a certain depth, and therefore it performs much better for end-bearing piles.

Alternatively, Dobry & Gazetas (1988) developed a wave-propagation-based expression for dynamic loading (discussed later) that for static conditions yields Eq. 5.46.

$$\psi(s) \approx \sqrt{\frac{D}{2s}} \qquad (5.46)$$

Note that, as expected for a sufficiently rigid pile, both are functions independent of depth, and therefore, simple analytical solutions for Eq. 5.44 are possible.

5.3.1 Homogeneous soil

For the simplest case of a single homogeneous soil layer ($k(z) = k$), Mylonakis (1995) and Mylonakis & Gazetas (1998a) solved Eq. 5.44 and applied the boundary conditions for both piles to give Eq. 5.47 for the interaction factor (Mylonakis & Gazetas 1998a):

$$\alpha = \psi(s)\zeta(\lambda L, \Omega) = \frac{\psi(s)}{2}\left[\frac{\left(\Omega^2 + 1\right)\sinh(2\lambda L) + 2\Omega\cosh(2\lambda L) + 2\Omega + 2\lambda L\left(\Omega^2 - 1\right)}{2\Omega\cosh(2\lambda L) + \left(\Omega^2 + 1\right)\sinh(2\lambda L)}\right]$$

$$= \frac{\psi(s)}{2}\left[1 - \frac{\left[\Omega + \lambda L\left(\Omega^2 - 1\right)\right]/\cosh^2(\lambda L)}{\left[\Omega + \tanh(\lambda L)\right]\left[1 + \Omega\tanh(\lambda L)\right]}\right] \tag{5.47}$$

where $\zeta(\lambda L, \Omega)$ is the so-called diffraction factor (Randolph 2003) representing the interaction between the pile and surrounding soil and is a function of both the dimensionless pile length and pile base conditions, as shown in Figure 5.15b. ζ attains simplified forms for the extreme values of the base conditions. For end-bearing piles, where $\Omega \to \infty$, Eq. 5.47 yields (Mylonakis and Gazetas 1998a):

$$\zeta = \frac{1}{2}\left[1 - \frac{2\lambda L}{\sinh(2\lambda L)}\right] \tag{5.48}$$

For fully floating piles with no base resistance $(\Omega = 0)$, Eq. 5.47 yields (Mylonakis & Gazetas 1998a)

$$\zeta = \frac{1}{2}\left[1 + \frac{2\lambda L}{\sinh(2\lambda L)}\right] \tag{5.49}$$

For a very long pile, as pile length approaches infinity, Eqs. 5.47–5.49 all converge to $\zeta = 1/2$. To investigate the validity of this result, Mylonakis & Gazetas (1998a) compared it to results from rigorous boundary element analyses. The interaction factor for long rigid piles, α_{rigid}, is approximately equal to the attenuation function, $\psi(s)$ (Randolph & Wroth 1979). Therefore, by dividing the interaction factor for a compressible pile with that for a rigid pile at the same pile length, a suitable approximation for the diffraction factor can be obtained. Table 5.4 shows this process for different pile lengths and pile-soil stiffness ratios. Approximate λL values are included for reference. The α_{rigid} values show good agreement with the attenuation function from Eq. 5.45, and it is apparent

Table 5.4 Approximate diffraction factor, ζ, as pile length increases from boundary element analyses ($v_s = 0.5$; $k/E_s = 0.6$, Roesset & Angelides (1980)) (Data from Mylonakis & Gazetas (1998a))

	L/D = 25	L/D = 50	L/D = 100
α_{rigid} (Poulos 1968)	0.64	0.68	0.74
$\psi(s)$ (Eq. 5.45)	0.66	0.68	0.75
$E_p/E_s = 1000$			
λL (Eq. 5.9)	0.7	1.4	2.8
α (Poulos & Davis 1980)	0.57	0.54	0.46
$\zeta \approx \alpha/\alpha_{rigid}$	0.89	0.80	0.62
$E_p/E_s = 100$			
λL (Eq. 5.9)	2.2	4.4	8.7
α (El-Sharnouby & Novak 1990)	0.43	0.39	0.41
$\zeta \approx \alpha/\alpha_{rigid}$	0.67	0.57	0.55

that the ζ value is approaching a value near 0.5 as pile length increases, in agreement with Eq. 5.47.

More recently, Randolph (2003) provided an alternative expression for the diffraction factor between piles of different diameters using the same Winkler approach.

So far, only the interaction between pile shafts has been considered. Following the same assumptions as for the single pile, the pile base can be modelled separately from the pile shaft as a rigid disc on an elastic half-space. The attenuation of displacement due to the pile base, $\psi_b(s)$, is therefore given by Eq. 5.50 (Davis & Selvadurai 1996). Employing St Venant's principle, an applied stress will appear as a point load at large distance from the source. Therefore, Randolph & Wroth (1979) recommend Eq. 5.51, which is shown in Figure 5.16a to match very well at reasonable pile spacings.

$$\psi_b(s) = \frac{2}{\pi} \sin^{-1}\left(\frac{D}{2s}\right) \tag{5.50}$$

$$\psi_b(s) \approx \frac{D}{\pi s} \tag{5.51}$$

Employing a similar three-step approach to that employed for the pile shaft interaction, Mylonakis & Gazetas (1998a) developed Eq. 5.52 for the pile-to-pile interaction factor due to a base load, α_b.

$$\alpha_b = \psi_b(s)\zeta_b(\lambda L, \Omega) = \psi_b(s)\frac{2\Omega}{2\Omega\cosh(2\lambda L)+\left(\Omega^2+1\right)\sinh(2\lambda L)}$$

$$= \psi_b(s)\frac{\Omega/\cosh^2(\lambda L)}{\left[\Omega+\tanh(\lambda L)\right]\left[1+\Omega\tanh(\lambda L)\right]} \tag{5.52}$$

where ζ_b is the base diffraction factor which is plotted in Figure 5.16b.

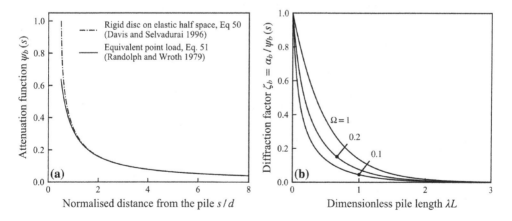

Figure 5.16 (a) Pile base displacement attenuation functions and (b) variation of base diffraction factor, ζ_b, with dimensionless pile length for different pile base conditions. (Adapted from Mylonakis & Gazetas (1998a).)

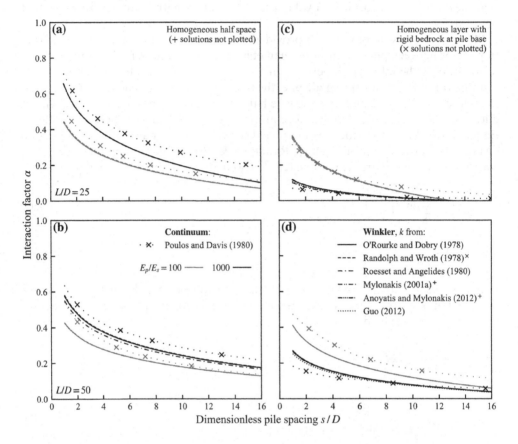

Figure 5.17 Interaction factor for piles embedded in a homogeneous half-space for (a) $L/D = 25$ and (b) $L/D = 50$, in a homogeneous soil layer with rigid bedrock at the pile base for (c) $L/D = 25$ and (d) $L/D = 50$ ($\psi(s)$ Eq. 5.45, r_m from Guo (2012), Table 5.1, $v_s = 0.5$). (Numerical data from Poulos & Davis (1980).)

Mylonakis & Gazetas (1998a) observed that, mostly due to the attenuation functions, α_b decays with distance from the pile at a much faster rate than α. As such, at most realistic pile spacings, α_b is negligible compared to α and can be neglected in most analyses.

Figure 5.17 shows a comparison between interaction factors calculated using boundary element methods and the Winkler method in Eqs. 5.47 and 5.45 for different Winkler stiffness equations (Table 5.1). All the expressions for the Winkler spring stiffness provide very similar results and perform well compared to the continuum solution by Poulos & Davis (1980).

5.3.2 Layered soil

Layered soil can be considered using a similar approach to that provided in Section 5.2.3 for single piles. For a pair of piles installed in M soil layers (as shown in Figure 5.5), the diffraction factor can be calculated iteratively starting from the bottom layer (layer M)

using Eq. 5.53. This directly follows from the solution in Mylonakis & Gazetas (1998b) for the interaction factor in layered soil and the assumption that $\psi(s)$ is independent of depth, z, and therefore the same for each soil layer.

$$\zeta_m = \zeta_{hom}(\lambda_m h_m, \Omega_m) + \zeta_{b,hom}(\lambda_m h_m, \Omega_m)\,\zeta_{m+1} \tag{5.53}$$

where λ_m and Ω_m are given in Eq. 5.12 and $\zeta_{hom}(h_m \lambda_m, \Omega_m)$ and $\zeta_{b,\ hom}(h_m \lambda_m, \Omega_m)$ are the diffraction factor (Eq. 5.47) and base diffraction factor (Eq. 5.52), respectively, for homogeneous soil with $\lambda_m h_m$ and Ω_m substituted in for λL and Ω, $\zeta_{M+1} = 0$ and the overall diffraction factor, $\zeta = \zeta_1$. For the simplest case of two soil layers, this yields (adapted from Mylonakis & Gazetas 1998a)

$$\zeta = \frac{1}{2} - \frac{\left[\lambda_1 h_1 \left(\Omega_1^2 - 1\right)\right]\big/\cosh^2\left(\lambda_1 h_1\right)}{2\left[\Omega_1 + \tanh\left(\lambda_1 h_1\right)\right]\left[1 + \Omega_1 \tanh\left(\lambda_1 h_1\right)\right]}$$
$$+ \frac{\Omega_1\left[\Omega_2 + \lambda_2 h_2 \left(\Omega_2^2 - 1\right)\right]\big/\left[\cosh^2\left(\lambda_1 h_1\right)\cosh^2\left(\lambda_2 h_2\right)\right]}{2\left[\Omega_1 + \tanh\left(\lambda_1 h_1\right)\right]\left[1 + \Omega_1 \tanh\left(\lambda_1 h_1\right)\right]\left[\Omega_2 + \tanh\left(\lambda_2 h_2\right)\right]\left[1 + \Omega_2 \tanh\left(\lambda_2 h_2\right)\right]} \tag{5.54}$$

This solution reduces to that for a homogenous soil ($\lambda = \lambda_2$, $\Omega = \Omega_m$) if the thickness of the first layer is set to zero ($\lambda_1 h_1 = 0$).

5.3.3 Inhomogeneous soil layers

As with the single pile case, in order to derive interaction factors for inhomogeneous soil using the method described above, a solution to Eq. (5.44) must be found for each family of inhomogeneity functions.

5.3.3.1 Linear and power-law stiffness profiles

Crispin & Leahy (2019) developed a closed form solution for Winkler spring stiffness varying according to Eq. 5.14. In this case, the diffraction factor is given by Eq. 5.55, which is plotted in Figure 5.18 for different base conditions and two soil profiles: linear ($n = 1$) and parabolic ($n = 1/2$), both with vanishing surface stiffness ($a = 0$). Alternative forms of Eq. 5.55 at the limits of a, Ω_L and L are shown in Table 5.5.

$$\zeta = \frac{1}{n+2} - \frac{\chi_L\left(\Omega_L^2 - 1\right) + 2v\Omega_L - \chi_L\chi_0^2\left[\left[S_1 + \Omega_L S_2\right]^2 - \left[S_3 + \Omega_L S_4\right]^2\right]}{\chi_L\chi_0\left[S_1 + \Omega_L S_2\right]\left[S_3 + \Omega_L S_4\right]} \tag{5.55}$$

where v, χ_0, χ_L, and Ω_L are from Eq. 5.16, and S_1–S_4 are given by:

$$S_1 = K_{v-1}(\chi_0)I_{v-1}(\chi_L) - I_{v-1}(\chi_0)K_{v-1}(\chi_L),$$

$$S_2 = K_{v-1}(\chi_0)I_v(\chi_L) + I_{v-1}(\chi_0)K_v(\chi_L),$$

$$S_3 = K_v(\chi_0)I_{v-1}(\chi_L) + I_v(\chi_0)K_{v-1}(\chi_L),$$

$$S_4 = K_v(\chi_0)I_v(\chi_L) - I_v(\chi_0)K_v \qquad (5.56)$$

Figure 5.19 shows a comparison of the performance of the Winkler solutions for inhomogeneous soil described in Eq. 5.55 against some existing boundary element solutions

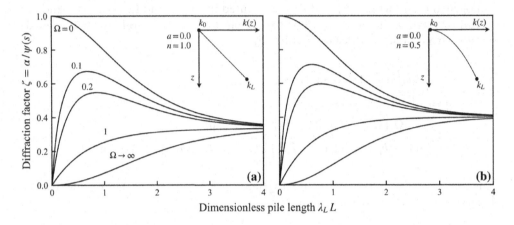

Figure 5.18 Diffraction factor, ζ, for piles embedded in (a) a linear soil profile with vanishing surface stiffness (Eq. 5.14, $n = 1$, $a = 0$) and (b) a parabolic soil profile with vanishing surface stiffness (Eq. 5.14, $n = 1/2$, $a = 0$). (Adapted from Crispin & Leahy (2019).)

Table 5.5 Summary of closed-form solutions for the diffraction factor, ζ, for piles embedded in an inhomogeneous soil (Eq. 5.14)

	General case $(a \neq 0)$	Vanishing surface stiffness $(a = 0)$
General case	Eq. 5.55	$\dfrac{1}{n+2} - \dfrac{\sin(\pi v)\left[\lambda_L L(\Omega^2 - 1) + \Omega\right]\big/\pi\lambda_L L}{\left[I_{1-v}(\chi_L) + \Omega I_{-v}(\chi_L)\right]\left[I_{v-1}(\chi_L) + \Omega I_v(\chi_L)\right]}$
Fully floating pile, $\Omega_L = 0$	$\dfrac{1}{n+2} + \dfrac{1 + \chi_0^2\left[S_1^2 - S_3^2\right]}{\chi_0 S_1 S_3}$	$\dfrac{1}{n+2} + \dfrac{\sin(\pi v)/\pi}{I_{1-v}(\chi_L)I_{v-1}(\chi_L)}$
End-bearing pile, $\Omega_L \to \infty$	$\dfrac{1}{n+2} - \dfrac{1 - \chi_0^2\left[S_2^2 - S_4^2\right]}{\chi_0 S_2 S_4}$	$\dfrac{1}{n+2} - \dfrac{\sin(\pi v)/\pi}{I_{-v}(\chi_L)I_v(\chi_L)}$
Infinitely long pile, $L \to \infty$	$\dfrac{1}{n+2} + \chi_D\left[\dfrac{K_{v-1}(\chi_D)}{K_v(\chi_D)}\right.$ $\left. - \dfrac{K_v(\chi_D)}{K_{v-1}(\chi_D)}\right]$	$\dfrac{1}{n+2}$

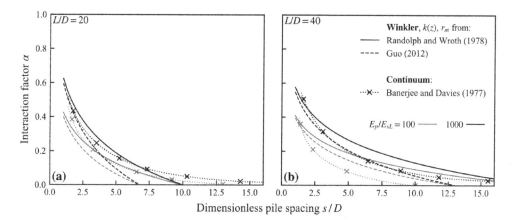

Figure 5.19 Interaction factor for piles embedded in an inhomogeneous half-space (Eq. 5.14, $n = 1$, $a = 0$) for (a) $L/D = 20$ and (b) $L/D = 40$ ($\psi(s)$ Eq. 5.45, $H/L = 2$, $v_s = 0.5$). (Numerical data from Banerjee & Davies (1977).)

for calculating interaction factors. In general, the Winkler solution shows good agreement with the continuum results; the largest divergences occur where the magnitude of the interaction factor is low and the impact on a predicted pile group response will be minimal.

5.3.3.2   Other inhomogeneous profiles

For more general soil profiles, a similar energy method to that used for single piles can be employed. By considering the virtual work done by a small virtual displacement, $w_v(z)$, an integral form of Eq. 5.44 can be obtained:

$$\int_0^L E_p A_p w_{21}''(z) w_v(z) dz - \int_0^L k(z) w_{21}(z) w_v(z) dz = -\psi(s) \int_0^L k(z) w_{11}(z) w_v(z) dz \quad (5.57)$$

The shape function, $\chi(z)$, in Eq. 5.19 can again be employed to describe the real and virtual pile displacement, yielding Eq. 5.58 for the diffraction factor, ζ. Note that the denominator of the fraction in this expression is the approximate pile head stiffness obtained in Eq. 5.18. In addition, both terms in the numerator are independent of the soil stiffness profile and were also evaluated in Eq. 5.18. Therefore, Eq. 5.58 can be calculated without evaluating any additional integrals and can be expressed in a simplified form in Eq. 5.59.

$$\zeta = \frac{w_{21}(0)}{\psi(s) w_{11}(0)} = 1 - \frac{\int_0^L E_p A_p \left[\chi'(z) \right]^2 dz + K_b \left[\chi(L) \right]^2}{\int_0^L E_p A_p \left[\chi'(z) \right]^2 dz + \int_0^L k(z) \left[\chi(z) \right]^2 dz + K_b \left[\chi(L) \right]^2} \quad (5.58)$$

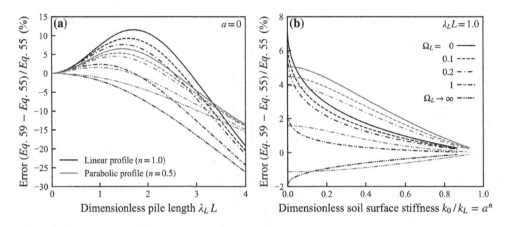

Figure 5.20 Error in energy method solution relative to the closed form expression from Eq. 5.55 with (a) dimensionless pile length and (b) dimensionless surface stiffness.

$$\zeta = 1 - \frac{1}{2}\left[1 + \frac{\left[\Omega_\mu + \mu L\left(\Omega_\mu^2 - 1\right)\right]\Big/\cosh^2\left(\mu L\right)}{\left[\Omega_\mu + \tanh\left(\mu L\right)\right]\left[1 + \Omega_\mu \tanh\left(\mu L\right)\right]}\right]\frac{\Omega_\mu + \tanh\left(\mu L\right)}{1 + \Omega_\mu \tanh\left(\mu L\right)}\Bigg/\frac{K_0}{E_p A_p \mu}$$

$$= 1 - \frac{K_{0,hom}\left(\mu L, \Omega_\mu\right)}{K_0}\left[1 - \zeta_{hom}\left(\mu L, \Omega_\mu\right)\right] \tag{5.59}$$

where $K_{0,hom}(\mu L, \Omega_\mu)$ and $\zeta_{hom}(\mu L, \Omega_\mu)$ are the equivalent homogeneous head stiffness and diffraction factor, respectively, obtained by substituting $\lambda L = \mu L$ and $\Omega = \Omega_\mu$ into Eqs. 5.3 and 5.47. For homogeneous soil, this expression reproduces the closed form solution in Eq. 5.47.

The performance of Eq. 5.59 is compared with the closed from solution for linear and parabolic soil profiles form Eq. 5.55 in Figure 5.20. This expression does not perform as well as the equivalent energy method for single pile response (shown in Figure 5.8). However, for $\lambda_L L < 2.5$, which encompasses most piles in practice and almost all piles not behaving as infinitely long, the absolute error is less than 15%. In addition, as shown in Figure 5.20b, this error decreases rapidly as the surface stiffness increases. Considering the uncertainty in the Winkler model parameters and the evident appeal of this simplified expression, Eq. 5.59 should be used where the more rigorous expressions presented earlier (Eq. 5.55 and Table 5.5) are not suitable or when analysing different soil profiles.

5.3.4 Dynamic response

Except perhaps at very low frequencies, static interaction factors cannot accurately predict the dynamic response of pile groups (Dobry & Gazetas 1988). This is because the cylindrical waves generated by each pile can arrive at neighbouring piles with a different, sometimes opposite, phase. This can result in sharp peaks in the dynamic stiffness of pile groups, resulting in some situations where the stiffness of a pile group is larger than the total stiffness of its component piles individually (Mylonakis 1995).

Kaynia (1982) and Kaynia & Kausel (1982) extended interaction factors to dynamic problems. In this case, the interaction factor has a complex value as shown in Eq. 5.60.

$$\alpha_{21}^* = \alpha_{12}^* = \frac{w_{21}^*(0)}{w_{11}^*(0)} = \left|\alpha^*\right| e^{i\varphi} \tag{5.60a}$$

$$\left|\alpha^*\right| = \sqrt{\left(\Re\left[\alpha^*\right]\right)^2 + \left(\Im\left[\alpha^*\right]\right)^2}, \quad \varphi = \tan^{-1}\left(\frac{\Im\left[\alpha^*\right]}{\Re\left[\alpha^*\right]}\right) \tag{5.60b}$$

where $w_{21}^*(0)$ and $w_{11}^*(0)$ are the complex-valued displacements of an unloaded pile (pile "2") and a harmonically loaded pile (pile "1"), respectively, due to the applied harmonic load and $\Re\left[\alpha^*\right]$, $\Im\left[\alpha^*\right]$, $\left|\alpha^*\right|$ and φ are the real component, imaginary component, magnitude (modulus), and phase (argument) of the complex interaction factor, respectively.

Kaynia (1982) and Kaynia & Kausel (1982) produced interaction factors for homogeneous soil using a rigorous boundary element type method. Additional solutions using the same formulation were provided for different pile base conditions and inhomogeneous soil profiles by Gazetas et al. (1991) and El-Marsafawi et al. (1992).

As a simplified alternative, Dobry & Gazetas (1988) suggested using the free field displacement due to the harmonically loaded pile at the location of the unloaded pile, normalised by the displacement at the pile-soil interface of the loaded pile. This is equivalent to assuming that the dynamic interaction factor is equal to a complex-valued dynamic attenuation function, $\psi^*(s)$. Gazetas & Makris (1991) developed Eq. 5.61 for $\psi^*(s)$ based on the solution by Novak (1974):

$$\psi^*(s) = \frac{K_0\left(\frac{2s}{D}s^*\right)}{K_0\left(s^*\right)}, \quad s^* = \frac{ia_0}{2\sqrt{1+2i\beta}} \tag{5.61}$$

Note that, unlike in the static case where the attenuated displacement is implicitly assumed to arrive at a neighbouring pile instantly, the time taken for the shear waves to reach the neighbouring pile is dependent on their velocity, V_s. As such, in inhomogeneous soils (where V_s is a function of depth), $\psi^*(s)$ can vary significantly with depth.

A simpler alternative to Eq. 5.61 can be obtained using the asymptotic expansions of the Bessel functions (Dobry & Gazetas 1988; modified by Mylonakis & Gazetas 1998b to give the correct result at the pile-soil interface):

$$\psi^*(s) \approx \sqrt{\frac{d}{2s}} \exp\left(-\frac{a_0}{2}\left[\beta_s + i\right]\left[\frac{2s}{d} - 1\right]\right) \tag{5.62}$$

Equations 5.61 and 5.62 are compared in Figure 5.21.

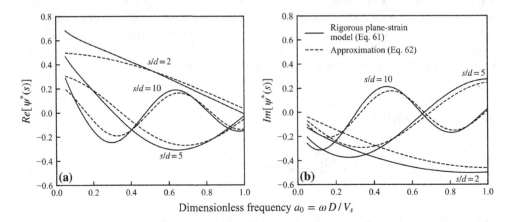

Figure 5.21 (a) Real and (b) imaginary components of the dynamic attenuation functions at different distances from the pile. (Adapted from Mylonakis & Gazetas (1998b).)

This model neglects the contribution of the second pile to the system stiffness. To correct this, Mylonakis & Gazetas (1998b) introduced a similar three-step model to that employed for the static case. This is shown in Figure 5.22 and results in the governing differential equation shown in Eq. 5.63.

$$E_p^* A_p \frac{d^2 w_{21}^*(z)}{dz^2} - \left[k^*(z) - \rho_p A_p \omega^2 \right] w_{21}^*(z) = -\psi^*(s) k^*(z) w_{11}^*(z) \tag{5.63}$$

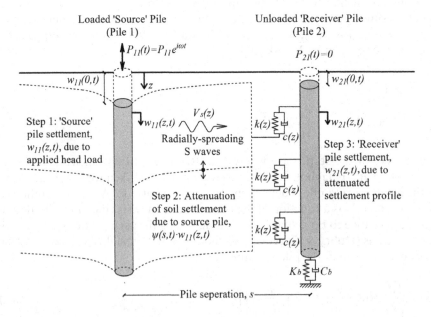

Figure 5.22 Three-step model for the calculation of interaction factors between two piles for dynamic loading. (Adapted from Mylonakis & Gazetas (1998b).)

5.3.4.1   Homogeneous soil

Similar to the dynamic single pile head stiffness, the dynamic interaction factor for homogeneous soil can be obtained by replacing the static terms in Eq. 5.47 with their dynamic counterparts. In addition, a term must be added to account for the mass of the receiver pile. This results in Eq. 5.64 (Mylonakis & Gazetas 1998b):

$$\alpha^* = \psi^*(s)\zeta^*\left(\lambda^* L, \Omega^*\right)$$

$$= \frac{\psi^*(s)}{2} \frac{k^*}{k^* - \rho_p A_p \omega^2}$$
$$\times \left[\frac{\left(\Omega^{*2} + 1\right)\sinh\left(2\lambda^* L\right) + 2\Omega^* \cosh\left(2\lambda^* L\right) + 2\Omega^* + 2\lambda^* L\left(\Omega^{*2} - 1\right)}{2\Omega^* \cosh\left(2\lambda^* L\right) + \left(\Omega^{*2} + 1\right)\sinh\left(2\lambda^* L\right)} \right]$$

$$= \frac{\psi^*(s)}{2}\left(\frac{\lambda^*_{massless}}{\lambda^*}\right)^2 \left[1 - \frac{\left[\Omega^* + \lambda^* L\left(\Omega^{*2} - 1\right)\right]/\cosh^2\left(\lambda^* L\right)}{\left[\Omega^* + \tanh\left(\lambda^* L\right)\right]\left[1 + \Omega^* \tanh\left(\lambda^* L\right)\right]} \right] \tag{5.64}$$

where $\lambda^*_{massless}$ is the value of λ^* calculated assuming a massless pile ($\rho_p = 0$).

Asymptotic forms of Eq. 5.64 for extreme values of the base stiffness can be obtained by inputting these conditions into Eq. 5.64 or repeating the above substitution with the relevant result from Eqs. 5.48 and 5.49. For end-bearing piles, where $\Omega^* \to \infty$, this yields (Mylonakis & Gazetas 1998b)

$$\zeta^* = \frac{1}{2}\frac{k^*}{k^* - \rho_p A_p \omega^2}\left[1 - \frac{2\lambda^* L}{\sinh\left(2\lambda^* L\right)} \right] \tag{5.65}$$

For fully floating piles with no base resistance $\left(\Omega^* = 0\right)$, this yields (Mylonakis & Gazetas 1998b)

$$\zeta^* = \frac{1}{2}\frac{k^*}{k^* - \rho_p A_p \omega^2}\left[1 + \frac{2\lambda^* L}{\sinh\left(2\lambda^* L\right)} \right] \tag{5.66}$$

For a very long pile, as pile length approaches infinity, Eqs. 5.64–5.66 all converge to the simple expression (Mylonakis & Gazetas 1998b)

$$\zeta^* = \frac{1}{2}\frac{k^*}{k^* - \rho_p A_p \omega^2} \tag{5.67}$$

One advantage of the analytical solution presented here is that it is possible to gain insight into the relative contributions of the two parts of the product that make up the dynamic interaction factor. Figure 5.23 shows the amplitude and phase angle of both

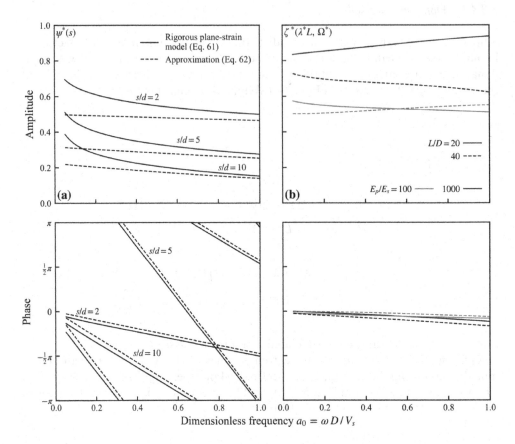

Figure 5.23 Amplitude and phase of the (a) dynamic attenuation function and (b) dynamic diffraction factor for different configurations of piles embedded in a homogeneous half-space $\left(v_s = 0.4, \beta_s = 0.05, v_p = 0.25, \ \beta_p = 0, \rho_p/\rho_s = 1.25\right)$.

the dynamic attenuation function, $\psi^*(s)$ and the dynamic diffraction factor ζ^*. The dynamic interaction factor calculated from the product of these results is compared to the rigorous method of Kaynia (1982) in Figure 5.24. The phase is wrapped (i.e., principal argument values are provided limited between $-\pi$ and π) to show the relative magnitudes of the two components more clearly. In addition, this highlights why the phase is particularly important for interaction factors; the displacement of a pile in a group is summed with the displacement of every other pile multiplied by the corresponding interaction factor; if the interaction factor is out of phase (near $-\pi$ or π), these values will destructively interfere resulting in a lower settlement amplitude. However, if the phase of the interaction factor is near zero, these values will constructively interfere resulting in much higher settlement amplitudes and a corresponding low group stiffness. The phase of the dynamic diffraction factor is approximately zero for all frequencies plotted; therefore, the phase of the interaction factor is mostly dependent on the attenuation function, and the stiffening effect of the pile only reduces the magnitude. For larger pile spacings, the phase of the dynamic attenuation function and as

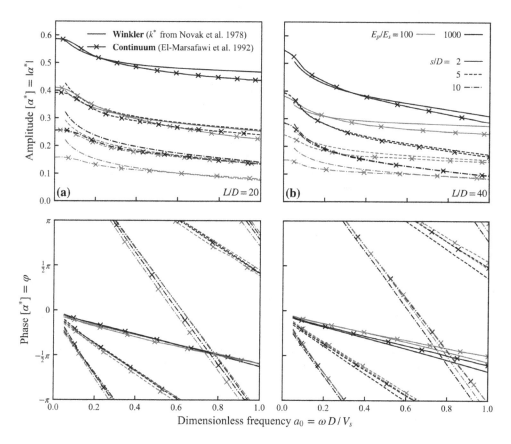

Figure 5.24 Dynamic interaction factor for piles embedded in a homogeneous half-space with (a) $L/D = 20$ and (b) $L/D = 40$ $\left(v_s = 0.4, \beta_s = 0.05, v_p = 0.25, \beta_p = 0, \rho_p/\rho_s = 1.25\right)$. (Numerical data from El-Marsafawi et al. (1992).)

a result, the dynamic interaction factor vary significantly with frequency. This means that the overall pile group response will mostly depend on whether the frequency of the applied load leads to constructive of destructive interference between the different pile responses. In general, the Winkler solution performs well compared to the rigorous results, providing accurate predictions of both the magnitude and phase of the dynamic interaction factor.

It should be noted that, in the presence of a stiff layer at the pile tip, the classical interaction factor approach (based on two piles) does not provide satisfactory results without accounting for the presence of the intermediate piles (El-Sharnouby & Novak 1990).

5.3.4.2 Layered soil

Layered soil can be analysed using the same method employed for static loads. However, as the attenuation function will be different for each layer, the interaction factor must be calculated for each stage in the iterative approach, rather than just the diffraction factor. This results in the following iterative expression for the dynamic interaction factor:

$$\alpha_m^* = \alpha_{hom}^*\left(\lambda_m^* h_m, \Omega_m^*\right) + \alpha_{m+1}^* \frac{\Omega_m^*/\cosh^2\left(\lambda_m^* h_m\right)}{\left[\Omega_m^* + \tanh\left(\lambda_m^* h_m\right)\right]\left[1 + \Omega_m^* \tanh\left(\lambda_m^* h_m\right)\right]} \quad (5.68)$$

where $\alpha_{hom}^*\left(\lambda_m^* h_m, \Omega_m^*\right)$ is the dynamic interaction factor for homogeneous soil with $\lambda_m^* h_m$ and Ω_m^* substituted for $\lambda^* L$ and Ω^*, respectively.

The overall complex interaction factor, $\alpha^* = \alpha_1^*$, for a pile in M soil layers can be obtained from Eq. 5.68 with the value below the base layer $\alpha_{M+1}^* = 0$.

5.3.4.3   Inhomogeneous soil layers

Closed-form solutions for interaction factors rely on solutions for the single pile case. As no closed form solutions are available for the harmonically loaded single pile embedded in commonly encountered inhomogeneous soil profiles, simplified solutions for dynamic interaction factors are limited to the approximate approaches developed for the static case. The soil profile can be split into many homogeneous layers, and the dynamic interaction factor can be calculated using Eq. 5.68 or, alternatively, an energy method similar to those presented earlier can be employed. The integral form of Eq. 5.63 is obtained as before:

$$\int_0^L E_p^* A_p w_{21}^{*\prime\prime}(z) w_v^*(z)\,dz - \int_0^L k^*(z) w_{21}^*(z) w_v^*(z)\,dz = -\int_0^L \psi^*(s) k^*(z) w_{11}^*(z) w_v^*(z)\,dz \quad (5.69)$$

Note that as the attenuation function varies with depth for the dynamic case, it must be included inside the integral on the right-hand side of the equation. Employing a complex-valued shape function χ^* as in the dynamic single pile case (such as Eq. 5.37), the above equation can be rearranged to give the dynamic interaction factor:

$$\alpha^* = \frac{1}{K_0^*} \int_0^L \psi^*(s,z) k^*(z)\left[\chi^*(z)\right]^2\,dz \quad (5.70)$$

The performance of this solution and an equivalent homogeneous layer solution is compared to rigorous results using the method of Kaynia (1982) in Figure 5.25. Both solutions give similar results to each other, and although the magnitude of the interaction factor is not predicted as well as for the homogeneous case, the phase shows very good agreement with the rigorous results.

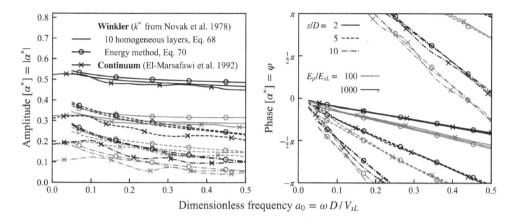

Figure 5.25 Dynamic interaction factor for piles embedded in an inhomogeneous half-space with a linear soil stiffness profile (Eq. 5.14, $a = 0$, $n = 1$); $\left(L/D = 20, v_s = 0.4, \beta_s = 0.05, v_p = 0.25, \beta_p = 0, \rho_p/\rho_s = 1.25\right)$. (Numerical data from El-Marsafawi et al. (1992).)

5.3.5 Applying interaction factors

Once the interaction factors between the different piles in the group have been calculated, the response of the whole group can be determined. Generally of interest are the overall stiffness of the pile group, K_G = total applied load on the group/group average settlement = $P_G/w_{0,avg}$; the distribution of load between different piles in the group; and the differential settlement between piles in the group. These can be calculated using the superposition method developed by Poulos (1968), who proposed that the additional settlement on each pile in the group due to the settlement of every other pile can be superimposed. This yields the following expression for the settlement of pile i, $w_{0,i}$ in a group of N piles (Poulos & Davis 1980):

$$w_{0,i} = \sum_{j=1}^{N} \frac{P_j \alpha_{ij}}{K_{0,j}} \tag{5.71}$$

where $K_{0,j}$ and P_j are the pile head stiffness and applied load at the head of pile j, respectively, and α_{ij} is the interaction factor between pile i and pile j, where the interaction factor of pile i with itself, $\alpha_{ii} = 1$. This expression generates N simultaneous equations for the pile group (although this can be reduced using symmetry) with two unknowns each, the load and displacement for each pile. These can be solved using the conditions connecting the piles at the pile cap.

For a perfectly rigid pile cap, the settlement of the group and therefore each pile is identical, and therefore, each of the equations generated using Eq. 5.71 is equated giving a set of $N-1$ equations for the pile head loads (N unknowns) which can be solved with the addition of Eq. 5.72 describing vertical (Poulos & Davis 1980). These can then be substituted back into Eq. 5.71 to yield the pile group displacement and stiffness.

$$P_G = \sum_{j=1}^{N} P_j \tag{5.72}$$

For a perfectly flexible pile cap, the load on the group is distributed evenly between the piles, i.e., $P_j = P_G/N$ (Poulos & Davis 1980). This can be inputted directly into Eq. 5.71 to calculate the settlement of each pile in the group. The average settlement of the group can then be calculated to yield the pile group stiffness.

In reality, the stiffness of a pile cap will lie between these extremes. However, these limits can be applied as simplifying assumptions. For example, a small pile group supporting one column will often have a sufficiently rigid cap to apply the perfectly rigid solution, whilst a piled raft may provide such little stiffness that the perfectly flexible solution can be applied (although the structure itself can add additional rigidity to the raft). Basile (2019) suggests bounds on the raft-soil stiffness ratio introduced by Horikoshi & Randolph (1997) inside which the above simplifying assumptions can be applied. Outside of these ranges, the problem can be considered as an elastic plate (the cap) supported by springs (the piles) with applied displacements and resolved numerically.

The importance of pile cap-soil interaction and the associated contact effects have been explored by Butterfield & Banerjee (1971) and more recently by Padrón et al. (2009, 2012). These effects are not considered here.

The performance of different configurations of a group of N identical piles can be compared using the settlement ratio, R_s, and a normalised pile group stiffness (Poulos & Davis 1980):

$$R_s = w_{0,avg} \bigg/ \frac{P_G}{NK_0} = \frac{NK_0}{K_G} \tag{5.73}$$

where $P_G/N\,K_0$ is the settlement of a single pile under the same average load of a pile in the group. For a single pile, $R_s = 1$, and as piles are added to the group, this value will increase. The lower the value, the more efficiently the piles in the group are being utilised.

For a designer comparing the efficiency of configurations with different numbers of piles, the group reduction factor, R_G, may be more appropriate (Poulos & Davis 1980):

$$R_s = w_{0,avg} \bigg/ \frac{P_G}{K_0} = \frac{K_0}{K_G} = \frac{R_s}{N} \tag{5.74}$$

where P_G/K_0 is the settlement of a single pile at the same total load as the group.

Both factors are functions of the interaction factors between the piles and the pile cap stiffness (which can be simplified with one of the two aforementioned assumptions). They can therefore be employed (using only the simple approaches presented in this section) to predict pile group response directly from analytical predictions of single pile response or single pile test results.

5.3.5.1   Illustrative example

Consider the group of six identical piles shown in Figure 5.26. The piles, with slenderness, $L/D = 25$, are embedded in a homogeneous soil with stiffness such that

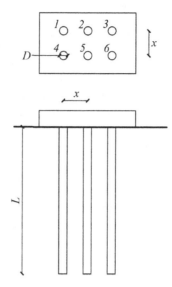

Figure 5.26 Illustrative example of pile group configuration.

$E_p/E_s = 2000$. The diffraction factor can be calculated from Eq. 5.47; employing $k = 0.6$ E_s (Roesset & Angelides 1980) and the simplified expressions in Eqs. 5.9 and 5.10 to give λL (= 0.49) and Ω (= 0.04), respectively, yields $\zeta = 0.85$. The piles are evenly spaced in a 3 by 2 grid with grid spacing $x/d = 5$. From symmetry, they can be split into two groups, the edge piles (1, 3, 4, and 6, Pile 1 used as the example) and the two centre piles (2 and 5, Pile 2 used as the example). Table 5.5 shows the values of the attenuation function (Eq. 5.45) and resulting interaction factors between each neighbouring pile. The results of Poulos & Davis (1980) who considered the same example are also shown.

Equation 5.71 can now be employed to generate two simultaneous equations based on the settlement of the centre and edge piles (see Table 5.6):

Table 5.6 Interaction factors for illustrative example (Numerical data from Poulos & Davis (1980))

Pile j	Pile 1 (edge pile)				Pile 2 (centre pile)			
	s/d	$\psi(s)$	α_{1j}	α_{1j} (Poulos & Davis 1980)	s/d	$\psi(s)$	α_{2j}	α_{2j} (Poulos & Davis 1980)
1	0	1.00	1.00	1.00	5	0.44	0.38	0.42
3	10	0.28	0.23	0.27	5	0.44	0.38	0.42
4	5	0.44	0.38	0.42	7.1	0.36	0.30	0.35
6	11.2	0.25	0.21	0.25	7.1	0.36	0.30	0.35
Σ edge	-	-	1.82	1.94	-	-	1.36	1.54
2	5	0.44	0.38	0.42	0	1.00	1.00	1.00
5	7.1	0.36	0.30	0.35	5	0.44	0.38	0.42
Σ centre	-	-	0.68	0.77	-	-	1.38	1.42

$$w_{0,edge} K_0 = 1.82 P_{edge} + 0.68 P_{centre}, \quad w_{0,centre} K_0 = 1.36 P_{edge} + 1.38 P_{centre} \qquad (5.75)$$

where $w_{0,edge}$ and $w_{0,centre}$, P_{edge} and P_{centre} are the head settlements head loads of the edge piles (1, 3, 4, and 6) and the centre piles (2 and 5), respectively.

For a rigid cap, $w_{0,edge} = w_{0,centre} = w_{0,avg}$; therefore, equating the two results in Eq. 5.75 yields $P_{edge} = 1.52 \ P_{centre}$. Considering vertical equilibrium for the group, 4 $P_{edge} + 2 \ P_{centre} = P_G$. These two equations can be used to calculate the proportion of the load carried by each pile: $P_{edge}/P_G = 0.19$, $P_{centre}/P_G = 0.12$, which can be inputted back into Eq. 5.75 to give the group reduction facto, $R_G = 0.43$ and the settlement ratio, $R_s = 2.56$. Poulos & Davis (1980) calculated corresponding values of $P_{edge}/P_G = 0.19$, $P_{centre}/P_G = 0.12$, $R_G = 0.46$, and $R_s = 2.77$.

For a flexible pile cap, $P_{edge} = P_{centre} = P_G/6$, which can be substituted into Eq. 5.75 to give $w_{0,centre} = 1.10 \ w_{0,edge}$ (Poulos & Davis 1980 obtained a value of 1.09). Inputting this into Eqs. 5.73 and 5.74 yields $R_G = 0.43$ and $R_s = 2.59$. Note that these values are extremely close to those calculated for the rigid cap assumption; Poulos & Davis (1980) suggest that suitable values of $w_{0,avg}$ and therefore R_s and R_G can be obtained with the flexible cap assumption regardless of the actual cap conditions.

5.3.5.2   Dynamic response

The response of pile groups to dynamic loading can also be analysed using the superposition method, substituting the quantities used in the static case with the complex-valued dynamic quantities discussed previously. To this end, the complex-valued settlement of pile i, $w_{0,i}^*$, is given by

$$w_{0,i}^* = \sum_{j=1}^{N} \frac{P_j^* \alpha_{ij}^*}{K_{0,j}^*} \qquad (5.76)$$

where P_j^* is the dynamic load carried by pile j, $K_{0,j}^*$ is the dynamic stiffness of pile j, and α_{ij}^* is the dynamic interaction factor between pile i and pile j.

This can be used (similarly to Eq. 5.71) to generate a set of simultaneous equations, which can be solved based on the pile cap conditions as before, yielding the average complex-valued settlement of the pile group, $w_{0,avg}^*$.

The efficiency of different group configurations can then be compared using the dynamic settlement ratio, R_s^*, and dynamic group reduction factor, R_G^*, defined in Eq. 5.77. Note that as the phase of the dynamic interaction factor is non-zero, R_s^* and R_G^* will have both an amplitude and phase (φ_s and φ_g, respectively). The amplitude is normally of most interest. The terms in the sum of Eq. 5.76 will interfere both constructively and destructively depending on different factors, particularly the frequency of the applied loading, spacing of the piles, and shear wave velocity of the soil. This can result in the amplitude of R_s^* varying significantly for the same pile group depending on the frequency of the applied load. If enough of the attenuated pile displacements interfere destructively, the amplitude of R_s^* can be lower than 1, indicating that the group is stiffer than the sum of its component piles. However, if the attenuated pile displacements interfere constructively, the amplitude of R_s^* can reach very large values.

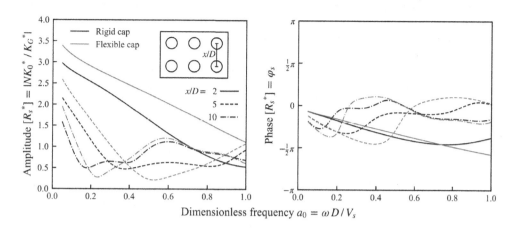

Figure 5.27 Dynamic settlement ratio, R_s^*, for the six-pile group (Figure 5.26) embedded in a homogeneous half-space; $\left(L/D = 25, E_p/E_s = 1000, \rho_p/\rho_s = 1.25, \beta_s = 0.05, \beta_p = 0, v_s = 0.4\right)$.

$$R_s^* = w_{0,avg}^* \bigg/ \frac{P_G^*}{NK_0^*} = \frac{NK_0^*}{K_G^*}, \quad R_G^* = w_{0,avg}^* \bigg/ \frac{P_G^*}{K_0^*} = \frac{K_0^*}{K_G^*} = \frac{R_s^*}{N} \tag{5.77}$$

where P_G^* is the applied dynamic group load, and K_G^* is the overall dynamic group stiffness.

Consider the example pile group shown in Figure 5.26. The same approach used in the static case can be employed to calculate the dynamic settlement ratios $\left(R_s^*\right)$ for different grid spacings (x/d) and soil, pile, and cap properties. The dynamic interaction factors can be calculated using Eq. 5.64 (for homogeneous soil; inhomogeneous soil is discussed in Section 5.3.4.2) and the attenuation function (Eqs. 5.61 or 5.62). Figure 5.27 shows the dynamic settlement ratio of the group shown in Figure 5.26 for different grid spacings $(x/d = 2, 5, 10)$, cap conditions (rigid or flexible), and applied load frequency (ω) when the piles of slenderness $L/D = 25$ are embedded in a homogeneous soil such that: $E_p/E_s = 1000, \rho_p/\rho_s = 1.25, \beta_s = 0.05$, and $\beta_p = 0$. The results are as expected, with R_s^* values below 1 for higher frequencies due to the destructive interference of the different piles. This happens at lower frequencies as the spacing increases.

5.4 SUMMARY AND CONCLUSIONS

A family of solutions of the Winkler type was reviewed for axially loaded single piles, pairs of piles, and pile groups in inhomogeneous soil, subjected to static and dynamic loads. The solutions were formulated in closed form or in terms of simple iterative formulae that can be implemented in design by means of spreadsheet (or even pocket calculator) analyses. Specifically

1   For single piles under static loading, exact solutions for homogeneous soil are presented in Eqs. 5.3–5.10, 5.19, 5.39 and in Table 5.1. Exact iterative solutions for

layered soil are presented in Eqs. 5.11–5.13. Exact solutions for smoothly inhomogeneous soil are presented in Eqs. 5.14–5.16 and in Table 5.2. Approximate energy solutions for inhomogeneous soil are presented in Eqs. 5.17–5.22.

2   For single piles under dynamic loading, exact solutions for homogeneous soil are presented in Eqs. 5.28–5.33 and in Table 5.3. Exact iterative solutions for layered soil are presented in Eqs. 5.34 and 5.35. Approximate energy solutions for inhomogeneous soil are presented in Eqs. 5.36–5.38.

3   A reduction in soil shear modulus to account for strain effects wisas presented in Eq. 5.42

4   For pairs of piles under static loading, exact solutions for homogeneous soil are presented in Eqs. 5.43–5.52 and in Table 5.4. Exact iterative solutions for layered soil are presented in Eqs. 5.53 and 5.54. Exact solutions for smoothly inhomogeneous soil are presented in Eqs. 5.54–5.56 and in Table 5.5. Approximate energy solutions for inhomogeneous soil are presented in Eqs. 5.57–5.59.

5   For pairs of piles under dynamic loading, exact solutions for interaction factors in homogeneous soil are presented in Eqs. 5.64–5.67. Exact iterative solutions for interaction factors in layered soil are presented in Eq. 5.68. Approximate energy solutions for inhomogeneous soil are presented in Eqs. 5.69 and 5.70.

6   For pile groups under static loading, solutions using the superposition method are presented in Eqs. 5.71–5.74.

7   For pile groups under dynamic loading, solutions using the superposition method are presented in Eqs. 5.76 and 5.77.

Satisfactory comparisons against rigorous continuum solutions (mostly of the boundary-element type) are presented for both static and dynamic loads.

ACKNOWLEDGEMENTS

The first author would like to thank the Engineering and Physical Sciences Research Council for their support (grant number EP/N509619/1). No new experimental data were collected during this study.

NOTATION

Latin

a, a_D	Soil inhomogeneity parameter
a_0	Dimensionless frequency
$a_{0,c}$	Dimensionless soil layer cut-off frequency
A	Soil displacement attenuation parameter
A_p	Pile cross sectional area
C_0	Pile head dashpot viscosity
C_b	Pile base dashpot viscosity
C_h	Pile head hysteretic dashpot viscosity
c_1	Strain-displacement proportionality coefficient
$c(z)$	Winkler dashpot viscosity

c_h	Winkler hysteretic dashpot viscosity
c_r	Winkler radiation dashpot viscosity
$\tilde{c}$	dynamic modifier of pile base damping constant
D	Pile diameter
E_p, E_p^*	Real and complex valued pile Young's modulus, respectively
E_s, E_s^*	Real and complex valued soil Young's modulus, respectively
E_{sL}, E_{sb}	Soil Young's modulus at and below the pile base, respectively
G_s, G_s^*	Real and complex soil shear modulus, respectively
H	Depth to underlying rigid layer
h_m	Thickness of soil layer m
$I_\mu()$	Modified Bessel function of the first kind of order μ
i	Imaginary unit
$J_\mu()$	Bessel function of the first kind of order μ
K_0, K_0^*	Real and complex valued pile head stiffness, respectively
$K_{0,hom}$	Pile head stiffness in homogeneous soil
$K_{0,j}, K_{0,j}^*$	Real and complex valued pile head stiffness of pile j, respectively
K_b, K_b^*	Real and complex pile base spring stiffness, respectively
K_G, K_G^*	Real and complex valued pile group stiffness, respectively
K_m	Pile stiffness at the top of layer m
$K_\mu()$	Modified Bessel function of the second kind of order μ
$k(z), k_0, k_L, k_D, k_m$	Winkler spring stiffness at depth z, the ground surface, the pile base, one diameter depth, and in soil layer m, respectively
k_{avg}	Average Winkler spring stiffness over the pile length
$k^*(z), k_0^*, k_L^*, k_m^*$	Complex Winkler spring stiffness at depth z, the ground surface, the pile base, and of soil layer m, respectively
$\tilde{k}$	Dynamic modifier of pile base stiffness constant
L	Pile length
N	Number of piles in the group
n	Soil inhomogeneity exponent
$P, P(t)$	Static and dynamic pile head load, respectively
P_{centre}	Load on the centre piles in a group
P_{edge}	Load on the edge piles in a group
P_G, P_G^*	Real and complex valued pile group load, respectively
P_j, P_j^*	Real and complex valued load on pile j, respectively
$P(z)$	Axial load in the pile at depth z
R_s, R_s^*	Real and complex valued pile group settlement ratio, respectively
R_G, R_G^*	Real and complex valued pile group reduction factor, respectively
r_m	Mobilised radius due to the pile movement
s	Distance between two piles
s^*	Complex dimensionless frequency
s_{st}	Dimensionless equivalent wavelength to determine static stiffness
t	Time
$V_s(z) V_s^*$	Real and complex soil shear wave velocity, respectively
V_{sL}	Soil shear wave velocity at the pile base
$Y_\mu()$	Bessel function of the second kind of order μ
$w(z), w^*(z)$	Real and complex valued pile displacement, respectively
$w_v(z)$	Virtual pile displacement

w_0	Pile head displacement
$w_{0,avg}, w_{0,avg}^*$	Real and complex valued average pile group displacement, respectively
$w_{0,i}, w_{0,i}^*$	Real and complex valued head displacement of pile i, respectively
$w_{0,centre}$	Displacement of the centre piles in a group
$w_{0,edge}$	Displacement of the edge piles in a group
$w_{ij}(z)$	Displacement of pile i at depth z due to a load on pile j
x	Pile group grid spacing
z	Depth below ground surface

Calligraphic

$\Im[\,]$	Imaginary component
$\Re[\,]$	Real component

Greek

$\alpha_{ij}, \alpha_{ij}^*$	Real and complex valued interaction factor, respectively, between pile i and pile j
α_m, α_m^*	Real and complex valued interaction factor, respectively, between pile i and pile j for soil layer m
α_{rigid}	Interaction factor between rigid piles
β_0	Pile head damping factor
β_p	Pile hysteretic damping factor
β_s	Soil hysteretic damping factor
$\Gamma()$	Gamma function
γ	Euler's number
$\gamma(z)$	Soil shear strain at depth z
ζ, ζ^*	Real and complex valued diffraction factor
ζ_b	Base diffraction factor
ζ_m	Diffraction factor of soil layer m
ζ_{hom}	Diffraction factor for homogeneous soil
$\zeta_{b,hom}$	Base diffraction factor for homogeneous soil
η_s	Dimensionless soil compressibility coefficient
θ	Phase of the complex pile head stiffness
$\lambda, \lambda_m, \lambda_0, \lambda_L, \lambda_D$	Load transfer parameter
λ^*	Complex load transfer parameter
$\lambda_{massless}^*$	Complex load transfer parameter assuming a massless pile
μ, μ^*	Real and complex valued load transfer parameter
v	Soil inhomogeneity parameter
v_p	Pile Poisson's ratio
v_s	Soil Poisson's ratio
ρ	Soil inhomogeneity parameter
ρ_p	Pile mass density
ρ_s	Soil mass density

φ	Phase of the complex interaction factor
φ_S	Phase of the complex pile group settlement ratio
$\chi(z), \chi^*(z)$	Real and complex valued shape function
χ_0, χ_L, χ_D	Arguments of Bessel functions
$\psi(s), \psi^*(s)$	Real and complex valued shaft displacement attenuation function
$\psi_b(s)$	Base displacement attenuation function
$\Omega, \Omega_L, \Omega_m, \Omega_\mu$	Dimensionless base stiffness
Ω^*, Ω_m^*	Complex valued dimensionless base stiffness
ω	Applied load angular frequency
ω_c	Cut-off soil layer frequency

REFERENCES

Akiyoshi, T. (1982). Soil-pile interaction in vertical vibration induced through a frictional interface. *Earthquake Engineering & Structural Dynamics*, 10 (1), 135–148.

Anoyatis, G. (2013) Contribution to kinematic and inertial analysis of piles by analytical and experimental methods. Ph.D. thesis, University of Patras, Patras, Greece.

Anoyatis, G. & Mylonakis, G. (2012) Dynamic Winkler modulus for axially loaded piles. *Géotechnique*, 62 (6), 521–536.

Anoyatis, G. & Mylonakis, G. (2020) Analytical solution for axially loaded piles in two-layer soil. *Journal of Engineering Mechanics, ASCE*, 146 (3), 04020003.

Anoyatis, G., Mylonakis, G. & Tsikas (2019) An analytical continuum model for axially loaded end-bearing piles in inhomogeneous soil. *International Journal for Numerical and Analytical Methods in Geomechanics*, 43 (6), 1162–1183.

Baguelin, F. & Frank, R. (1980). Theoretical studies of piles using the finite element method. In: Smith, I.M., George, P. & Rigden, W.J. (eds.) *Numerical Methods in Offshore Piling*, Institution of Civil Engineers, London, UK, pp. 83–91.

Banerjee, P.K. & Davies, T.G. (1977) Analysis-of pile groups embedded in Gibson soil. In: *Proceedings of the* 13th *International Conference on Soil Mechanics and Foundation Engineering*, Tokyo, July 11–15, 1, pp. 381–386.

Banerjee, P.K. & Davies, T.G. (1978) The behaviour of axially and laterally loaded single piles embedded in nonhomogeneous soils. *Géotechnique*, 28 (3), 43–60.

Baranov, V.A. (1967) On the calculation of excited vibrations of an embedded foundation (in Russian). *Voprosy Dynamiki i Prochnoeti*, (14), Polytechnic Institute, Riga, Latvia, 195–209.

Basile, F. (2019) The role of cap flexibility in pile group design. In: 17th European Conference on Soil Mechanics and Geotechnical Engineering, Reykjavík, Iceland.

Bateman, A.H., Crispin, J.J. & Mylonakis, G. (2021) A simplified analytical model for developing "t-z" curves for axially loaded piles. *Under review*.

Blaney, G.W., Kausel, E. & Roesett, J.M. (1975). Dynamic stiffness of piles. In: *Proceedings of the 2nd International Conference on Numerical Methods in Geomechanics*, Blacksburg, VA, 2, pp. 1010–1012.

Blaney, G.W., Muster, G.L. & O'Neill, M.W. (1987) Vertical vibration test of a full-scale pile group. In: Nogami, T. (ed.) *Dynamic Response of Pile Foundations – Experiment, Analysis and Observation*, Geotechnical Special Publication No. 11, ASCE, New York, NY, USA, pp. 149–165.

Butterfield, R. & Banerjee (1971) The elastic analysis of compressible piles and pile groups. *Géotechnique*, 21 (1), 43–60.

Cairo, R., Dente, G. & Troncone, A. (1999) Cone model for a pile foundation embedded in a soil layer. In: Brebbia, C.A. & Oliveto, G. (eds.) *Earthquake Resistant Structures II*, Transactions on the Built Environment, 41, WIT Press, Southampton, UK, pp. 565–574.

Comodromos, E.M., Anagnostopoulos, C.T. & Georgiadis, M.K. (2003) Numerical assessment of axial pile group response based on load test. *Computers and Geotechnics*, 30 (6), 505–515.

Cooke, R.W. (1974) The settlement of friction pile foundations. *Building Research Station*, Current Paper CP 12/75, 1–16.

Coyle, H.M. & Reese, L.C (1966). Load transfer for axially loaded piles in clay. *Journal of the Soil Mechanics and Foundation Division, ASCE*, 92 (2), 1–26.

Crispin, J.J., Leahy, C.P. & Mylonakis, G. (2018). Winkler Model for axially-loaded piles in inhomogeneous soil. *Géotechnique Letters*, 8 (4), 290–297.

Crispin, J.J. & Leahy, C.P. (2019) Settlement of axially loaded pile groups in inhomogeneous soil. *DFI Journal – Journal of the Deep Foundation Institute*, 12 (3), 163–170.

Crispin, J.J., Su, H., Round, K. & Mylonakis, G. (2019) Energy formulation for axial pile head stiffness in inhomogeneous soil. In: El-Naggar, H., Abdel-Rahman, K., Fellenius, B. & Shehata, H. (eds.) *Sustainability Issues for the Deep Foundations*. GeoMEast 2018. Sustainable Civil Infrastructures, Springer, Cham, Switzerland, pp. 181–190.

Darendeli, M.B. (2001) Development of a new family of normalized modulus reduction and material damping curves. Ph.D. thesis, University of Texas at Austin, Austin, TX.

Davies, T.G., Sen, R. & Banerjee, P.K. (1985) Dynamic behaviour of pile groups in inhomogeneous soil. *Journal of Geotechnical Engineering, ASCE*, 111 (12), 1365–1379.

Davis, R.O. & Selvadurai, A.P.S. (1996) *Elasticity and Geomechanics*, Cambridge University Press, Cambridge, UK.

Dezi, F., Carbonari, S. & Leoni, G. (2009) A model for the 3D kinematic interaction analysis of pile groups in layered soils. *Earthquake Engineering & Structural Dynamics*, 38 (11), 1281–1305.

Dezi, F., Carbonari, S. & Morici, M. (2016) A numerical model for the dynamic analysis of inclined pile groups. *Earthquake Engineering and Structural Dynamics*, 45(1), 45-68.

Di Laora, R., Grossi, Y. & Viggiani, G.M.B. (2017) An analytical solution for the rotational component of the Foundation Input Motion induced by a pile group. *Soil Dynamics and Earthquake Engineering*, 97, 424–438.

Dobry, R. & Gazetas, G. (1988) Simple method for dynamic stiffness and damping of floating pile groups. *Géotechnique*, 38 (4), 557–574.

Dobry, R., Vicente, E., O'Rourke, M.J. & Roesset, J.M. (1982) Horizontal stiffness and damping of single piles. *Journal of Geotechnical Engineering, ASCE*, 108 (3), 439–459.

El-Marsafawi, H., Kaynia, A.M. & Novak, M. (1992) Interaction factors and the superposition method for pile group dynamic analysis. Geotechnical Research Center, Department of Civil Engineering, University of Western Ontario, Canada, Report number GEOT-1-92.

El Naggar, M.H. & Novak, M. (1994a) Non-linear model for dynamic axial pile response. *Journal of Geotechnical Engineering, ASCE*, 120 (2), 308–329.

El Naggar, M.H. & Novak, M. (1994b) Nonlinear axial interaction in pile dynamics. *Journal of Geotechnical Engineering, ASCE*, 120 (4), 678–696.

El-Sharnouby, B. & Novak, M. (1990). Stiffness constants and interaction factors for vertical response of pile groups. *Canadian Geotechnical Journal*, 27 (6), 813–822.

Fleming, K., Weltman, A., Randolph, M. & Elson, K. (2009). *Piling Engineering*, 3rd edition, Taylor & Francis, Abingdon, UK.

Gazetas, G. & Dobry, R. (1984) Horizontal response of piles in layered soil. *Journal of Geotechnical Engineering, ASCE*, 110(1), 20–40.

Gazetas, G., Fan, K. & Kaynia, A. (1993) Dynamic response of pile groups with different configurations. *Soil Dynamics and Earthquake Engineering*, 12 (4), 239–257.

Gazetas, G., Fan, K., Kaynia, A. & Kausel, E. (1991) Dynamic interaction factors for floating pile groups. *Journal of Geotechnical Engineering, ASCE*, 117 (10), 1531–1548.

Gazetas, G. & Makris, N. (1991) Dynamic pile-soil-pile interaction. Part 1: Analysis of axial vibration. *Earthquake Engineering and Structural Dynamics*, 20 (2), 115–132.

Gazetas, G. & Mylonakis, G. (1998) Seismic soil-structure interaction: New evidence and emerging issues state of the art paper. In: Dakoulas, P., Yegian, M. & Holtz, R.D. (eds.) Geotechnical Earthquake Engineering and Soil Dynamics III, 3–6 August 1998, Seattle, WA, USA. Geotechnical Special Publication, GSP75, ASCE, Reston, VA. Vol. 2, pp. 1119–1174.

Guo, W.D. (2012) *Theory and Practice of Pile Foundations*, CRC Press, Boca Raton, FL.

Guo, W.D. & Randolph, M.F. (1997) Vertically loaded piles in non-homogeneous media. *International Journal for Numerical and Analytical Methods in Geomechanics*, 21 (8), 507–532.

Guo, W.D. & Randolph, M.F. (1998) Rationality of load transfer approach for pile analysis. *Computers and Geotechnics*, 23 (1–2), 85–112.

Gupta, B.K. & Basu, D. (2018) Dynamic analysis of axially loaded end-bearing pile in homogeneous viscoelastic soil. *Soil Dynamics and Earthquake Engineering*, 111, 31–40.

Horikoshi, K. & Randolph, M.F. (1997) On the definition of raft-soil stiffness ratio for rectangular rafts. *Géotechnique*, 47 (5), 1055–1061.

Ishihara, K. (1996) *Soil Behaviour in Earthquake Geotechnics*, Clarendon Press, Oxford, UK.

Jia, J. (2018) *Soil Dynamics & Foundation Modelling: Offshore and Earthquake Engineering*, Springer, Cham, Switzerland.

Kagawa, T. & Kraft, L.M. (1981) Lateral pile response during earthquakes. *Journal of the Geotechnical Engineering Division, ASCE*, 107 (12), 1713–1731.

Kanellopoulos, K. & Gazetas, G. (2020) Vertical static and dynamic pile-to-pile interaction in non-linear soil. *Géotechnique*, 70 (5), 432–447.

Karatzia, X. & Mylonakis, G. (2017) Horizontal stiffness and damping of piles in inhomogeneous soil. *Journal of Geotechnical and Geoenvironmental Engineering, ASCE*, 143 (4), 04016113.

Kaynia, A.M. (1980) Dynamic stiffness and seismic response of sleeved piles. Department of Civil Engineering, Massachusetts Institute of Technology, Cambridge, MA, Research report no. R80-12.

Kaynia, A.M. (1982) Dynamic stiffness and seismic response of pile groups. Ph.D. thesis, Massachusetts Institute of Technology, Cambridge, MA.

Kaynia, A.M. & Kausel, E. (1982) Dynamic stiffness and seismic response of pile groups. Department of Civil Engineering, Massachusetts Institute of Technology, Cambridge, MA, Research report no. R82-03.

Kaynia, A.M. & Kausel, E. (1991) Dynamics of piles and pile groups in layered soil media. *Soil Dynamics and Earthquake Engineering*, 10 (8), 386–401.

Kraft, L.M., Ray, R.P. & Kagawa, T. (1981). Theoretical t-z Curves. *Journal of the Geotechnical Engineering Division, ASCE*, 107 (11), 1543–1561.

Kuhlemeyer, R.L. (1979) Vertical vibration of piles. *Journal of the Geotechnical Engineering Division, ASCE*, 105 (2), 273–287.

Madabhushi, G., Knappett, J. & Haigh, S. (2010) *Design of Pile Foundations in Liquefiable Soils*, Imperial College Press, London, UK.

Maheshwari, B.K., Truman, K.Z., El Naggar, M.H. & Gould, P.L. (2004) Three-dimensional nonlinear analysis for seismic soil–pile-structure interaction. *Soil Dynamics and Earthquake Engineering*, 24 (4), 343–356.

Makris, N. & Gazetas, G. (1993) Displacement phase differences in a harmonically oscillating pile. *Géotechnique*, 43 (1), 135–150.

Mamoon, S.M., Kaynia, A.M. & Banerjee, P.K. (1990) Frequency domain dynamic analysis of piles and pile groups. *Journal of Engineering Mechanics, ASCE*, 116 (10), 2237–2257.

Mandolini, A. & Viggiani, C. (1997) Settlement of piled foundations. *Géotechnique*, 47 (4), 791–816.

Matlock, H. (1970) Correlations for design of laterally loaded piles in soft clay. In: *Proceedings of the 2nd Offshore Technology Conference, Houston 22–24 April 1970*, Offshore Technology Conference, Houston, TX, USA, OTC paper 1204, pp. 577–588.

Michaelides, O., Gazetas, G., Bouckovalas, G. & Chrysikou, E. (1998) Approximate non-linear dynamic axial response of piles. *Géotechnique*, 48 (1), 33–53.

Mizuno, H. (1987) Pile damage during earthquake in Japan. In: Nogami, T. (ed.) *Dynamic Response of Pile Foundations – Experiment, Analysis and Observation, Geotechnical Special Publication No. 11*, ASCE, New York, NY, USA, pp. 53–78.

Motta, E. (1994) Approximate elastic-plastic solution for axially loaded piles. *Journal of Geotechnical Engineering, ASCE*, 120 (9), 1616–1624.

Mylonakis, G. (1995) Contributions to static and seismic analysis of piles and pile-supported bridge piers. Ph.D. thesis, State University of New York at Buffalo, Buffalo, NY.

Mylonakis, G. (2000) A new elastodynamic model for settlement analysis of a single pile in multi-layered soil, Discussion to Paper by K.M. Lee and Z.R. Xiao. *Soils and Foundations*, 40 (4), 163–166.

Mylonakis, G. (2001a) Winkler model for axially loaded piles. *Géotechnique*, 51 (5), 455–461.

Mylonakis, G. (2001b) Elastodynamic model for large-diameter end-bearing shafts. *Soils and Foundations*, 41 (3), 31–44.

Mylonakis, G. & Gazetas, G. (1998a) Settlement and additional internal forces of grouped piles in layered soil. *Géotechnique*, 48 (1), 55–72.

Mylonakis, G. & Gazetas, G. (1998b) Vertical vibrations and additional distress of grouped piles in layered soil. *Soils and Foundations*, 38 (1), 1–14.

NEHRP (2012) Soil-structure interaction for building structures. National Institute of Standards and Technology (NIST). Report number: GCR 12-917-21.

Nogami, T. (1983) Dynamic group effect in axial responses of grouped piles. *Journal of Geotechnical Engineering, ASCE*, 109 (2), 228–243.

Nogami, T. & Konagai, K. (1987) Dynamic response of vertically loaded nonlinear pile foundations. *Journal of Geotechnical Engineering, ASCE*, 113 (2), 147–160.

Novak, M. (1974) Dynamic stiffness and damping of piles. *Canadian Geotechnical Journal*, 11 (4), 574–598.

Novak, M. (1991) Piles under dynamic loads. In: *Proceedings of the 2nd International Conference on Recent Advances in Geotechnical Earthquake Engineering and Soil Dynamics*, 11–15 March, St. Louis, MO, (2), 2433–2456.

Novak, M., Nogami, T. & Aboul-Ella, F. (1978). Dynamic soil reactions for plane strain case. *Journal of Engineering Mechanics, ASCE*, 104 (4), 953–959.

Novak, M., Sheta, M., El-Hifnawy, L., El-Marsafawi, H. & Ramadan, O. (1993) DYNA4 – A computer program for foundation response to dynamic loads. User's Manual, Geotechnical Research Centre, University of Western Ontario, London, Canada.

O'Neill, M.W. (2001) Side resistance in piles and drilled shafts. *Journal of Geotechnical and Geoenvironmental Engineering, ASCE*, 127 (1), 1–16.

O'Neill, M.W., Hawkins, R.A. & Audibert, J.M.E. (1982) Installation of pile group in overconsolidated clay. *Journal of the Geotechnical Engineering Division, ASCE*, 108 (11), 1369–1386.

O'Rourke, M.J. & Dobry, R. (1978). Spring and dash pot coefficients for machine foundation on piles. American Concrete Institute, Detroit, MI. Report number SP-10, pp. 177–198.

Osman, A.S. & Bolton, M.D. (2005) Simple plasticity-based prediction of the undrained settlement of shallow circular foundations on clay. *Géotechnique*, 55 (6), 435–447.

Ottaviani, M. (1975) Three-dimensional finite element analysis of vertically loaded pile groups. *Géotechnique*, 25 (2), 159–174.

Padrón, L.A., Aznárez, J.J. & Maeso, O. (2007) BEM–FEM coupling model for the dynamic analysis of piles and pile groups. *Engineering Analysis with Boundary Elements*, 31 (6), 472–484.

Padrón, L.A., Mylonakis, G. & Beskos, D.E. (2009) Importance of footing-soil separation on dynamic stiffness of piled embedded footings. *International Journal of Numerical and Analytical Methods in Geomechanics*, 33 (11), 1439–1448.

Padrón, L.A., Mylonakis, G. & Beskos, D.E. (2012) Simple superposition approach for dynamic analysis of piled embedded footings. *International Journal of Numerical and Analytical Methods in Geomechanics*, 36 (12), 1523–1534.

Pender, M. (1993). Aseismic pile foundation design analysis. *Bulletin of the New Zealand National Society for Earthquake Engineering*, 26 (1), 49–160.

Pipkin, A.C. (1972) *Lectures on Viscoelastic Theory*, Springer, New York, NY.

Poulos, H.G. (1968) Analysis of the settlement of pile groups. *Géotechnique*, 18 (4), 449–471.

Poulos, H.G. (1989) Pile behaviour – Theory and application. *Géotechnique*, 39 (3), 365–415.

Poulos, H.G. (2017) *Tall Building Foundation Design*, CRC Press, Boca Raton, FL.

Poulos, H.G. & Davis, E.H. (1980) *Pile Foundation Analysis and Design*, John Wiley & Sons, Inc., New York, NY.

Prakash, S. & Sharma, H.D. (1990) *Pile Foundations in Engineering Practice*, John Wiley & Sons, Inc., New York, NY.

Rajapakse, R.K.N.D. (1990) Response of an axially loaded elastic pile in a Gibson soil. *Géotechnique*, 40 (2), 237–249.

Randolph, M.F. (1977) A theoretical study of the performance of piles. Ph.D. thesis, University of Cambridge, Cambridge, UK.

Randolph, M.F. (2003) Science and empiricism in pile foundation design. *Géotechnique*, 53 (10), 847–875.

Randolph, M.F. & Wroth, C.P. (1978) Analysis of deformation of vertically loaded piles. *Journal of Geotechnical and Geoenvironmental Engineering, ASCE*, 104 (12), 1465–1488.

Randolph, M.F. & Wroth, C.P. (1979) An analysis of vertical deformation of pile groups. *Géotechnique*, 29 (4), 423–439.

Reese, L.C., Isenhower, W.M. & Wang, S-T. (2005) *Analysis and Design of Shallow and Deep Foundations*, John Wiley & Sons, Inc., Hoboken, NJ.

Roesset, J.M. (1980) Stiffness and damping coefficients of foundations. In: O'Neil, M. & Dobry, R. (eds.) *Dynamic Response of Pile Foundations*, ASCE, New York, NY, USA, pp. 1–29.

Roesset, J.M. & Angelides, D. (1980) Dynamic stiffness of piles. In: Smith, I.M., George, P. & Rigden, W.J. (eds.) *Numerical Methods in Offshore Piling*, Institution of Civil Engineers, London, UK, pp. 75–81.

Rollins, K.M., Clayton, R.J., Mikesell, R.C. & Blaise, B.C. (2005) Drilled shaft side friction in gravelly soils. *Journal of Geotechnical and Geoenvironmental Engineering, ASCE*, 131 (8), 987–1003.

Salgado, R. (2008) *The Engineering of Foundations*, International edition, McGraw-Hill, New York, NY.

Sanchez-Salinero, I. (1982). Static and dynamic stiffness of single piles. Geotechnical Engineering Report, University of Texas, Austin, TX. Report number: GR82-31.

Sanchez-Salinero, I. (1983). Dynamic stiffness of pile groups: Approximate solutions. Geotechnical Engineering Report, University of Texas, Austin, TX. Report number: GR83-5.

Scott, R.F. (1981) *Foundation Analysis*, Prentice Hall, Englewood Cliffs, NJ.

Seed, H.B. & Reese, L.C. (1957) The action of soft clay along friction piles. *Transactions of the American Society of Civil Engineers*, 122, 731–764.

Syngros, K. (2004) Seismic response of piles and pile-supported bridge piers evaluated through case histories. Ph.D. thesis, The City College of the City University of New York, New York, NY.

Tajimi, H. (1969) Dynamic analysis of a structure embedded in an elastic stratum. In: *Proceeding of the 4th World Conference on Earthquake Engineering*, 13–18 January 1969, Santiago, Chile. Vol. 3, A6-53-70.

Vardanega, P.J. & Bolton, M.D. (2011) Strength mobilization in clays and silts. *Canadian Geotechnical Journal*, 48 (10), 1485–1503.

Vardanega, P.J., Williamson, M.G. & Bolton, M.D. (2012) Bored pile design in stiff clay II: Mechanisms and uncertainty. *Proceedings of the ICE – Geotechnical Engineering*, 165 (4), 233–246; corrigendum 166 (5), 518.

Veletsos, A.S. & Verbic, B. (1973) Vibration of viscoelastic foundations. *Earthquake Engineering and Structural Dynamics*, 2 (1), 87–102.

Viggiani, C., Mandolini, A. & Russo, G. (2011) *Piles and Pile Foundations*, CRC Press, Boca Raton, FL.

Vlasov, V.Z. & Leontiev, N.N. (1966) Beams, plates and shells on elastic foundations. Translated from Russian, Israel program for scientific translations, Washington, DC. NIST No. N67-14238.

Voyagaki, E., Crispin, J., Gilder, C., Nowak, P., O'Riordan, N., Patel, D. & Vardanega, P. (2019) Analytical approaches to predict pile settlement in London clay. In: El-Naggar, H., Abdel-Rahman, K., Fellenius, B. & Shehata, H. (eds.) *Sustainability Issues for the Deep Foundations, GeoMEast 2018*. Sustainable Civil Infrastructures, Springer, Cham, Switzerland, pp. 162–180.

Voyagaki E., Crispin J.J., Gilder C.E.L., Ntassiou K., O'Riordan N., Nowak P., Sadek T., Patel D., Mylonakis G., and Vardanega P.J. 2021. The DINGO Database of Axial Pile Load Tests for the UK: Settlement prediction in fine-grained soils. *Under review*.

Vucetic, M. & Dobry, R. (1991) Effect of soil plasticity on cyclic response. *Journal of Geotechnical Engineering, ASCE*, 117 (1), 89–107.

Wolf, J.P. (1994) *Foundation Vibration Analysis Using Simple Physical Models*, Prentice-Hall Inc., Englewood Cliffs, NJ.

Wolf, J.P., Meek, J.W. & Song, C. (1992) Cone models for a pile foundation. In: Prakash, S. (ed.) *Piles Under Dynamic Loads*, 13–17 September 1992, New York, NY. ASCE Geotechnical Special Publication, GSP34, pp. 1–27.

Chapter 6

Simplified models for lateral static and dynamic analysis of pile foundations

George E. Mylonakis
University of Bristol
Khalifa University
University of California at Los Angeles

Jamie J. Crispin
University of Bristol

CONTENTS

6.1   INTRODUCTION

Analysis of flexural pile response to lateral loads is of fundamental importance for structural/geotechnical design against earth pressures, earthquakes, landslides, scour, floods, collision of ships on bridge piers, machine vibrations, wind, and blasts. The problem has been thoroughly investigated for more than 70 years, and valuable knowledge has been acquired as to the physics of the response and solution methods.

Analysis approaches for laterally loaded piles can be roughly classified into the following three main groups:

A   Rigorous numerical methods, mainly finite element and boundary element formulations which treat the soil around and underneath the pile as a continuum. Specifically
  • Finite element methods (FEMs) are general-purpose continuum formulations capable of handling a wide range of pile-related problems including elastic analyses (Blaney et al. 1976; Randolph 1977, 1981; Wolf & Von Arx 1978; Kuhlemeyer 1979; Kaynia 1980; Roesset & Angelides 1980; Wolf et al. 1981; Dobry et al. 1982; Sanchez-Salinero 1982, 1983; Velez et al. 1983; Waas & Hartmann 1984; Gazetas 1984; El-Sharnouby & Novak 1985; Syngros 2004; Di Laora et al. 2012; Shadlou & Bhattacharya 2014; Di Laora & Rovithis 2015) and non-linear analyses involving yielding materials for the soil and the pile (Angelides & Roesset 1981; Trochanis et al. 1991; Maheshwari et al. 2004; Radhima et al. 2021);
  • On the other hand, boundary-element methods are best suited for linear elastostatic/elastodynamic analyses and can better handle infinite domains (especially in three dimensions) and dynamic loads (Poulos 1971a, b; Banerjee & Davies 1978; Kaynia 1982; Kaynia & Kausel 1982, 1991; Tyson & Kausel 1983; Rajapakse & Shah 1989; Mamoon et al. 1990; Fan et al. 1991; El-Marsafawi et al. 1992b; Gazetas et al. 1991, 1993; Kaynia & Mahzooni 1996; Guin 1997; Padrón et al. 2007a, b, 2009, 2012);
  • Some hybrid boundary-element based non-linear formulations are also available (Davies & Budhu 1986; Budhu & Davies 1987, 1988).
B   Experimental methods. These can be further classified as follows:
  • Laboratory testing of small-scale models in centrifuges, shaking table devices, or fixed soil containers using idealised soil – mainly sand (Gohl 1991; Meymand 1998; Boulanger et al. 1999; Bhattacharya & Adhikari 2011; Chidichimo et al. 2014; Durante et al. 2016; Goit et al. 2016; Fiorentino et al. 2021);
  • Field testing of full- or reduced-scale models, or monitoring of piles supporting real structures, or analysis of case histories (Reese et al. 1974, 1975; Ting 1978; Scott et al. 1982; Ohira et al. 1984; Tazoh et al. 1987; El-Marsafawi et al. 1992a; Makris et al. 1994; Rollins et al. 1998; Weaver et al. 1998; Nikolaou et al. 2001; Syngros 2004; Mylonakis et al. 2006; Gerber & Rollins 2008).
C   Approximate analytical models, mainly Winkler formulations substituting the soil by a bed of distributed independent springs (or springs and dashpots attached in parallel). These can be further classified as follows:

- Linear elastic Winkler models developed for static (Hetenyi 1946; Barber 1953; Matlock & Reese 1960; Francis 1964; Franklin & Scott 1979; Mylonakis 1995) or dynamic conditions (Baranov 1967; Yoshida & Yoshinaka 1972; Novak 1974; Berger et al. 1977; Novak et al. 1978; O'Rourke & Dobry 1978; Kaynia 1980; Novak & Sheta 1980; Flores-Berrones & Whitman 1982; Dobry et al. 1982; Dobry & O'Rourke 1983; Barghouthi 1984; Gazetas & Dobry 1984a, b; Nogami 1985; Makris & Gazetas 1992, 1993; Mylonakis 1995; Mylonakis & Gazetas 1999; Tabesh & Poulos 2000, 2001; Mylonakis & Roumbas 2001; Nikolaou et al. 2001; Varun et al. 2009; Dezi et al. 2009, 2016; Karatzia & Mylonakis 2012, 2017; Rovithis et al. 2013; Karatzia et al. 2014, 2015; Di Laora & Rovithis 2015; Sextos et al. 2015);
- Non-linear Winkler models encompassing classic "p-y" formulations for static and cyclic loads (McClelland & Focht 1956; Matlock 1970; Reese et al. 1974, 1975; O'Neill et al. 1977; Haiderali & Madabhushi 2016; Byrne et al. 2019), or special "p-y" formulations for dynamic loads (Parmelee et al. 1964; Penzien et al. 1964; Penzien 1970; Kagawa & Kraft 1981; Trochanis et al. 1991; Nogami et al. 1992; El Naggar & Novak 1995, 1996; Badoni & Makris 1996; Boulanger et al. 1999; Gerolymos & Gazetas 2005; Papadopoulou & Comodromos 2014);
- Higher-order Winkler formulations employing rotational springs attached in parallel to the translational spring bed, following the spatial condensation model originally proposed for surface footings by Vlasov & Leontiev (1966) (Mylonakis 2000; Guo & Lee 2001; Basu & Salgado 2008; Basu et al. 2009; Shadlou & Bhattacharya 2014), as well as modifications of the above model (Mylonakis 2000, 2001; Agapaki et al. 2018);
- Semi-continuum formulations following the constrained elasticity models originally suggested by Matsuo & Ohara (1960) for walls and Tajimi (1966, 1969) for piles (Nogami & Novak 1977; Saitoh & Watanabe 2004; Saitoh 2005; Anoyatis 2013; Anoyatis et al. 2016; Anoyatis & Lemnitzer 2017)
- Other approximate solutions based on theory of elasticity and wave propagation (Biot 1937; Vesić 1961; Dobry & Gazetas 1988; Wolf 1994; Cairo et al. 1999; Mylonakis et al. 2001).

Review papers on the subject (Tajimi 1977; Roesset 1980; Hadjian et al. 1992; Novak 1991; Gazetas 1991; Pender 1993; Gazetas & Mylonakis 1998; Finn 2005) and reports (API 1982; Reese 1986; Lam & Martin 1986; NEHRP 2012) are available. More complete summaries focusing on static and cyclic loads are included in several books (Hetenyi 1946; Poulos & Davis 1980; Scott 1981; Prakash & Sharma 1990; Reese et al. 2005; Salgado 2008; Fleming et al. 2009; Reese & Van Impe 2011; Viggiani et al. 2011; Guo 2012; Poulos 2017). Books focusing on dynamic and earthquake loads are also available (Wolf 1994; Madabhushi et al. 2010; Jia 2018; Bhattacharya 2019).

The methods in Group C are deemed particularly advantageous for practical applications given their conceptual simplicity and ability to reduce the dimensionality of the problem from three to a single dimension (or a superposition of a series of two-dimensional "horizontal soil slice" problems and one-dimensional beam problems), which requires significantly smaller computational effort than rigorous continuum formulations. Despite this simplification, these methods are still able to

- Yield predictions that are in satisfactory agreement with more rigorous solutions (both in the elastic and inelastic regime) after calibration of the governing parameters
- Incorporate variation of soil properties with load amplitude (non-linearity) and depth (in-homogeneity)
- Incorporate contact effects such as slippage and gapping at the pile-soil interface
- Model response to biaxial and triaxial loads
- Incorporate advanced modelling options, such as rotational springs attached in parallel to the classical translational springs, to better capture soil-pile interaction effects
- Model geometric non-linearity such as "P-Δ" effects, buckling, and large deformations
- Incorporate material non-linearity in the pile
- Model kinematic effects, such as the action of impinging seismic waves, or static soil displacements due to excavation or slope movements
- Be extended to the dynamic regime by adding pertinent distributed dampers to the spring bed, accounting for wave radiation and material energy dissipation
- Incorporate pile group effects through pertinent pile-soil-pile interaction models
- Be easily implemented into finite difference, finite element, and boundary element codes

With reference to the last item in the above list, this work recognises that rigorous numerical methods such as FEM are not ideally suited for routine design as they are time-consuming and limited by the lack of general-purpose constitutive models for soil behaviour (i.e., for cases deviating from those involving clean sands and idealised clays) under combined compression and shearing, especially under cyclic loads and undrained conditions. In addition, rigorous continuum models typically suffer from various theoretical and computational complexities associated with the underlying numerical procedures, which make them unappealing to practicing engineers.

In this chapter, simplified pile-soil interaction models based on the concept of a beam on a Winkler foundation (BWF) are reviewed and extended. Both static and dynamic loads at the pile head are considered, for single piles and pile groups (kinematic loads are treated in Chapter 8). The physics and salient features of these models are discussed, together with pertinent modifications for inhomogeneous soil and dynamics loads. An advance over existing solutions is proposed, by introducing pertinent complex-valued shape functions for pile response in a virtual work formulation of the problem. A number of models for deriving or calibrating Winkler springs and dashpots are discussed. Spring and dashpot moduli are compared to available numerical data from the literature (methods in Group A above).

6.2   SINGLE PILE RESPONSE

The problem examined in this chapter is shown in Figure 6.1. The vertical cylindrical pile, modelled as a beam with stiffness E_p, length L, diameter D, and second moment of area I, is supported by distributed Winkler springs with stiffness $k(z)$. This parameter is known as the *modulus of subgrade reaction*, defined as force per unit pile length

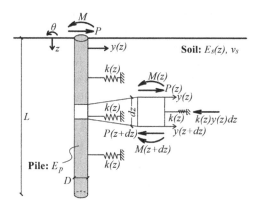

Figure 6.1 The Winkler model for a pile under lateral load and moment at the pile head.

over lateral displacement, and is measured in units of force per length squared.[1] This stiffness is typically assumed proportional to the soil stiffness, E_s, and approaches to select it are discussed later in this chapter. A load, P, and an anticlockwise moment, M, are applied at the pile head, resulting in a pile head deflection, y_0, and rotation, θ_0. Employing the moment-curvature relationship from Euler-Bernoulli beam theory yields the well-known governing differential equation for a BWF in Eq. 6.1 (Biot 1937; Hetenyi 1946; Poulos & Davis 1980; Scott 1981). Note that in this notation, positive bending moments induce negative curvatures and vice versa.

$$E_p I \, \frac{d^4 y(z)}{dz^4} + k(z)\, y(z) = 0 \tag{6.1}$$

where $y(z)$ is the lateral deflection of the pile at depth z. For a solid cylindrical pile, $I = \pi D^4 / 64$.

6.2.1   Homogeneous soil

For the simplest case of homogeneous soil [$k(z) = k$], the Winkler parameter λ (units of length^{-1}) can be introduced to simplify Eq. 6.1:

$$\frac{d^4 y(z)}{dz^4} + 4\lambda^4 y(z) = 0 \tag{6.2a}$$

$$\lambda = \sqrt[4]{\frac{k}{4 E_p I}} \tag{6.2b}$$

The emergence of this parameter (which can be interpreted either as a characteristic length or a wavenumber controlling the attenuation of pile settlement with depth) is

---

1 The specific definition of k differs from the so-called *coefficient of subgrade reaction* (usually denoted by k_s), which is defined as pressure over displacement and is measured in units of force per length cubed.

a fundamental property of Winkler models which allows direct normalisation of the independent variable z. The product λL can be viewed as a dimensionless "mechanical pile length" encompassing both geometric slenderness, L/D, and pile-soil relative stiffness, E_p/k, thus reducing the number of dimensionless parameters governing pile response by one (Anoyatis et al. 2013). As will be shown later in this chapter, this characteristic length is not an artifact of the model since it is inherently linked to the concept of "active pile length" which is of paramount importance for interpreting the behaviour of laterally loaded piles (Di Laora & Rovithis 2015).

By solving the equilibrium equation (Eq. 6.1a) and imposing the boundary conditions at the pile head and tip, the stiffness at the pile head can be expressed as a 2×2 matrix in the form

$$\begin{bmatrix} P \\ M \end{bmatrix} = K u = \begin{bmatrix} K_{11} & K_{12} \\ K_{21} & K_{22} \end{bmatrix} \begin{bmatrix} y_0 \\ \theta_0 \end{bmatrix} \tag{6.3}$$

where K_{11}, K_{22}, and K_{12} ($= K_{21}$) are the swaying, rocking, and cross-swaying-rocking stiffnesses, respectively (Mylonakis 1995; Mylonakis & Gazetas 1999). The element K_{ij} of this matrix is defined as the force or moment along the degree-of-freedom i under a unit displacement or rotation along the degree-of-freedom j when all other degrees of freedom are fixed. The diagonal terms of the matrix are positive, whilst the off-diagonal may be positive or negative depending the choice of the coordinate system. By virtue of reciprocity associated with the linear elastic nature of Eq. 6.3, the stiffness matrix is symmetric, i.e., $K_{12} = K_{21}$. A relevant discussion on the topic is provided by Banerjee (1994) and Achenbach (2004). Moreover, the stiffness matrix is positive definite as the scalar $u^T K u$ (proportional to the energy stored in the pile) is strictly positive for every non-zero column vector u.

For a pile of finite length, these stiffness terms naturally depend on the boundary conditions at the pile base. If the pile base is unrestrained (floating condition), these are given by (Mylonakis 1995; Mylonakis & Gazetas 1999)

$$K_{floating} = \begin{bmatrix} 4E_pI\lambda^3 \dfrac{\sin(2\lambda L)+\sinh(2\lambda L)}{2+\cos(2\lambda L)+\cosh(2\lambda L)} & 2E_pI\lambda^2 \dfrac{-\cos(2\lambda L)+\cosh(2\lambda L)}{2+\cos(2\lambda L)+\cosh(2\lambda L)} \\ sym & 2E_pI\lambda \dfrac{-\sin(2\lambda L)+\sinh(2\lambda L)}{2+\cos(2\lambda L)+\cosh(2\lambda L)} \end{bmatrix}$$

$$\tag{6.4a}$$

where sym indicates that the matrix is symmetric. If the pile base is fixed against displacement and rotation (clamped condition), the stiffnesses are given by (Mylonakis 1995; Mylonakis & Gazetas 1999)

$$K_{fixed} = \begin{bmatrix} 4E_pI\lambda^3 \dfrac{\sin(2\lambda L)+\sinh(2\lambda L)}{-2+\cos(2\lambda L)+\cosh(2\lambda L)} & 2E_pI\lambda^2 \dfrac{-\cos(2\lambda L)+\cosh(2\lambda L)}{-2+\cos(2\lambda L)+\cosh(2\lambda L)} \\ sym & 2E_pI\lambda \dfrac{-\sin(2\lambda L)+\sinh(2\lambda L)}{-2+\cos(2\lambda L)+\cosh(2\lambda L)} \end{bmatrix}$$

$$\tag{6.4b}$$

Finally, if the base is fixed against displacement, but free to rotate (hinged condition), the stiffness matrix is given by (Mylonakis 1995)

$$
\boldsymbol{K}_{hinged} = \begin{bmatrix} 4E_p I\lambda^3 \dfrac{\cos(2\lambda L)+\cosh(2\lambda L)}{-\sin(2\lambda L)+\sinh(2\lambda L)} & 2E_p I\lambda^2 \dfrac{\sin(2\lambda L)+\sinh(2\lambda L)}{-\sin(2\lambda L)+\sinh(2\lambda L)} \\[2ex] sym & 2E_p I\lambda \dfrac{-\cos(2\lambda L)+\cosh(2\lambda L)}{-\sin(2\lambda L)+\sinh(2\lambda L)} \end{bmatrix}
$$

(6.4c)

Note that in the sign convention chosen (see Figure 6.1), positive applied head moment is anticlockwise (the same as positive rotation), and all head stiffness terms are positive.

Alternatively, the pile behaviour can be expressed as a flexibility matrix, the inverse of the stiffness matrix:

$$
\begin{bmatrix} y_0 \\ \theta_0 \end{bmatrix} = \boldsymbol{F}\,\boldsymbol{P} = \begin{bmatrix} F_{11} & F_{12} \\ F_{21} & F_{22} \end{bmatrix} \begin{bmatrix} P_0 \\ M_0 \end{bmatrix} = \boldsymbol{K}^{-1}\,\boldsymbol{P}
$$

(6.5)

In this formulation (which is often preferred by geotechnical engineers as it allows direct estimates of pile response to a given head load), the element F_{ij} is defined as the displacement or rotation at the pile head along the degree of freedom i for a unit force or moment along the degree-of-freedom j, when all the other degrees of freedom are not loaded. Like the stiffness matrix, the flexibility matrix is symmetric and has positive diagonal terms. The off-diagonal terms can be positive or negative depending on the choice of coordinate system and carry the opposite sign of the corresponding stiffness terms. In the notation adopted in this paper, shown in Figure 6.1, $K_{12} > 0$, $F_{12} < 0$.

Of particular interest is the inverse of the F_{11} term, which is equal to the head stiffness of a pile unrestrained against rotation at the pile head (free-head condition), which is referred to as the free-head stiffness, K_{free}. This can also be obtained by direct application of the pertinent boundary conditions or from static condensation of the stiffness matrix via

$$
K_{free} = 1/F_{11} = K_{11} - K_{12}^2/K_{22}
$$

(6.6)

In this way, the following free-head stiffnesses can be obtained for different pile base conditions:

$$
K_{free,floating} = 2E_p I\lambda^3 \frac{-2 + \cos(2\lambda L) + \cosh(2\lambda L)}{-\sin(2\lambda L) + \sinh(2\lambda L)}
$$

(6.7a)

$$
K_{free,fixed} = 2E_p I\lambda^3 \frac{2 + \cos(2\lambda L) + \cosh(2\lambda L)}{-\sin(2\lambda L) + \sinh(2\lambda L)}
$$

(6.7b)

$$
K_{free,hinged} = 2E_p I\lambda^3 \frac{\left[\sin(2\lambda L) + \sinh(2\lambda L)\right]^2}{\left[\sin(2\lambda L) - \sinh(2\lambda L)\right]\left[\cos(2\lambda L) - \cosh(2\lambda L)\right]}
$$

(6.7c)

It will be shown that the algebraic operation in Eq. 6.6 can be used to establish a number of constraints for the stiffness coefficients of long piles – a property that remained unidentified in previous studies.

Figure 6.2 shows the pile head stiffness terms K_{11}, K_{free}, K_{12}, and K_{22} as a function of dimensionless pile length, λL. As pile length increases, the elements of the stiffness matrix (and consequently the flexibility matrix) asymptotically approach a specific value, which was used to normalise the results. For the terms plotted (Hetenyi 1946; Poulos & Davis 1980; Scott 1981; Mylonakis 1995; Guo 2012),

$$\mathbf{K}_{L\to\infty} = \begin{bmatrix} 4E_pI\lambda^3 & 2E_pI\lambda^2 \\ sym & 2E_pI\lambda \end{bmatrix}, \quad K_{free,\,L\to\infty} = 2E_pI\lambda^3 \tag{6.8}$$

which is naturally independent of tip boundary conditions. Comparison of the stiffness term K_{11} in the above matrix with the term K_{free} suggests that the moment release at the pile head reduces stiffness by a factor of 2. In reality, however, the reduction in stiffness is smaller as the Winker stiffness, k, for free-head conditions is higher than under fixed head conditions. (In other words, the λ terms are different in the two expressions and do not cancel out). This is discussed later in this chapter.

It should be noted that the ratio K_{21}/K_{11} (= $1/2\lambda$) defines an auxiliary length – which provides a good approximation of the depth where the overall soil reaction acts, under fixed-head conditions. This property is discussed in Maravas et al. (2007, 2014).

The length beyond which the pile head response is no longer affected by base conditions is known as the "active pile length", L_a. The exact value of this length is discussed in detail later in this chapter, as it depends on the specific definition chosen (e.g., tolerance of head stiffness value from Eq. 6.8, curvature below a certain value) and the relationship between the Winkler spring stiffness and soil elastic properties. A traditional definition by Kuhlemeyer (1979) is that the active pile length coincides with the elevation down the pile, where deflection is 1/1000 of that at the head.

With reference to Figure 6.2, it is evident that for a mechanical length λL in the region of approximately 2.5, all stiffness terms have become asymptotic regardless of tip conditions. [Alternatively, considering the depth where pile displacement is zero for fixed-head conditions according to the Winkler model (Eq. 6.13a) yields the similar value $3\pi/4 \approx 2.35$.] For instance, for a pile-soil stiffness ratio $E_p/E_s = 1000$, this yields a pile slenderness $L_a/D \approx 9$. Under dynamic loading, this value can increase, as discussed later in this chapter. This length is much shorter than that of many piles installed in practice; therefore, this chapter will focus on long piles that can be analysed with the simpler solutions for infinitely long piles. Note that from this point, the notation $L \to \infty$ is omitted for brevity. Except where specifically otherwise indicated, all results presented in the following refer to infinitely long piles.

6.2.2 Selection of k(z)

Application of these simplified solutions requires careful selection of a suitable Winkler spring stiffness. In the absence of exact analytical solutions, this is often carried out using calibrations against numerical continuum solutions, by dividing the soil reaction per unit pile length by the resulting displacement at each depth. A problem

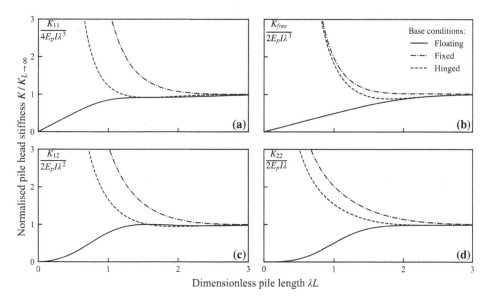

Figure 6.2 Normalised head stiffness of a pile of finite length embedded in a homogeneous soil, Eqs. 6.3 and 6.6, (a) K_{11}, (b) K_{free}, (c) K_{21}, and (d) K_{22}. (Adapted from Mylonakis (1995), Mylonakis & Gazetas (1999).)

associated with this direct approach is that k is not a soil property and depends on depth even in homogeneous soil[2] (Anoyatis 2013; Anoyatis & Lemnitzer 2017). Alternatively, depth-independent "average" values can be obtained by comparing a key response parameter between the Winkler model and a more rigorous solution. Suitable parameters could include one of the head stiffness terms in Eq. 6.8 (notably the K_{11} or K_{free} terms), the pile head displacement or rotation, or even the peak bending moment at depth. It should be noted that as only a single parameter is available for calibration, it is impossible to perfectly match all the above parameters by a single k.

Table 6.1 includes a number of calibrations available in the literature; similar tables have been provided by various authors including Shadlou & Bhattacharya (2014) and Anoyatis & Lemnitzer (2017). Early solutions were based on solutions to the beam-on-elastic-foundation problem (i.e., with soil resisting foundation movement only on one side of the beam). Using this model, Biot (1937) obtained a simplified analytical solution for a beam on a homogeneous half-space, with which to match the peak bending moment against the Winkler model. This solution was extended by Vesić (1961) who matched the variation with depth of the pile bending moment between the two solutions. The resulting expression has been employed by many authors including Broms (1964a, b) for fixed-head laterally loaded piles. Francis (1964) and later Pender

---

2 A simple way to demonstrate this is through the Boussinesq solution, according to which stresses in a homogeneous half-space due to a point load P acting at the surface attenuate in proportion to P/z^2, whilst displacements attenuate in proportion to $P/(G_s z)$. Accordingly, Winkler modulus (i.e., stresses over displacements) varies with depth in proportion to $1/z$.

Table 6.1 Static Winkler spring stiffness equations

Source	Winkler spring stiffness, k
Biot (1937)	$k = 1.26 E_s \left(1 - v_s^2\right)^{-1.11} \left(\dfrac{E_p}{E_s}\right)^{-0.11}$
Vesić (1961)	$k = 0.65 \dfrac{E_s}{1 - v_s^2} \sqrt[12]{\dfrac{E_s D^4}{E_p I}} = 0.84 \dfrac{E_s}{1 - v_s^2} \left(\dfrac{E_p}{E_s}\right)^{-1/12}$
Francis (1964), Yoshida & Yoshinaka (1972)	$k = 1.30 \dfrac{E_s}{1 - v_s^2} \sqrt[12]{\dfrac{E_s D^4}{E_p I}} = 1.67 \dfrac{E_s}{1 - v_s^2} \left(\dfrac{E_p}{E_s}\right)^{-1/12}$
Roesset (1980)	$k = 1.2 \, E_s$
Dobry et al. (1982)	$\dfrac{k}{E_s} = 1.67 \left(\dfrac{E_p}{E_s}\right)^{-0.053}$
Dobry & O'Rourke (1983)	$k = 3 \, G_s$
Gazetas & Dobry (1984a)	Fixed head. $k = 1.0 \, E_s$ to $1.2 \, E_s$, Free head. $k = 1.5 \, E_s$ to $2.5 \, E_s$
Syngros (2004)	Fixed head. $\dfrac{k}{E_s} = 2.0 \left(\dfrac{E_p}{E_s}\right)^{-0.075}$, Free head. $\dfrac{k}{E_s} = 3.5 \left(\dfrac{E_p}{E_s}\right)^{-0.11}$
Karatzia & Mylonakis (2017), Mylonakis (2001)	$k = \dfrac{4 \pi G_s \eta_u^2}{\left(1 + \eta_u^2\right)\left[\ln(4/\alpha_c) - \gamma\right] + \ln(\eta_u)}$, $\eta_u^2 = \dfrac{2 - v_s}{1 - v_s}$, $\alpha_c = c_1 \left(\dfrac{E_p}{E_{s,D}}\right)^{c2}$, where c_1 and c_2 are constants that depend on the pile head conditions and soil stiffness profile. Tabulated values are provided in the paper.

(1993) observed that in the pile problem, soil is present on both sides of the beam, and therefore, the Winkler spring stiffness should be double that calculated for a beam on an elastic foundation problem.

Roesset (1980), Dobry et al. (1982), and Syngros (2004) provide solutions calibrated with finite element analyses for a linearly elastic half-space. The first two solutions are calibrated against results for piles fixed at the head obtained with the elastodynamic formulation by Blaney et al. (1976). Gazetas & Dobry (1984a) and later Syngros (2004) noted that different Winkler moduli must be selected for each problem in order to match both the fixed-head and free-head stiffness.

A physically motivated analytical model was proposed by Mylonakis (2001) who integrated the governing equation describing soil response around a pile over the vertical coordinate using a preselected shape function. This technique is analogous to the one employed by Vlasov & Leontiev (1966) for spread footings (in the former study the integration was performed in the direction perpendicular to the soil-foundation interface) and yields an improved version of Baranov's plane-strain model where the interslice shear stresses are accounted for, leading to finite static stiffness. Mylonakis (2001) provided a simplified expression for k using asymptotic forms of the pertinent Bessel functions. A sinusoidal shape function was employed based on soil response for finite soil layers, as well as one selected based on the pile response from the Winkler model. Karatzia & Mylonakis (2017) extended the solution to inhomogeneous soil and provided simple fitted expressions for key parameters for different pile head fixities and soil profiles.

The aforementioned expressions for Winkler spring stiffness are compared in Figure 6.3 for both fixed- and free-head piles. In general, there is limited variation between the different expressions, except that those for free head piles are notably higher than those for fixed head piles. The dependence of the Winkler spring stiffness on pile-soil stiffness contrast is negligible. As there is no clear physical justification for the nearly zero values of the exponents in the (E_p/E_s) terms in Table 6.1, these values are suspect since they might be associated with imperfect regressions (curve fitting) of the underlying numerical data. Also, there is no dependence of the Winkler spring stiffness on the diameter of pile (or, equivalently, on λ). Evidently, such a dependence would not be compatible with the physics of the problem, as k and E_s (or soil shear modulus, G_s) carry the same units, and the presence of pile diameter would violate the dimensionally homogeneous nature of the solutions in Table 6.1.

Some authors (e.g., Guo & Lee 2001) suggest employing an improved two-parameter Winkler model to better match continuum solutions. The second parameter can either be interpreted as a membrane under tension connecting the base of the springs (Hetenyi 1946; Guo 2012) or as distributed rotational springs acting in parallel to the translational Winkler springs (Kaynia 1980; Sanchez-Salinero 1982; Scott 1981; Agapaki et al. 2018). These can be calibrated against continuum solutions in a similar way to the one-parameter model; alternatively, analytical expressions are obtained when using the Vlasov & Leontiev (1966) approach and integrating in the radial direction. Agapaki et al. (2018) suggest the introduction of a third parameter, an additional soil-structure-interaction-induced bending stiffness, to allow matching of the solution directly to all three parameters in the stiffness matrix simultaneously. These higher-order Winkler

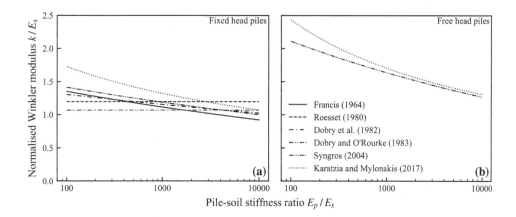

Figure 6.3 Static Winkler modulus (Table 6.1) for (a) fixed head piles and (b) free head piles $(v_s = 0.4)$.

solutions can better match continuum solutions. However, they lead to more complex results, particularly for inhomogeneous soils, dynamic loading, and pile group interaction effects. This chapter focusses on the one-parameter model.

6.2.3   Comparison with numerical results

Recasting Eq. 6.8 into Eq. 6.9 provides some basic expressions which allow for convenient comparison between Winkler solutions and numerical continuum counterparts:

$$\frac{K_{11}}{E_s D} = \left(\frac{\pi}{16}\right)^{1/4} \left(\frac{k}{E_s}\right)^{3/4} \left(\frac{E_p}{E_s}\right)^{1/4} \tag{6.9a}$$

$$\frac{K_{free}}{E_s D} = \frac{1}{2}\left(\frac{\pi}{16}\right)^{1/4} \left(\frac{k}{E_s}\right)^{3/4} \left(\frac{E_p}{E_s}\right)^{1/4} \tag{6.9b}$$

$$\frac{K_{12}}{E_s D^2} = \frac{1}{2}\left(\frac{\pi}{16}\right)^{1/2} \left(\frac{k}{E_s}\right)^{1/2} \left(\frac{E_p}{E_s}\right)^{1/2} \tag{6.9c}$$

$$\frac{K_{22}}{E_s D^3} = \frac{1}{2}\left(\frac{\pi}{16}\right)^{3/4} \left(\frac{k}{E_s}\right)^{1/4} \left(\frac{E_p}{E_s}\right)^{3/4} \tag{6.9d}$$

The advantage of Eq. 6.9 over Eq. 6.8 lies in the monomial nature of the (k/E_s) and (E_p/E_s) terms which provide insight into the nature of the solution and associated powers.

Randolph (1981), Davies & Budhu (1986), Gazetas (1991), and Syngros (2004) have provided the simplified expressions for pile head stiffness shown in Table 6.2. These were fitted against continuum finite element solutions and provide a simple way to employ numerical results without repeating the analysis. However, they are often limited to homogeneous soil and not generalisable to other soil profiles. Note that Randolph (1981) and Davies & Budhu (1986) provided results in the form of a flexibility matrix that has been inverted to provide the stiffness matrix. Also, Gazetas (1991) did not consider free-head piles. Note that the ratio K_{free}/K_{11} is in excess of 1/2 in all solutions, which highlights that Winkler spring stiffness, k, is higher in free-head than in fixed-head piles. This difference in the values of the Winkler constants can be established by taking the ratio $[2K_{free}/K_{11}]^{4/3}$. Using the available expressions in Table 6.2 yields (k_{free}/k_{fixed}) values ranging between 1.2 and 1.4. In addition, in order to ensure the monomial character in (E_p/E_s) of both the flexibility matrix and K_{free}, the exponents, n_{ij}, of the various terms should be related such that $n_{11} = n_{free} = (2\,n_{12} - n_{22})$ or, equivalently, n_{12} is the arithmetic mean of n_{11} and n_{22}. Evidently, this property holds for the expressions from Randolph (1981) and Davies & Budhu (1986) (and the Winkler expressions in Eq. 6.9 assuming the same k expression is employed for each term) but not in those by Gazetas (1991) and Syngros (2004).

Figure 6.4 shows a comparison of the performance of the Winkler solutions described in Eq. 6.9 and Table 6.1 against some existing continuum solutions and the fitted functions in Table 6.2. In general, there is good agreement between the Winkler solution and the more rigorous results (including the fitted expressions). The lowest

Table 6.2 Fitted head stiffness functions for homogeneous soil

Source	Randolph (1981)[*]	Davies & Budhu (1986)[*]	Gazetas et al. (1991)	Syngros (2004)
$K_{11}/E_s D$	$3.147\eta\left(\dfrac{E_p}{E_s\eta}\right)^{1/7}$	$1.292\left(\dfrac{E_p}{E_s}\right)^{2/11}$	$1.00\left(\dfrac{E_p}{E_s}\right)^{0.21}$	$1.24\left(\dfrac{E_p}{E_s}\right)^{0.18}$
$K_{free}/E_s D$	$2.000\eta\left(\dfrac{E_p}{E_s\eta}\right)^{1/7}$	$0.769\left(\dfrac{E_p}{E_s}\right)^{2/11}$		$0.72\left(\dfrac{E_p}{E_s}\right)^{0.18}$
$K_{12}/E_s D^2$	$0.531\eta\left(\dfrac{E_p}{E_s\eta}\right)^{3/7}$	$0.309\left(\dfrac{E_p}{E_s}\right)^{5/11}$	$0.22\left(\dfrac{E_p}{E_s}\right)^{0.50}$	$0.21\left(\dfrac{E_p}{E_s}\right)^{0.50}$
$K_{22}/E_s D^3$	$0.246\eta\left(\dfrac{E_p}{E_s\eta}\right)^{5/7}$	$0.183\left(\dfrac{E_p}{E_s}\right)^{8/11}$	$0.15\left(\dfrac{E_p}{E_s}\right)^{0.75}$	$0.15\left(\dfrac{E_p}{E_s}\right)^{0.75}$

Where $\eta = \dfrac{4+3v_s}{8(1+v_s)}$, [*]Obtained by inverting the flexibility matrix.

discrepancy is observed for the K_{22} term (where the beam rotational stiffness dominates), then the K_{12} term, the K_{11} term and finally the K_{free} term (where the soil resistance dominates). This corresponds to the power of the dependence of the terms in Eq. 6.9 on E_p/E_s. A higher power indicates that the overall stiffness is influenced more by the pile response than the soil response, the variation in predictions for the latter part being much higher. In addition, the k expressions that were not calibrated for free head piles expectedly performed poorly for predicting K_{free}.

6.2.4 Inhomogeneous soil

Piles are often installed in soils with properties that vary with depth. A very limited number of analytical solutions are available for inhomogeneous soils. Both employing the Winkler model, Hetenyi (1946) and Franklin & Scott (1979) developed closed-form solutions for soils with stiffness varying linearly with depth. The former employs hypergeometric series and the latter a contour integral solution, for which tabulated results are provided in Scott (1981). However, both solutions are difficult to employ and limited to linear variation of Winkler stiffness with depth. Alternatively, numerical solutions are available for inhomogeneous soils including soils with stiffness varying according to a power-law function of depth and a finite difference approach by Matlock & Reese (1960) (see also Lam & Martin 1986). Of particular interest is the approximate solution by Karatzia & Mylonakis (2017), an energy solution based on earlier work by Mylonakis (1995), Mylonakis & Gazetas (1999), and Mylonakis & Roumbas (2001). By considering the work done by a small virtual displacement, $y_v(z)$, an integral form of Eq. 6.1 can be obtained:

$$\int_0^L E_p I y''''(z)\,y_v(z)\,dz + \int_0^L k(z)\,y(z)\,y_v(z)\,dz = 0 \tag{6.10}$$

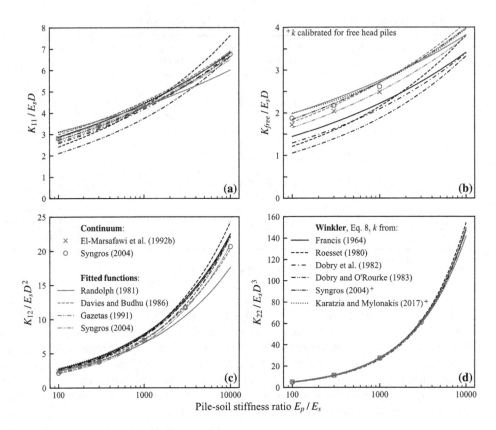

Figure 6.4 Variation of normalised head stiffness with pile-soil stiffness ratio for piles embedded in a homogeneous half-space; (a) K_{11}, (b) K_{free}, (c) K_{21}, and (d) K_{22} $(v_s = 0.5)$. (Numerical data from Syngros (2004).)

Through integration by parts, the weak form of this equation can be obtained and, using Euler-Bernoulli beam theory, related to the load and moment applied at the pile head. For long piles, the upper integration limit has a negligible effect on the results, and therefore, this can be expressed as an improper integral in the form

$$Py_v(0) + My_v'(0) = \int_0^\infty E_p I y''(z) y_v''(z) dz + \int_0^\infty k(z) y(z) y_v(z) dz \qquad (6.11)$$

Note that the two terms on the right-hand side express the contributions of the pile flexural stiffness and soil stiffness, respectively, to the overall stiffness at the pile head. Dimensionless shape functions, $\chi_i(z)$ and $\chi_j(z)$, can then be selected to describe the variation with depth of the real and virtual displacements, respectively.

$$\frac{P}{y_i}\chi_j(0) + \frac{M}{y_i}\chi_j'(0) = K_{ij} = \int_0^\infty E_p I \chi_i''(z) \chi_j''(z) dz + \int_0^\infty k(z) \chi_i(z)\chi_j(z) dz \qquad (6.12)$$

where $y_i = y(0)/\chi_i(0)$. It is evident that this expression produces a pile head stiffness term, K_{ij}, that depends on the shape functions chosen. If χ_i is chosen such that $\chi_i(0) = 1$ and either $\chi_i'(0) = 0$ or $\chi_i''(0) = 0$, the stiffness term will relate to head displacement. If, instead, χ_i is chosen such that $\chi_i'(0) = 1$ and $\chi_i(0) = 0$, the stiffness term will relate to head rotation. Similarly, if $\chi_j(0) = 1$ and either $\chi_i'(0) = 0$ or $\chi_i''(0) = 0$, the stiffness term will relate to the applied load and if $\chi_j'(0) = 1$ and $\chi_j(0) = 0$, the stiffness term will relate to the applied moment. Note that for long piles, the shape functions must approach zero at large depths. Suitable shape functions can be obtained from the solution for homogeneous soil (Mylonakis 1995; Mylonakis & Gazetas 1999; Mylonakis & Roumbas 2001; Karatzia & Mylonakis 2012, 2017):

$$\chi_1(z) = e^{-\mu z}\left[\sin(\mu z) + \cos(\mu z)\right] \tag{6.13a}$$

$$\chi_2(z) = \frac{e^{-\mu z}}{\mu}\sin(\mu z) \tag{6.13b}$$

$$\chi_{free}(z) = e^{-\mu z}\cos(\mu z) \tag{6.13c}$$

where the parameter μ is analogous to λ for homogeneous soil. For Eqs. 6.13a and b, the subscripts of the shape functions chosen for χ_i and χ_j correspond to the component of the stiffness matrix, K_{ij}, obtained [e.g., $\chi_i(z) = \chi_1(z)$ and $\chi_j(z) = \chi_2(z)$ yield K_{12}].

Using Eq. 6.13c $\left[\chi_i(z) = \chi_j(z) = \chi_{free}(z)\right]$ yields K_{free}.

Application of these shape functions requires selection of a suitable value for μ. For homogeneous soils, selecting $\mu = \lambda$ yields the analytical results in Eq. 6.8. For inhomogeneous soils, Mylonakis (1995) and Mylonakis & Gazetas (1999) recommend treating Eq. 6.2 as a function of depth and finding the average value over the active pile length, L_a:

$$\mu = \frac{1}{L_a}\int_0^{L_a}\lambda(z)dz, \quad \lambda(z) = \sqrt[4]{\frac{k(z)}{4E_pI}} \tag{6.14}$$

A variety of expressions for L_a are available in the literature based on a number of different definitions of active length, some of which are compared in Figure 6.5. In general, these are provided in the form shown in Eq. 6.15:

$$L_a/D = \chi_L\left(\frac{E_p}{E_s}\right)^{n_L} \tag{6.15}$$

where χ_L and n_L are empirical constants that must be specifically calibrated for each soil profile based on the chosen definition.

Note that for this purpose, L_a is a model parameter rather than a property of the solution. Therefore, perfectly matching a chosen active length definition will not necessarily provide better head stiffness results. Instead, Di Laora & Rovithis (2015) suggest selecting a value of μL_a and then employing Eq. 6.14 and an iterative procedure to calculate μ and L_a, a process that can be conducted for any arbitrary soil stiffness profile. Di Laora & Rovithis (2015) recommended choosing $\mu L_a = 2.5$ based on the

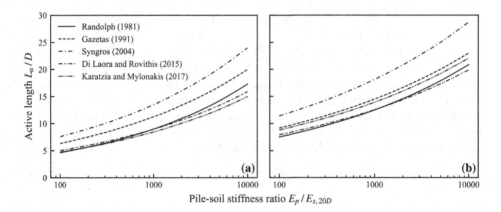

Figure 6.5 Active length expressions for piles embedded in (a) homogeneous soil and (b) soil with a linear stiffness variation with depth and zero surface stiffness.

value for homogeneous soil, which is compared to other active length definitions in Figure 6.5. In general, the head stiffness results are not particularly sensitive to the value of μL_a chosen.

Equations 6.12 and 6.13 can be resolved numerically for arbitrary soil stiffness profiles. In addition, they have been resolved analytically for a range of specific soil stiffness profiles, including linear, power-law, and exponential functions of depth as well as layered soil profiles (Mylonakis & Gazetas 1999; Mylonakis & Roumbas 2001; Karatzia & Mylonakis 2017).

Consider the linear Winkler stiffness profile shown in Figure 6.6a, solved analytically by Franklin and Scott (1979). The Winkler spring stiffness variation can be expressed in the form shown in Eq. 6.16:

$$k(z) = k_0 + k_{sl}z \tag{6.16}$$

where k_0 is the Winkler spring stiffness at the pile head, and k_{sl} is the increase in Winkler spring stiffness per unit depth (slope of the $k(z)$ function, units of force/length^3). Inputting this function into Eq. 6.12 with the corresponding shape functions in Eq. 6.13 yields the following pile head stiffness equations (adapted from Mylonakis & Roumbas 2001; Karatzia & Mylonakis 2017):

$$\frac{K_{11}}{E_p I \mu^3} = 1 + \frac{16}{\pi} \frac{k_{sl}D}{E_p} \frac{1}{(\mu D)^4} \left(\frac{3}{2\mu D} + 3 \frac{k_0}{k_{sl}D} \right) \tag{6.17a}$$

$$\frac{K_{free}}{E_p I \mu^3} = \frac{1}{2} + \frac{16}{\pi} \frac{k_{sl}D}{E_p} \frac{1}{(\mu D)^4} \left(\frac{1}{2\mu D} + \frac{3}{2} \frac{k_0}{k_{sl}D} \right) \tag{6.17b}$$

$$\frac{K_{12}}{E_p I \mu^2} = 1 + \frac{16}{\pi} \frac{k_{sl}D}{E_p} \frac{1}{(\mu D)^4} \left(\frac{3}{4\mu D} + \frac{k_0}{k_{sl}D} \right) \tag{6.17c}$$

$$\frac{K_{22}}{E_p I \mu} = \frac{3}{2} + \frac{16}{\pi} \frac{k_{sl} D}{E_p} \frac{1}{(\mu D)^4} \left(\frac{1}{2\mu D} + \frac{1}{2} \frac{k_0}{k_{sl} D} \right) \tag{6.17d}$$

Note that if $k_{sl} = 0$, Eq. 6.17 reproduces the homogeneous results in Eq. 6.8 with $k = k_0$ and therefore $\mu = \lambda$. If a value of μL_a is selected (as discussed previously), Eqs. 6.14 and 6.16 can be employed to develop Eq. 6.18, an analytical expression for the active length that is in good agreement with numerical results (Karatzia & Mylonakis 2016).

$$L_a / D = \left[\frac{5}{8} \mu L_a \left(\frac{\pi E_p}{k_{sl} D} \right)^{1/4} + \left(\frac{k_0}{k_{sl} D} \right)^{5/4} \right]^{4/5} - \frac{k_0}{k_{sl} D} \tag{6.18}$$

The performance of Eq. 6.17 is compared with the closed-form solution provided by Franklin & Scott (1979) in Figure 6.6b. The approximate expressions perform very well compared to the analytical solution, with a negligible margin of error relative to the likely confidence in the input parameters in practice. In addition, this figure demonstrates the different dependence of each stiffness term on the soil response that was also observed in Figure 6.4. The K_{11} and K_{free} terms are most affected by the change in soil stiffness near the surface, corresponding to the E_p/E_s exponent of 1/4 in Eq. 6.9, whilst the K_{22} term is almost completely unaffected, corresponding to an E_p/E_s exponent of 3/4.

As with the homogeneous soil case, simplified functions fitted to numerical continuum solutions are available for a linear soil stiffness variation with zero surface stiffness (shown in Table 6.3). These functions are limited to specific soil profiles for which they were calibrated, and therefore, few additional soil profiles have been considered. Randolph (1981) introduced a simple process for interpolating between results for homogenous soils and the aforementioned linear soil stiffness profile to produce results for a linear soil stiffness profile with non-zero surface stiffness. Gazetas (1991) also

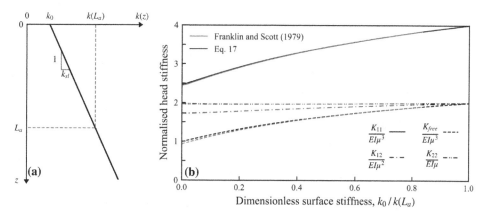

Figure 6.6 (a) Winkler spring stiffness profile, $k(z)$, and (b) normalised pile head stiffness of a pile embedded in an inhomogeneous soil with stiffness varying linearly with depth $\left(\mu L_a = 2.5 \right)$.

Table 6.3 Fitted head stiffness functions for soil with stiffness varying linearly with depth with zero surface stiffness

Source	Randolph (1981)[*]	Budhu & Davies (1988)[*]	Gazetas et al. (1991)	Syngros (2004)
$K_{11}/E_{s,D}D$	$1.129\eta\left(\dfrac{2E_p}{E_{s,D}\eta}\right)^{3/9}$	$0.734\left(\dfrac{E_p}{E_{s,D}}\right)^{3/9}$	$0.6\left(\dfrac{E_p}{E_{s,D}}\right)^{0.35}$	$0.65\left(\dfrac{E_p}{E_{s,D}}\right)^{0.35}$
$K_{free}/E_{s,D}D$	$0.473\eta\left(\dfrac{2E_p}{E_{s,D}\eta}\right)^{3/9}$	$0.313\left(\dfrac{E_p}{E_{s,D}}\right)^{3/9}$		$0.28\left(\dfrac{E_p}{E_{s,D}}\right)^{0.35}$
$K_{12}/E_{s,D}D^2$	$0.300\eta\left(\dfrac{2E_p}{E_{s,D}\eta}\right)^{5/9}$	$0.270\left(\dfrac{E_p}{E_{s,D}}\right)^{5/9}$	$0.17\left(\dfrac{E_p}{E_{s,D}}\right)^{0.60}$	$0.17\left(\dfrac{E_p}{E_{s,D}}\right)^{0.60}$
$K_{22}/E_{s,D}D^3$	$0.135\eta\left(\dfrac{2E_p}{E_{s,D}\eta}\right)^{7/9}$	$0.173\left(\dfrac{E_p}{E_{s,D}}\right)^{7/9}$	$0.15\left(\dfrac{E_p}{E_{s,D}}\right)^{0.85}$	$0.13\left(\dfrac{E_p}{E_{s,D}}\right)^{0.80}$

Where $\eta = \dfrac{4+3v_s}{8(1+v_s)}$, [*]obtained by inverting the flexibility matrix.

provided results for a parabolic soil stiffness profile with zero surface stiffness. Note that the expressions by Randolph (1981), Budhu & Davies (1988), and Gazetas (1991) retain their monomial character in (E_p/E_s) in both the flexibility matrix form and in K_{free} due to the relationship between the exponents (discussed previously for the results in Table 6.2).

Figure 6.7 shows a comparison of the performance of the approximate Winkler energy solution in Eq. 6.17 against some existing continuum solutions and the fitted functions in Table 6.3. This figure shows similar trends to Figure 6.4 with the Winkler solution performing well compared to the more rigorous results.

6.2.4.1 Layered soil

Piles are often embedded in soil that can be discretised into piece-wise layers. Alternatively, more complex soil stiffness profiles can be approximated with a number of homogeneous layers to enable easier computation. An example of a pile installed in M soil layers is shown in Figure 6.8.

Mylonakis (1995) provided two implicit analytical solutions for the response of piles in layered soil, a transfer matrix solution and an iterative approach. Both methods utilise the solutions for short piles embedded in homogeneous soils with different base boundary conditions applied, resulting in relatively long unwieldy expression. Instead, applying the energy method in Eq. 6.12 yields the following simple expressions for the pile head stiffness terms (adapted from Mylonakis & Gazetas 1999 and Karatzia & Mylonakis 2017), which were shown to be in good agreement with the closed-form solution:

$$\frac{K_{11}}{E_p I \mu^3} = 1 + \sum_{m=1}^{M}\left(\frac{\lambda_m}{\mu}\right)^4 \left[e^{-2\mu z}\left(2 + \cos(2\mu z) + \sin(2\mu z)\right)\right]_{z=z_m+h_m}^{z=z_m} \tag{6.19a}$$

$$\frac{K_{free}}{E_pI\mu^3} = \frac{1}{2} + \frac{1}{2}\sum_{m=1}^{M}\left(\frac{\lambda_m}{\mu}\right)^4\left[e^{-2\mu z}\left(2 + \cos(2\mu z) - \sin(2\mu z)\right)\right]_{z=z_m+h_m}^{z=z_m} \tag{6.19b}$$

$$\frac{K_{12}}{E_pI\mu^2} = 1 + \sum_{m=1}^{M}\left(\frac{\lambda_m}{\mu}\right)^4\left[e^{-2\mu z}\left(1 + \sin(2\mu z)\right)\right]_{z=z_m+h_m}^{z=z_m} \tag{6.19c}$$

$$\frac{K_{22}}{E_pI\mu} = \frac{3}{2} + \frac{1}{2}\sum_{m=1}^{M}\left(\frac{\lambda_m}{\mu}\right)^4\left[e^{-2\mu z}\left(2 - \cos(2\mu z) + \sin(2\mu z)\right)\right]_{z=z_m+h_m}^{z=z_m} \tag{6.19d}$$

where z_m, h_m, and λ_m are the top depth, thickness, and Winkler parameter of layer m, respectively. The latter can be calculated from Eq. 6.2b with $k = k_m$, the Winkler spring stiffness of layer m. These expressions reproduce the values for homogeneous soil in Eq. 6.8 when one layer is employed with $k_1 = k$ and $h_1 \rightarrow \infty$.

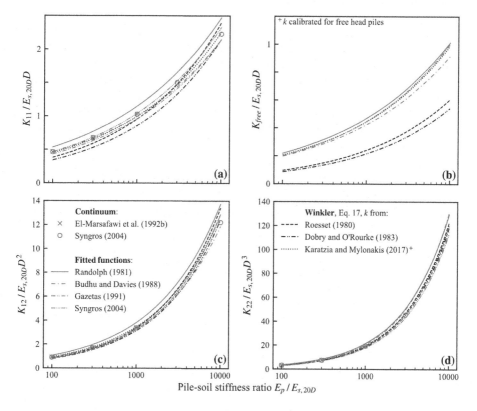

Figure 6.7 Variation of normalised head stiffness with pile-soil stiffness ratio for piles embedded in an inhomogeneous soil with stiffness varying linearly with depth from a value of zero at the surface; (a) K_{11}, (b) K_{free}, (c) K_{21}, (d) K_{22}, where $E_{s,20D}$ is the soil stiffness at a depth of 20 pile diameters $(v_s = 0.5)$. (Numerical data from Syngros (2004).)

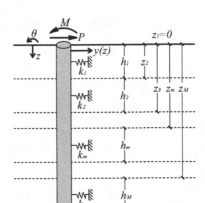

Figure 6.8 Winkler model for a pile embedded in M soil layers.

6.2.5 Dynamic impedance

Dynamic loading can be handled using a slightly enhanced version of the classical Winkler model. Consider the pile shown in Figure 6.9, a load, $P(t)$ and/or moment $M(t)$ is applied at the pile head at a particular angular frequency, ω. This is resisted by distributed springs and dashpots of stiffness $k(z)$ and viscosity $c(z)$, respectively. In this representation, the springs account for soil stiffness and the dashpots for energy dissipation due to friction and radiation damping (NEHRP 2012). It is often convenient to express these impedances as a complex valued stiffness, $k^* = k + i\omega c$. k^* is dependent on the frequency of the load applied, the soil stiffness, $E_s(z)$, shear wave velocity, $V_s(z)$, Poisson's ratio, v_s, mass density, ρ_s and hysteretic (material) damping factor, β_s, defined through the complex soil stiffness $E_s^* = E_s(1 + 2i\beta_s)$. The pile itself is modelled as an elastic beam with second moment of area I, mass density ρ_p, and complex stiffness $E_p^* = E_p(1 + 2i\beta_p)$, where β_p is the hysteretic material damping coefficient for the pile.

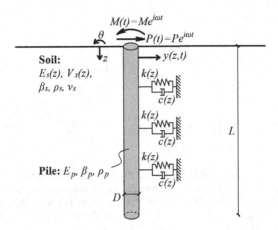

Figure 6.9 Winkler model for a pile under a dynamic lateral load.

Considering steady state harmonic oscillations, the governing differential equation is given by

$$E_p^* I \frac{d^4 y^*(z)}{dz^4} + \left[k(z) + i\omega c(z) - \rho_p A_p \omega^2 \right] y^*(z) = 0 \tag{6.20}$$

where $y^*(z) = y^*(z, t) e^{-i\omega t}$ and $y^*(z, t)$ is the time dependent displacement profile down the pile, and A_p is the pile cross sectional area. Using this complex (phasor) representation of harmonic pile displacements instead of elementary sinusoidal functions allows the dynamic pile stiffness and damping to be established in a compact, time-independent form as shown in the following.

As with the static case, the overall behaviour of the pile can be expressed as a stiffness matrix, K^*, which describes the pile head response and is, in this case, complex valued.

$$K^* = \begin{bmatrix} K_{11}^* & K_{12}^* \\ K_{21}^* & K_{22}^* \end{bmatrix} = K + i\,\omega\, C = \begin{bmatrix} K_{11} & K_{12} \\ K_{21} & K_{22} \end{bmatrix} + i\,\omega \begin{bmatrix} C_{11} & C_{12} \\ C_{21} & C_{22} \end{bmatrix} \tag{6.21}$$

where $K = \Re\left[K^* \right]$ is the stiffness matrix for the pile head response, and $C = \frac{1}{\omega} \Im\left[K^* \right]$ is the damping (viscosity) matrix.

6.2.5.1 Homogeneous soil

When $k^*(z) = k^*$, Eq. 6.20 can be reduced to a form similar to Eq. 6.2:

$$\frac{d^4 y^*(z)}{dz^4} + 4\lambda^{*4}\, y^*(z) = 0 \tag{6.22a}$$

$$\lambda^* = \sqrt[4]{\frac{k^* - \rho_p A_p \omega^2}{4 E_p^* I}} \tag{6.22b}$$

where λ^* is a complex wavenumber analogous to λ for the static case. Therefore, the static solutions presented in Section 6.2.1 can simply be modified by substituting the dynamic terms E_p^* and λ^* for the corresponding static terms, E_p and λ, respectively. In this way, the dynamic pile head stiffness of a long pile can be obtained from Eq. 6.8:

$$K^* = \begin{bmatrix} 4 E_p^* I \lambda^{*3} & 2 E_p^* I \lambda^{*2} \\ sym & 2 E_p^* I \lambda^* \end{bmatrix} \tag{6.23}$$

A decomposition of Eqs. 6.22b and 6.23 into real and imaginary parts is discussed in Makris & Gazetas (1993) and in Maravas et al. (2007, 2014).

6.2.5.2 Selection of $k^*(z)$

Various approaches are available to select the dynamic Winkler impedances, some of which are shown in Table 6.4. Note that these expressions are for fixed head piles and inertial loading (i.e., forces and moments) applied at the pile head.

Table 6.4 Dynamic Winkler impedance equations (fixed head piles, inertial interaction) (Adapted from Shadlou & Bhattacharya (2014), Anoyatis & Lemnitzer (2017))

Source	Real component	Imaginary component
Baranov (1967), Novak (1974), Novak et al. (1978)	$k^* = \pi G_s^* s^2 \dfrac{4K_1(q)K_1(s)+sK_1(q)K_0(s)+qK_0(q)K_1(s)}{sK_1(q)K_0(s)+qK_0(q)K_1(s)+qsK_0(q)K_0(s)}$, where $s = \dfrac{ia_0}{2\sqrt{1+2i\beta_s}}$, $q = \dfrac{s}{\eta_s}$, $a_0 = \omega D/V_s$ and $\eta_s = \sqrt{\dfrac{2(1-v_s)}{1-2v_s}}$	
O'Rourke & Dobry (1978)	$k = 2.3G_s$ to $4.2G_s$	$\dfrac{c_r}{\rho_s V_s D} = 2\left[1 + \dfrac{V}{V_s}\right]$, where $V_s \le V \le V_c$
Roesset & Angelides (1980), Roesset (1980)	$k = 1.2E_s$	$c_r = 5\rho_s V_s D$
Kaynia (1980)	$k = 3.5G_s$	$c_r = 5\rho_s V_s D$
Dobry et al. (1982)	$\dfrac{k}{E_s} = 1.67\left(\dfrac{E_p}{E_s}\right)^{-0.053}$	$\dfrac{c_r}{\rho_s V_s D} = 1.55(1+v_s)\left(\dfrac{E_p}{E_s}\right)^{0.124}$
Gazetas & Dobry (1984a)	$k = 1.0E_s$ to $1.2E_s$	$z \le 2.5D.$ $\dfrac{c_r}{\rho_s V_s D} = 4\left(\dfrac{\pi}{4}\right)^{\frac{3}{4}} a_0^{-\frac{1}{4}}$ $z \ge 2.5D.$ $\dfrac{c_r}{\rho_s V_s D} = 2\left(\dfrac{\pi}{4}\right)^{\frac{3}{4}} a_0^{-\frac{1}{4}}\left[1+\left(\dfrac{V_{La}}{V_s}\right)^{\frac{5}{4}}\right]$ where $\dfrac{V_{La}}{V_s} = \dfrac{3.4}{\pi(1-v_s)}$
Makris & Gazetas (1993)	$k = 1.2E_s$	$\dfrac{c_r}{\rho_s V_s D} = 2a_0^{-\frac{1}{4}}\left[1+\left(\dfrac{V_{La}}{V_s}\right)^{\frac{5}{4}}\right] \approx 6\,a_0^{-\frac{1}{4}}$
Mylonakis (2001)	$k^* = \pi G_s^* s^2 \dfrac{4K_1(q^*)K_1(s^*)+s^*K_1(q^*)K_0(s^*)+q^*K_0(q^*)K_1(s^*)}{s^*K_1(q^*)K_0(s^*)+q^*K_0(q^*)K_1(s^*)+q^*s^*K_0(q^*)K_0(s^*)}$, where: $s^* = \dfrac{1}{2}\sqrt{a_{0,c}^2 - a_0^2/(1+2i\beta_s)}$, $q^* = \dfrac{s^*}{\eta_u}$, $\eta_u = \sqrt{\dfrac{2-v_s}{1-v_s}}$ and $a_{0,c} = \dfrac{\pi}{2}\left(\dfrac{H}{D}\right)^{-1}$	

(Continued)

Table 6.4 (Continued) Dynamic Winkler impedance equations (fixed head piles, inertial interaction) (Adapted from Shadlou & Bhattacharya (2014), Anoyatis & Lemnitzer (2017))

Source	Real component	Imaginary component
Anoyatis et al. (2016), Anoyatis & Lemnitzer (2017)	$k^* = \pi G_s^* s_m^* \left[s_m^* + 4 \dfrac{K_1\left(s_m^*\right)}{K_0\left(s_m^*\right)} \right] + 2.2 G_s i \sqrt{a_0^2 - a_{0,c}^2}$ where $s_m^* = \dfrac{1}{2(\eta_s)^\chi} \sqrt{a_{0,c}^2 - a_0^2 / (1 + 2i\beta_s)}$, and χ is an empirical correction factor (dependent on v_s). Tabulated results for χ are provided in the paper, and these follow the function. $\chi = 6 - 10 v_s$.	
Karatzia & Mylonakis (2017)	$k = \dfrac{4\pi G_s \eta_u^2}{\left(1 + \eta_u^2\right)\left[\ln\left(4/\alpha_c\right) - \gamma\right] + \ln\left(\eta_u\right)}$ See Table 6.1 for α_c definition	$\dfrac{c_r}{\rho_s V_s D} = \pi a_0^{-0.4} \left[\dfrac{1}{4} + \dfrac{4}{5} \dfrac{V_c}{V_s} \right]$, where $\dfrac{V_c}{V_s} = \sqrt{\dfrac{2}{1 - v_s}}$

Baranov (1967) and Novak (1974) presented an analytical solution considering the plane-strain response of a horizontal soil slice under zero vertical normal strain and corresponding shear strain in an undamped medium. Novak et al. (1978) employed the correspondence principle of viscoelasticity (Pipkin 1972) to extend the analytical solution to a damped medium. This involved substituting the material stiffnesses in the elastic solution with a corresponding complex modulus as follows:

$$G_s \to G_s^* = G_s\left(1 + 2i\beta_s\right), \quad E_s \to E_s^* = E_s\left(1 + 2i\beta_s\right), \quad V_s \to V_s^* = V_s\sqrt{1 + 2i\beta_s} \qquad (6.24)$$

Whilst this solution is rigorous for the plane-strain case, it results in a complex stiffness that cannot be conveniently separated into real and imaginary parts in closed form, as the Bessel functions of complex argument are not separable. In addition, for very low frequencies, this solution yields zero static stiffness. Novak et al. (1993) resolves this by applying an empirical correction to the solution by assuming a constant spring value below a certain threshold frequency. However, various alternative simplified expressions are available instead.

Regarding the spring stiffness, k, O'Rourke & Dobry (1978), Roesset (1980), Roesset & Angelides (1980), Kaynia (1980), Dobry et al. (1982), Gazetas & Dobry (1984a), and Makris & Gazetas (1993) all provide simple (and similar) expressions that have been fitted against more rigorous numerical continuum results.

With reference to the dashpot viscosity, c, Roesset & Angelides (1980), Roesset (1980), Kaynia (1980), and Dobry et al. (1982) also provide similarly fitted results for the radiation damping component, c_r.

In an earlier paper, O'Rourke & Dobry (1978) instead employ a simplified analytical model by Berger el al. (1977) that considers the dashpots required to dissipate the energy due to both a P-wave (travelling at the soil compression-extension wave velocity, V_c) and an S-wave (travelling at the soil shear wave velocity, V_s) travelling along rods parallel and perpendicular to the axis of loading, respectively (the stripe model shown in Figure 6.10). This results in a radiation dashpot that is independent

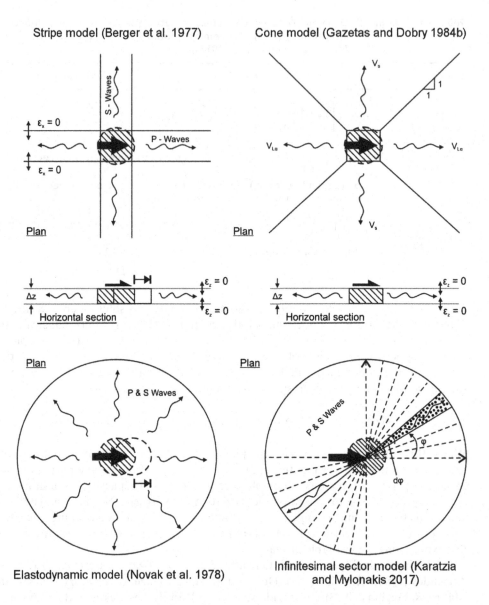

Figure 6.10 Radiation damping models. (Reproduced from Karatzia & Mylonakis (2017), © ASCE.

of frequency. O'Rourke & Dobry (1978) suggest selecting a lower velocity for the rods along the loading axis.

In a follow-up paper, Gazetas & Dobry (1984b) developed an alternative approximation for the plane-strain (horizontal soil "slice") solution by splitting the domain into four independent quarter planes, split at 45° to the direction of loading such that the loading vector and the perpendicular axis pass along the centreline of two quarters each (the cone model shown in Figure 6.10). To this end, a circular pile had to be replaced by a square one with the same perimeter. Compression waves travelling at Lysmer's analog wave velocity, V_{La}, are assumed to propagate in the two quarters centred along the axis of loading and shear waves in the other two quarters. This results in a frequency-dependent expression, against which Gazetas & Dobry (1984a) fitted a simplified monomial expression using regression analysis. For shallow depths, shear waves were assumed to propagate in all four quarters instead, to simulate surface waves. Makris & Gazetas (1993) also utilised this model with some minor simplifications in the resulting expression.

Karatzia & Mylonakis (2017) developed a more refined model employing infinitesimal independent sectors through which both shear and compression waves propagate (the infinitesimal sector model shown on Figure 6.10). In this way, there is no need to replace the circular pile by a square one as in the original model. Summing the response of all sectors through integration over the polar angle φ yields a solution in terms of Bessel functions. A simple (frequency-dependent) monomial expression has been fit to these results using non-linear regression.

The aforementioned approximate solutions only consider the radiation damping component of the dashpot viscosity coefficient, c_r. For a damped medium, the correspondence principle of viscoelasticity can be employed as before. In this context, Dobry et al. (1982) and Gazetas & Dobry (1984a) suggest an approximate treatment of simply adding a hysteretic damping term, c_h, to the radiation dashpot as shown:

$$c \approx c_h + c_r, \quad c_h \approx 2\beta_s \frac{k}{\omega} \tag{6.25}$$

For a homogeneous half-space, the aforementioned solutions can be employed without modification. However, a finite soil layer resting on a stiff deposit exhibits a cut-off frequency, ω_c, below which shear waves do not propagate and therefore, radiation damping does not exist. This cut-off frequency is practically equal to the natural frequency of the soil layer in shear oscillations. For a homogeneous soil layer over rock, this frequency is approximately given by Eq. 6.26 (Gazetas & Dobry 1984a), expressed as a dimensionless cut-off frequency, $a_{0,c}$.

$$a_{0,c} = \frac{\omega_c D}{V_s} \approx \frac{\pi}{2} \left(\frac{H}{D} \right)^{-1} \tag{6.26}$$

For frequencies below the cut-off frequency $(\omega \leq \omega_c)$, no radiation damping is generated, and therefore, the Winkler impedances can be obtained using the simplified expressions in Eq. 6.27 (using the same assumptions as for Eq. 6.25): Gazetas & Dobry (1984a) also suggest further modifying the frequency-dependent dashpot using a straight line connecting $c = c_h$ at $\omega = \omega_c$ and the value of c at $\omega = 2\omega_c$.

$$k = \Re\left[k^* \right], \quad c = c_h \approx 2\beta_s \frac{k}{\omega} \tag{6.27}$$

Mylonakis (2001) introduced the third dimension (depth) to the horizontal slice model by integrating the governing equation over the vertical coordinate using a preselected shape function. This approach is analogous to the technique introduced by Vlasov & Leontiev (1966) for surface footings and yields an improved version of the plane-strain model with more realistic behaviour near the cut-off frequency (therefore, introducing Eq. 6.27 is not required for this solution). Also, unlike the original plane-strain approach of Baranov and Novak, this solution yields a finite stiffness at zero frequency for finite soil thickness and can also predict the cut-off frequency. In addition, for the limit case of a homogeneous half-space and using a pertinent shape function, this solution directly reproduces the plane-strain results, and the static stiffness approaches zero. Anoyatis et al. (2016) and Anoyatis & Lemnitzer (2017) suggested an empirical modification to the solution by Mylonakis (2001) resulting in a simplified function which matches well with an analytical solution. However, Eq. 6.27 should be used for $\omega \le \omega_c$.

A number of dynamic Winkler impedance expressions are plotted in Figure 6.11 for a homogeneous half-space. The empirical modification proposed by Anoyatis et al. (2016) is shown to be in good agreement with the more rigorous plane-strain solution by Novak et al. (1978). For higher frequencies, both solutions provide a stiffness component, k, in a similar range to the empirical expressions. However, in the plane-strain model, this component approaches zero at low frequencies, which is unrealistic. The simplified analytical models for the radiation damping coefficient $\beta_r = \Im[K]/(2\,\Re[K])$ by Gazetas & Dobry (1984b) and Karatzia & Mylonakis (2017) give similar results to the plane-strain model by Novak et al. (1978). It is worth mentioning that in terms of the damping coefficient, β_r, the plane-strain model is not singular at low frequencies. Indeed, whilst both $\Re(K)$ and $\Im(K)$ attain unrealistically low values in the zero-frequency limit, their ratio exhibits a much more stable behaviour and attains similar values to those of more rigorous solutions.

At higher frequencies (a_o of about 1), the empirical expressions based on frequency-independent dashpots c_r also match well. However, their behaviour at low frequencies is noticeably different as β_r varies linearly with a_o which does not faithfully reproduce wave radiation near the static limit.

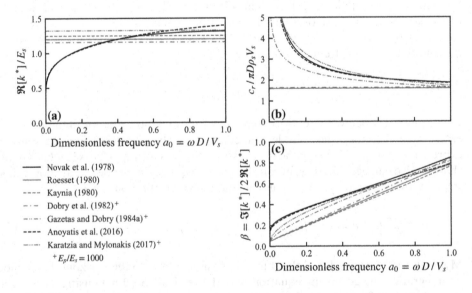

Figure 6.11 Dynamic Winkler modulus for a pile embedded in a homogeneous half-space $\left(v_s = 0.4,\ \beta_s = 0.05\right)$.

6.2.5.3 Comparison with numerical results

Figure 6.12 shows the performance of the Winkler solution (Eq. 6.23) for different impedance values (Table 6.4) against existing continuum solutions for piles embedded in a homogeneous half-space by El-Marsafawi et al. (1992b) and Syngros (2004). In general, the Winkler solution performs well and gives results in meaningful agreement with the more rigorous solutions. However, at low frequencies, the impedances calculated using Novak et al. (1978) approach zero. Therefore, the empirical correction suggested by Novak et al. (1993) (selecting a minimum stiffness, k, for frequencies below a certain threshold) is recommended for the near-zero frequency limit.

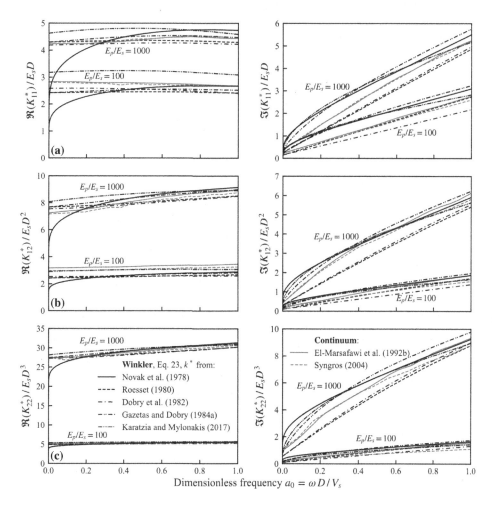

Figure 6.12 Normalised dynamic lateral head stiffness of piles embedded in a homogeneous half-space; (a) K_{11}, (b) K_{21}, and (c) K_{22} $\left(v_s = 0.4, \beta_s = 0.05, v_p = 0.25, \beta_p = 0.01, \rho_p/\rho_s = 1.25\right)$. (Numerical data from Syngros (2004).)

6.2.5.4   Inhomogeneous soil

No closed form solutions to the dynamic Winkler governing equation (Eq. 6.20) are known to the authors for inhomogeneous soil.[3] The different dependence on depth for each of the three terms (i.e., stiffness, damping, and inertia) that multiply displacement in the governing equation prohibits the simple modification of static results employed for homogeneous soils. Instead, an approximate approach must be employed. One option is to split the problem into a number of homogeneous layers and then employ the transfer matrix or iterative solutions available in Mylonakis (1995). This is an easy-to-implement, yet tedious approach. Alternatively, the energy method employed for static loading earlier in this chapter can be modified for dynamic loading (Mylonakis 1995; Mylonakis & Gazetas 1999; Mylonakis & Roumbas 2001; Karatzia & Mylonakis 2017).

Following the same approach as that employed to derive Eq. 6.12, the following equation can be obtained from Eq. 6.20:

$$K_{ij}^* = \int_0^\infty E_p^* I \chi_i^{*\prime\prime}(z) \chi_j^{*\prime\prime}(z) dz + \int_0^\infty \left[k^*(z) - \rho_p A_p \omega^2 \right] \chi_i^*(z) \chi_j^*(z) dz \tag{6.28}$$

where K_{ij}^* is the dynamic head stiffness term corresponding to the choice of $\chi_i^*(z)$ and $\chi_j^*(z)$, the dimensionless shape functions analogous to $\chi_i(z)$ and $\chi_j(z)$.

Previous solutions have taken these functions to be real-valued and employed the static functions in Eq. 6.13 to obtain the complex pile head stiffness (Mylonakis 1995; Mylonakis & Gazetas 1999; Mylonakis & Roumbas 2001; Karatzia & Mylonakis 2012, 2017) or just the dashpot coefficients for the translational mode (O'Rourke & Dobry 1978; Dobry et al. 1982; Gazetas & Dobry 1984a). Simplified expressions for the pile head dashpot coefficients obtained using this approach for a number of inhomogeneous soil profiles are provided by Mylonakis & Roumbas (2001) and Karatzia & Mylonakis (2012, 2017). However, employing a real-valued shape function neglects the phase difference between the response at different pile elevations and may lead to significant errors in the predictions of the method for certain types of inhomogeneous media. Instead, the complex-valued shape functions from Eq. 6.29 can be employed. These are based on the dynamic response of a pile embedded in a homogeneous soil and inherently account for phase differences in the response of different points.

$$\chi_1^*(z) = e^{-\mu^* z} \left[\sin\left(\mu^* z\right) + \cos\left(\mu^* z\right) \right] \tag{6.29a}$$

---

3 This is related to the fourth-order nature of the governing equation, which drastically limits the number of inhomogeneous media that can be tackled by exact methods. This lies beyond the scope of this chapter.

$$\chi_2^*(z) = \frac{e^{-\mu^* z}}{\mu^*} \sin\left(\mu^* z\right)$$
(6.29b)

$$\chi_{free}^*(z) = e^{-\mu^* z} \cos\left(\mu^* z\right)$$
(6.29c)

where the parameter μ^* is analogous to λ^* in the same way as μ is analogous to λ and can be obtained using a similar approach to Eq. 6.14:

$$\mu^* = \frac{1}{L_a} \int_0^{L_a} \lambda^*(z)\,dz, \quad \lambda^*(z) = \sqrt[4]{\frac{k^*(z) - \rho_p A_p \omega^2}{4 E_p^* I}}$$
(6.29d)

The active length for a pile under dynamic load has been reported as longer than for the static case (Velez et al. 1983). However, for the purpose of this energy solution, L_a is a model parameter (rather than a property of the solution), and the resulting head stiffness values are not particularly sensitive to the value chosen. Therefore, it can be obtained using the simple fitted functions for statically loaded piles available for certain soil stiffness profiles (e.g., those in Figure 6.5). For a linear soil stiffness profile with zero surface stiffness Syngros (2004) recommends Eq. 6.15 with $\chi_L = 2.5$ and $n_L = 0.2$. For arbitrary soil profiles, the iterative approach for static loading (Di Laora & Rovithis 2015) can be employed to get $L_a = 2.5/\mu$ from Eq. 6.15 and $k(z) = \Re\left[k^*(z)\right]$.

The performance of this solution is compared to rigorous numerical results from El-Marsafawi et al. (1992b) and Syngros (2004) in Figure 6.13, by comparing the predicted head stiffness of a pile embedded in an inhomogeneous soil with stiffness varying linearly with depth and zero surface stiffness. The approximate approach in Eq. 6.28 performs well and captures the shape and magnitude of the variation with frequency for all pile head impedances. As in the homogeneous case (Figure 6.12), the stiffnesses from Novak et al. (1978) approach zero at low frequencies, and therefore, an empirical correction is required if these Winkler springs are employed. Note that perfect absorbing boundaries are not available for the inhomogeneous soil profile considered, and therefore, numerical solutions must impose a rigid boundary (or a homogenous half-space) at a certain depth. This is evident in the oscillations present in the more rigorous results and the behaviour at low frequencies, particularly below the (quite evident) cut-off frequency for the soil layer. This limitation is not present in the simplified Winkler model.

6.2.6 Displacement, rotation, bending moment, and shear force variation with depth

To design a piled foundation, engineers often require more information than just the response at the pile head. Of particular interest are the pile displacement and rotation with depth, the bending moment and shear force profile with depth, and the peak bending moment in the pile. From Euler-Bernoulli beam theory, based on the sign convention in Figure 6.1, the rotation, $\theta(z)$, bending moment, $M(z)$, and shear force,

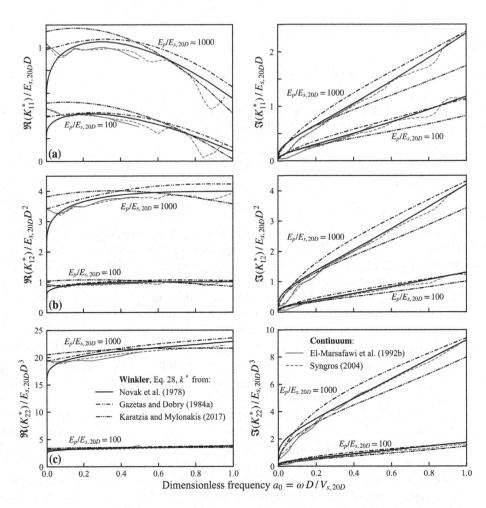

Figure 6.13 Normalised dynamic lateral head stiffness of piles embedded in an inhomogeneous soil with stiffness varying linearly with depth from a value of zero at the surface; (a) K_{11}, (b) K_{21}, and (c) K_{22} $\left(v_s = 0.4, \beta_s = 0.05, v_p = 0.25, \beta_p = 0.01, \rho_p/\rho_s = 1.25\right)$. ($L_a$ and numerical data from Syngros (2004).)

$P(z)$, in the pile can be obtained from successive differentiation of the displaced shape of the pile, $y(z)$, using Eq. 6.31:

$$\theta(z) = +\frac{dy(z)}{dz} \tag{6.30a}$$

$$M(z) = -E_p I \frac{d^2 y(z)}{dz^2} \tag{6.30b}$$

$$P(z) = +E_p I \frac{d^3 y(z)}{dz^3} \qquad\qquad (6.30c)$$

The displaced shape of the pile obtained from the Winkler model for long piles embedded in homogeneous soil has previously been employed to generate the shape functions in Eq. 6.13. Recalling their definition leads to the following expression for $y(z)$:

$$y(z) = y_0 \chi_1(z) + \theta_0 \chi_2(z) \qquad\qquad (6.31)$$

where the head displacement, y_0, and rotation, θ_0, can be obtained from the applied load P and moment, M, using Eq. 6.3 and the stiffness terms for the pile head response.

The shape functions and their derivatives are presented in Table 6.5 and plotted in Figure 6.14. Inputting these into Eqs. 6.30 and 6.31 with $\mu = \lambda$ yields the exact solutions (in the Winkler domain) for homogeneous soil.

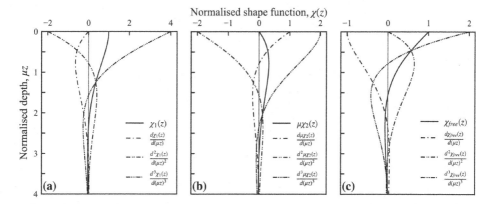

Figure 6.14 Normalised shape functions and their derivatives (a) $\chi_1(z)$, (b) $\chi_2(z)$, and (c) $\chi_{free}(z)$.

Table 6.5 Shape functions and their derivatives (Hetenyi 1946, Pender 1993, Karatzia 2017)

$\dfrac{d}{dz}$	$\chi_1(z)$	$\chi_2(z)$	$\chi_{free}(z)$
$\chi_i(z)$	$e^{-\mu z}\left[\cos(\mu z)+\sin(\mu z)\right]$	$\dfrac{e^{-\mu z}}{\mu}\sin(\mu z)$	$e^{-\mu z}\cos(\mu z)$
$\dfrac{d\chi_i(z)}{dz}$	$-2\mu e^{-\mu z}\sin(\mu z)$	$e^{-\mu z}\left[\cos(\mu z)-\sin(\mu z)\right]$	$-\mu e^{-\mu z}\left[\cos(\mu z)+\sin(\mu z)\right]$
$\dfrac{d^2\chi_i(z)}{dz^2}$	$2\mu^2 e^{-\mu z}\left[\sin(\mu z)-\cos(\mu z)\right]$	$-2\mu e^{-\mu z}\cos(\mu z)$	$2\mu^2 e^{-\mu z}\sin(\mu z)$
$\dfrac{d^3\chi_i(z)}{dz^3}$	$4\mu^3 e^{-\mu z}\cos(\mu z)$	$2\mu^2 e^{-\mu z}\left[\cos(\mu z)+\sin(\mu z)\right]$	$2\mu^3 e^{-\mu z}\left[\cos(\mu z)-\sin(\mu z)\right]$

Evidently, as the solutions to the Winkler equation in Eq. 6.2 involve products of exponential and trigonometric functions of depth, the pile always displaces (and rotates) in the direction opposite to that of the applied head load beyond a certain depth.

For fixed head piles ($\theta_0 = 0$), the peak bending moment, M_{max}, occurs at the pile head and is the moment required to ensure the head fixity condition, M_{fix}:

$$M_{max} = M_{fix} = K_{21}y_0 = P\frac{K_{21}}{K_{11}} \tag{6.32}$$

For homogeneous soils, this yields (Pender 1993): $M_{max} = P/2\lambda$. For inhomogeneous soil, simply replacing λ by μ is not possible, as evident from the stiffness in Eq. 6.17 and Figure 6.6.

For peak bending moments at depth, the critical elevation, z_{Mmax}, can be obtained from Eq. 6.33 for homogeneous soils (Pender 1993).

$$z_{Mmax} = \frac{1}{\lambda}\tan^{-1}\left(\frac{1}{1-2\lambda M/P}\right) \tag{6.33}$$

Substituting this back into Eq. 6.30b yields M_{max}.

For the special case of a free head pile ($M = 0$), the displaced shape is instead given by Eq. 6.34, the derivatives of which are presented in Table 6.5. In this case, z_{Mmax} and M_{max} are given by Eq. 6.35.

$$y(z) = y_0\chi_{free}(z) \tag{6.34}$$

$$z_{Mmax} = \frac{\pi}{4\lambda}, \quad M_{max} = -\frac{Pe^{-\pi/4}}{\lambda\sqrt{2}} \approx -0.322\frac{P}{\lambda} \tag{6.35}$$

For piles embedded in inhomogeneous soil, these solutions (Eqs. 6.30–6.32) can be employed with μ selected using Eq. 6.14, as before. However, as the displaced shape is approximate, results will be less and less accurate upon successive differentiation. Instead, a more accurate approach would be to approximate the soil reaction acting on the pile, $p(z) = k(z)y(z)$, and obtain the shear force and bending moment profile using equilibrium (using the notation in Figure 6.1):

$$P(z) = P - \int_0^z k(x)y(x)dx \tag{6.36}$$

$$M(z) = M - Pz + \int_0^z k(x)y(x)(z-x)\,dx \tag{6.37}$$

where x is a dummy depth variable.

For piles under dynamic loads, Eqs. 6.30 and 6.31 with Table 6.5 can be employed with the static terms replaced with the corresponding complex-valued dynamic terms (e.g., substituting the dynamic terms E_p^* and μ^* for the corresponding static terms,

E_p and μ, respectively). The resulting displacement, rotations, bending moments, and shear forces will be complex valued. These can be expressed as a magnitude and a phase difference from the applied load and moment at the pile head.

6.2.7 Applying these solutions

Similar to the corresponding simplified solutions for axially loaded piles discussed in the previous chapter, the solutions presented in this chapter require selection of an appropriate soil stiffness variation with depth, $E_s(z)$. Ideally, this would be based on in-situ measurements of soil stiffness supported by high-quality laboratory testing with small-strain measurement. Small-strain shear modulus, $G_{s,0}$, can be evaluated from the corresponding shear wave velocity, $V_{s,0}$, using Eq. 6.40. Young's modulus for soil can be related to shear modulus using Eq. 6.41.

$$G_s = \rho_s V_s^2 \tag{6.38}$$

$$E_s(z) = 2\left(1 + v_s\right) G_s(z) \tag{6.39}$$

For very small strains, $G_{s,0}$ may be sufficient. However, soil shear modulus should be reduced relative to $G_{s,0}$ for larger strain effects. Using results from some of the earliest full-scale tests on instrumented laterally loaded piles, McClelland & Focht (1956) observed similarity in shape between the load-displacement response at the pile interface (the "p-y" curve) and the axial stress-strain curves from triaxial tests in clay at similar confining pressures. Therefore, they suggested that soil strains around a laterally loaded pile can be evaluated in an approximate manner as

$$\gamma(z) = \left(1 + v_s\right) y(z) / c_1 D \tag{6.40}$$

where $\gamma(z)$ denotes an average soil strain at depth z, $y(z)$ the corresponding horizontal pile displacement and c_1 a pertinent proportionality coefficient. By comparing the strains and displacements at failure between the two curves, McClelland & Focht (1956) suggested a value of $c_1 = 1/2$. However, based on a similar approach by Skempton (1951) for pad foundations, Matlock (1970) and Kagawa & Kraft (1981) suggest much higher values ranging between 2 and 3. The method employed to obtain this factor is discussed in Reese & Van Impe (2011), involving selecting a factor to match the elastic stiffness of the two curves when normalised by their corresponding strengths. However, the solution does not account for the presence of soil on both sides of the pile (i.e., as opposed to a beam on an elastic foundation, modelling a flexible surface footing). Francis (1964) suggested that a more appropriate stiffness for laterally loaded piles is double that for the beam-on-elastic-foundation problem, shown to be in good agreement with more modern solutions in Figure 6.3. Therefore, a more appropriate choice for laterally loaded piles may be at around $c_1 = 5$. On the basis of Eq. 6.40, a strain-compatible soil shear modulus can be obtained through conventional modulus reduction curves (e.g., Darendeli 2001; Ishihara 1996; Vucetic & Dobry 1991; Vardanega & Bolton 2011) or directly from laboratory soil tests on samples from the site. These corrections can be employed in an iterative manner until a pre-described tolerance condition is fulfilled.

For larger strains, various simple extension to the static solutions presented in this chapter are available in the literature (e.g., Scott 1981; Guo 2012) involving linear elastic perfectly plastic Winkler springs to account for soil non-linearity. Corresponding dynamic solutions are much harder to obtain, and hence, the available information is limited (Parmelee et al. 1964; Penzien 1970; Trochanis et al. 1991; Nogami et al. 1992; Badoni & Makris 1996; El Naggar & Novak 1995, 1996; Maheshwari et al. 2004; Gerolymos & Gazetas 2006; Radhima et al. 2021).

6.3   PILE GROUP RESPONSE

Grouped piles are subjected to two types of loading: (i) loads transmitted onto the pile heads from the superstructure through the cap and (ii) additional loading imposed along their length by the displacement field generated by the neighbouring piles. Following the original work by Poulos (Poulos 1971b; Poulos & Davis 1980), such effects can be described using interaction factors, which account for the interplay between two piles at a time. The interaction factor, α, between two piles can be defined based on the response of a pile carrying no load at its head (termed pile 2 or "receiver" pile) due to a head load (force or moment) applied on a neighbouring pile (termed pile 1 or "source" pile). The interaction factor between two identical piles is then defined as the response (translation or rotation) atop the receiver pile, $\Delta_{21}(0)$, normalised by the corresponding response of the source pile, $\Delta_{11}(0)$, due to its own loading (force or moment), as shown in Eq. 6.41a:

$$\alpha_{ij}(s,\phi) = \Delta_{21}(0)/\Delta_{11}(0) \tag{6.41a}$$

where i is the degree of freedom corresponding to a displacement or rotation atop pile 2, j is the degree of freedom corresponding to a force or moment atop pile 1, s is the distance between the pile centre lines (axis-to-axis distance), and ϕ is the aperture angle between the direction of loading and the line connecting the pile centres.

Poulos introduced five interaction factors and employed a BEM-like approach to derive them. Tabulated results are presented in Poulos & Davis (1980). These factors are

α_{uP} – receiver pile displacement due to free head load
$\alpha_{\theta P}$ – receiver pile rotation due to free head load
α_{uM} – receiver pile displacement due to free head moment
$\alpha_{\theta M}$ – receiver pile rotation due to free head moment
α_{uF} – receiver pile displacement due to fixed head load (interaction between fixed-head piles)

Accordingly, under free-head conditions and for a specific set of s, ϕ values, the interaction between two piles is represented by a 2×2 matrix in the form

$$\alpha = \begin{bmatrix} \alpha_{uP} & \alpha_{uM} \\ \alpha_{\theta P} & \alpha_{\theta M} \end{bmatrix} \tag{6.41b}$$

Naturally, all interaction factors in Eq. 6.41b have magnitude of less than one. Due to reciprocity, $|\alpha_{uM}| = |\alpha_{\theta P}|$. Also, due to the sign convention employed here (Figure 6.1)

factors, α_{uM} and $\alpha_{\theta M}$ are negative. An updated set of interaction factors for lateral static loads was derived by El-Sharnouby & Novak (1985) using finite element analysis.

Randolph (1977, 1981) employed a combination of finite element analyses and collections of boundary-element-like results by Poulos (1971b) to propose a set of fitted functions for each of the interaction factors. Using the notation adopted in this chapter, these functions can be expressed in the following forms:

$$\alpha_{uP} \approx \frac{1}{2}\left[\frac{8(1+v_s)}{4+3v_s}\frac{E_p}{E_s}\right]^{\frac{1}{7}}\left(\frac{D}{2s}\right)\left[1+\cos^2(\phi)\right] \le 0.5 \qquad (6.42a)$$

$$\alpha_{uM} = -\alpha_{\theta P} \approx -\alpha_{uP}^2 \qquad (6.42b)$$

$$\alpha_{\theta M} \approx -\alpha_{uP}^3 \qquad (6.42c)$$

$$\alpha_{uF} \approx \frac{6}{5}\alpha_{uP} \qquad (6.42d)$$

Note that α_{uF} is always larger (by 20% or so) than α_{uP}, whilst α_{uM} and $\alpha_{\theta M}$ are much smaller than α_{uP}. Also, the interaction factor between two piles aligned in the same direction to that of the applied load is twice as large as that in the direction perpendicular to the applied load, i.e., $\alpha_{uP}(s,0) \approx 2\,\alpha_{uP}(s,\pi/2)$. These properties are further discussed later in this section.

To ensure bounded behaviour near the source pile, when α_{uP} exceeded 0.5, Randolph proposed replacing α_{uP} in Eq. 6.42a with $1-1/(4\alpha_{uP})$ using the value of α_{uP} calculated from the original expression.

A simplified theoretical formulation was proposed by Dobry & Gazetas (1988), who suggested that $\alpha_{uP} = \alpha_{uF} = \psi(s,\phi)$ and $\alpha_{uM} = \alpha_{\theta P} = \alpha_{\theta M} = 0$. The function $\psi(s,\phi)$ corresponds to the attenuation in the far field of a cylindrical wave field and for static conditions can be approximated by

$$\psi(s,\phi) \approx \sqrt{\frac{D}{2s}} \qquad (6.43a)$$

Note that this simplified method makes no distinction between free-head and fixed piles, i.e., $\alpha_{uP} = \alpha_{uF}$. Also, this solution is independent of the aperture angle ϕ, which implies that the interaction factors between two piles aligned in the same direction to the applied load $(\phi = 0)$ and perpendicular to the direction of the applied load $(\phi = \pi/2)$ are approximately equal, in contrast to the factor of 2 difference obtained by Randolph (1981).

Theoretical solutions for the function $\psi(s,\phi)$ under static conditions were derived by Baguelin et al. (1977), Novak & Sheta (1980), and Karatzia et al. (2014) using the horizontal soil "slice" model of Baranov. These solutions are analogous to that based on the "shearing of concentric cylinders" model and exhibit a logarithmic variation with radial distance, thereby requiring introducing a rigid boundary at a certain distance from the active pile (at a different distance than in the axial mode). Ways to incorporate 3D effects into these solutions are discussed in Baguelin et al. (1977), Mylonakis (2001), and Karatzia et al. (2014).

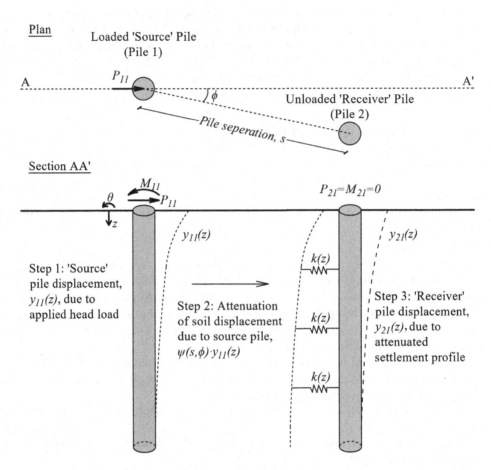

Figure 6.15 Three-step model for the calculation of interaction factors between two piles. (Adapted from Mylonakis & Gazetas (1999).)

A more rational theoretical formulation in the form of a simple, three-step model was introduced by Makris & Gazetas (1992) and later expanded by Mylonakis (1995) and Mylonakis & Gazetas (1999), as shown in Figure 6.15. The model is formulated based on the following steps:

Step 1: The deflected shape of the source pile, hereafter referred to as $y_{11}(z)$, to a unit force or moment applied at its head is determined using a pertinent analytical approach (Winkler, FEM, or BEM) such as that employed in Section 6.2 of this chapter.

Step 2: A displacement field is generated at the periphery of the source pile with an initial amplitude equal to the deflected pile shape, $y_{11}(z)$. It is assumed that the displacement field attenuates in a horizontal manner, depending on radial distance and aperture angle (s, ϕ) (as will be discussed later in this section, this corresponds to an assumption of horizontal wave propagation in the dynamic regime.) Therefore, the free-field soil displacement at a distance s and aperture angle ϕ from the direction of loading is given by $U(s, z, \phi) = \psi(s, \phi) \, y_{11}(z)$, which is a "separation of variables" approximation. Eq. 6.43a can be used to approximate the function $\psi(s, \phi)$. However, in order to give more rational results for different aperture angles, an empirical correction based on the fitted results from Randolph (1981) can be employed:

$$\psi(s, \phi) \approx \sqrt{\frac{D}{2s}} \frac{1 + \cos^2(\phi)}{2} \tag{6.43b}$$

Step 3: Key to the present analysis is that the receiver pile does not follow exactly the free field motion described in step 2. Its inertial and flexural resistance would give rise to an "inclusion effect", i.e., an interaction between the receiver pile and the surrounding soil, leading to a reduction of the displacement amplitude. Thereby, the pile displacement will be different than $U(s, z, \phi)$. Moreover, a rotation will be generated at the head of the receiver pile, which cannot be calculated directly from function $U(s, z, \varphi)$. To account in a simple yet rational way for this interaction, the receiver pile is modelled as a beam on a Winkler foundation in which the excitation $U(s, z, \phi)$ is applied at as a support motion at the base of the distributed springs attached to the pile. The mechanics of this loading is in a sense the reverse of step 1. In step 1, the source pile induces displacements on soil through its "reacting" springs, whereas in step 3, the soil induces displacements on the receiver pile through its "transmitting" springs.

Considering the equilibrium of an infinitesimal piece of the receiver piles, this interaction can be expressed by the following ordinary differential equation:

$$E_p I \frac{d^4 y_{21}(z)}{dz^4} + k(z) \, y_{21}(z) = k(z) \, U(s, z, \phi) = k(z) \, \psi(s, \phi) \, y_{11}(z) \tag{6.44}$$

in which $y_{21}(z)$ stands for the unknown displacement of the receiver pile.

6.3.1 Homogeneous soil

A closed-form solution to Eq. 6.44 for homogeneous soil $[k(z) = k]$ is available in Mylonakis (1995) and further discussed in Mylonakis & Gazetas (1999). As the attenuation function is assumed to be independent of depth, the resulting interaction factors can be expressed as a product of the attenuation function and a "diffraction" factor, ζ_{ij}, representing the inclusion (stiffening) effect of the receiver pile on the overall system.

$$\alpha_{ij}(s, \phi) = \psi(s, \phi) \, \zeta_{ij} \tag{6.45a}$$

For long piles, the inclusion factors corresponding to each interaction factor are given in Eq. 6.45b (Mylonakis 1995; Mylonakis & Gazetas 1999):

$$\zeta = \begin{bmatrix} \zeta_{uP} & \zeta_{uM} \\ \zeta_{\theta P} & \zeta_{\theta M} \end{bmatrix} = \begin{bmatrix} 3/4 & -1/2 \\ 1/2 & -1/4 \end{bmatrix}, \quad \zeta_{uF} = 3/4 \tag{6.45b}$$

Of particular interest is the result for fixed head piles. This indicates that the boundary condition at the pile head has no effect on the diffraction factor for long piles. This is contrary to the 20% increase in the value of ζ for fixed head piles in the correlations by Randolph (1981). The stronger interaction between fixed-head piles relative to free-head ones can be justified in light of the higher displacements developing near the head of the source pile, $y_{11}(z)$, under fixed-head conditions, which can lead to higher imposed displacements on the receiver pile.

Interaction factors can also be also obtained for piles of finite length using this solution. The corresponding expressions are complicated and are provided in Mylonakis (1995) and Mylonakis & Gazetas (1999).

Figure 6.16 shows how the inclusion factor varies with pile length for piles unrestrained at the tip (floating condition). For zero pile length, all inclusion functions tend to unity, which is anticipated because a floating pile of zero length has no flexural resistance and follows exactly the free-field motion. For $\lambda L > 0$, the factors ζ_{uP}, ζ_{uM}, and $\zeta_{\theta M}$ decrease monotonically with pile length, reaching their asymptotic values 3/4, 1/2, and 1/4, respectively, at $\lambda L \approx 3$, which is larger than the active length value of the corresponding solitary piles (Figure 6.2). This suggests that two interacting piles mobilise soil reaction at greater depths than a solitary pile. On the other hand, the inclusion factor ζ_{uF} of a fixed-head pile exhibits a more complicated behaviour, with undulations before reaching the asymptotic value 3/4 at $\lambda L \approx 3$.

The interaction factors obtained for long piles from Eqs. 6.43b and 6.45 are shown in Figure 6.17. These are compared with the fitted equations from Randolph (1981) as well as the rigorous numerical results by El-Marsafawi et al. (1992b). The Winkler solutions for α_{uP} and α_{uF} provide good approximations of the interaction factors relative to the most rigorous results by El-Marsafawi et al (1992b). On the other hand, the rotational components α_{uM} and $\alpha_{\theta M}$ are overpredicted by the Winkler model, especially at distances s/D higher than approximately 3.

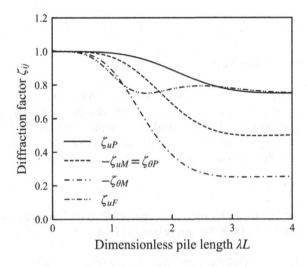

Figure 6.16 Variation of diffraction factors, ζ_{ij}, with pile length for floating piles in homogeneous soil. (Adapted from Mylonakis and Gazetas (1999).)

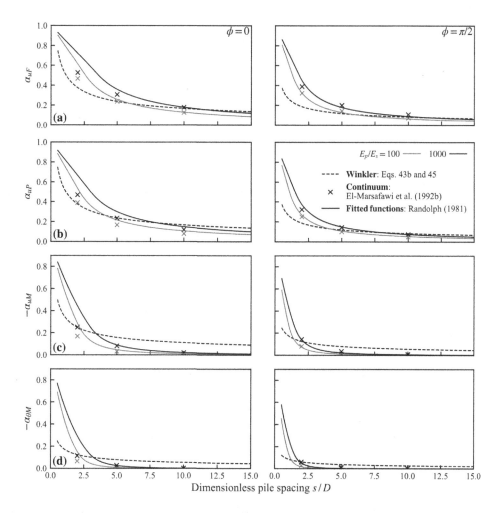

Figure 6.17 Interaction factors between piles in homogeneous soil. (Data from El-Marsa-fawi et al. (1992b).)

6.3.2 Dynamic response

Kaynia (1982) and Kaynia & Kausel (1982) extended the concept of an interaction factor to the dynamic regime. Dynamic pile-to-pile interaction is described by complex-valued interaction factors which account for the dynamic interplay between neighbouring piles, including phase differences between the response at the pile heads.

In the realm of the herein adopted Winkler analysis, the three-step model (shown in Figure 6.18) is conceptually similar to that for static load, with the following key differences: *First*, functions y_{11}, y_{21}, and ψ are complex valued. This is particularly important for the function ψ which describes a horizontally propagating wave, emitted from the source and arriving at the location of the receiver pile at a different phase depending

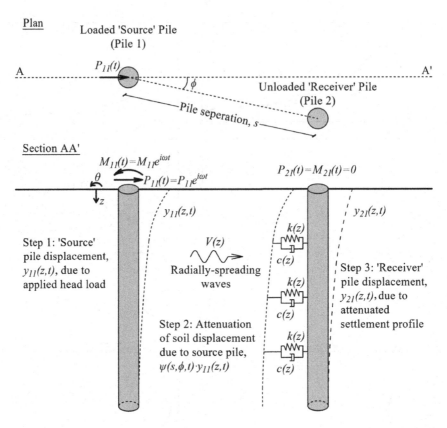

Figure 6.18 Three-step model for the calculation of interaction factors between two piles under dynamic loading. (Adapted from Makris & Gazetas (1992) and Mylonakis & Gazetas (1999).)

on distance, frequency, and wave propagation velocity. *Second*, the coefficients ζ_{ij} can now be interpreted as "diffraction" coefficients accounting for the diffraction of the impinging waves due to the presence of the receiver pile. *Third*, as wave propagation velocity is different in the direction parallel and perpendicular to the applied load, the function $\psi(s, \phi)$ is dependent on the aperture angle according to the theoretical expression (Novak & Sheta 1980; Kaynia 1982):

$$\psi(s,\phi) = \psi(s,0)\cos^2(\phi) + \psi\left(s,\frac{\pi}{2}\right)\sin^2(\phi) \tag{6.46}$$

where (from Makris & Gazetas 1992, modified from Dobry & Gazetas 1988)

$$\psi(s,0) \approx \left(\frac{D}{2s}\right)^{1/2} \exp\left[-(\beta_s + i)\left(\frac{s}{D} - \frac{1}{2}\right)\frac{V_s}{V_c}\, a_0\right] \tag{6.47a}$$

$$\psi(s,\pi/2) \approx \left(\frac{D}{2s}\right)^{1/2} \exp\left[-(\beta_s + i)\left(\frac{s}{D} - \frac{1}{2}\right)a_0\right] \tag{6.47b}$$

in which V_s and V_c are the propagation velocities of s waves and compression waves, respectively, at the particular elevation, β_s is the hysteretic damping coefficient, i is the complex unit corresponding to a phase angle of $\pi/2$, and a_0 is the dimensionless frequency from Section 6.2.5. In the original model V_c was taken equal to the so-called "Lysmer analogue" wave velocity $V_{La} = 3.4\ V_s\ [(1-v)\ \pi]$ (Gazetas & Dobry 1984a). Alternative formulations for V_c are available in Anoyatis et al. (2016).

Equation 6.47 is plotted against the solution for function $\psi(s, \phi)$ from the rigorous 2D plane-strain model of Baranov & Novak in Figure 6.19. Despite the differences in the formulations (see Nogami 1985; Mylonakis 1995), the two models provide quite similar results, which implies that they both can be used for predicting interaction effects between neighbouring piles.

With reference to Step 3 in the Winkler model at hand, the governing differential equation is written as

$$E_p I \frac{d^4 y_{21}(z)}{dz^4} + \left[k^*(z) - \rho_p A_p \omega^2 \right] y_{21}(z) = k^*(z)\ \psi(s,\phi)\ y_{11}(z) \tag{6.48}$$

in which $k^* = k + i\omega c$ is the complex-valued Winkler impedance discussed in Section 6.2.5.

For long piles in homogeneous soil, the analytical solution can be obtained with the following simple correction to the static values of the diffraction factors in Eq. 6.45b (Makris & Gazetas 1992; Mylonakis 1995; Mylonakis & Gazetas 1999):

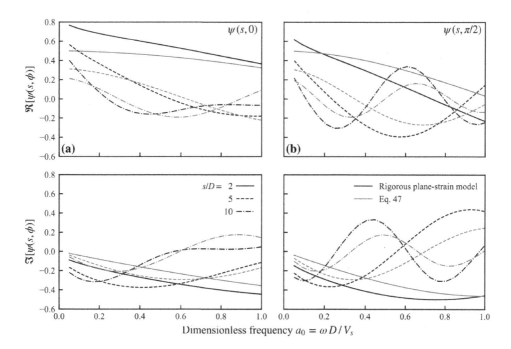

Figure 6.19 Real and imaginary components of the dynamic attenuation functions at different distances from the pile for (a) $\phi = 0$ and (b) $\phi = \pi/2$. (Adapted from Mylonakis (1995)), $\left(v_s = 0.4, \beta_s = 0.05 \right)$.

$$\zeta_{ij}^* = \frac{k^*}{k^* - \rho_p A_p \omega^2} \zeta_{ij} = \left(\frac{\lambda_{massless}^*}{\lambda^*}\right)^4 \zeta_{ij} \qquad (6.49)$$

where $\lambda_{massless}^*$ is the value of λ^* calculated from Eq. 6.22b assuming a massless pile ($\rho_p = 0$).

The interaction factors for long *fixed-head* piles are shown in Figure 6.20, plotted for three pile spacings ($s/D = 2$, 5, and 10). These are compared with the predictions of the simplified three-step model based on the Winkler impedances k^* of the Baranov-Novak model (see Table 6.4) and attenuation functions ψ from Eq. 6.47 as well as the rigorous numerical results by El-Marsafawi et al. (1992b). The increasing phase angle with frequency suggests an oscillatory behaviour of the interaction factors with frequency. This can be attributed mainly to the attenuation function $\psi(s, \phi)$ in Eq. 6.47 and to a lesser extent on pile-soil interaction (Eq. 6.49). The phase is wrapped (i.e., principal argument values are provided limited between $-\pi$ and π). This highlights why the phase is particularly important for interaction factors; the deflections of a pile in a group are summed with the deflection of every other pile multiplied by the corresponding interaction factor, and if the interaction factor is out of phase (near $-\pi$ or π), these values will destructively interfere resulting in a lower settlement amplitude. However, if the phase of the interaction factor is near zero, these values will constructively interfere resulting in much higher settlement amplitudes and a corresponding low group stiffness. The satisfactory performance of the simplified three-step method is evident in a wide range of frequencies and pile separations examined.

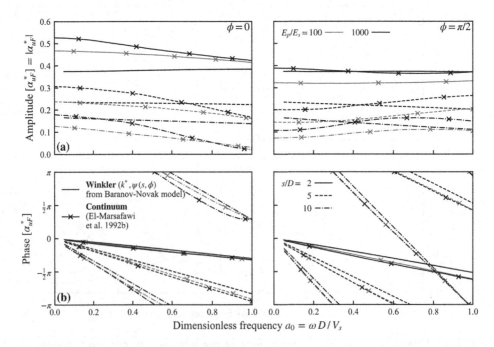

Figure 6.20 Interaction factors between fixed head piles in homogeneous soil. (Data from El-Marsafawi et al. (1992b)), $\left(v_s = 0.4, \beta_s = 0.05\right)$.

Corresponding interaction factors for long *free-head* piles are shown in Figure 6.21. These, again, are compared with the existing simplified solutions using the Baranov-Novak model and the rigorous numerical results by El-Marsafawi et al. (1992b). The Winkler values are equal to the ones for fixed-head piles in Figure 6.20 (see Eq. 6.47). The performance of the three-step model is, again, satisfactory for a wide range of frequencies and pile separations examined.

From Eqs. 6.45 and 6.47, the following relations can be derived between the interaction factors for long piles:

$$\alpha_{\theta P} = 2/3 \; \alpha_{uP} \tag{6.50a}$$

$$\alpha_{\theta M} = 1/3 \; \alpha_{uP} \tag{6.50b}$$

A comparison of these expressions against the empirical static formulae of Randolph in Eq. 6.42, i.e., $\alpha_{\theta P} \approx \alpha_{uP}^2, \alpha_{\theta M} \approx \alpha_{uP}^3$ is performed graphically in Figure 6.22. Also plotted on the graph are numerical results from the rigorous Green's function formulation of Kaynia & Kausel (1982) for nine different frequencies and three pile separations. Evidently, at low frequencies ($a_0 = 0.1$), Randolph's curves fit better to the rigorous results which explains the better performance of those formulae against the BEM results in Figure 6.17c and d. With increasing frequency, however, the rigorous results move closer to the derived expressions in Eq. 6.50. To explain this pattern, recall that with increasing frequency, waves emitted from a pile tend to follow horizontal paths (i.e., perpendicular to the pile longitudinal axis) – a result of wave interference (Gazetas 1987),

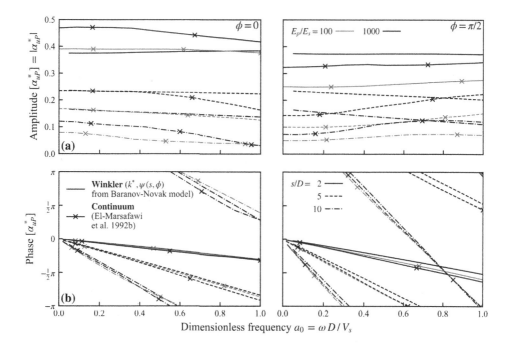

Figure 6.21 Interaction factors between free head piles in homogeneous soil. (Data from El-Marsafawi et al. (1992b)), $\left(v_s = 0.4, \beta_s = 0.05\right)$.

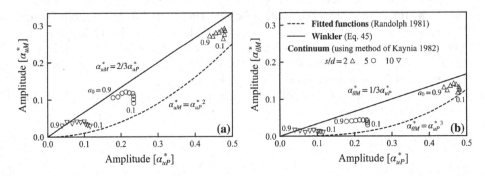

Figure 6.22 Relationship between free-head interaction factors for long piles in homo-
geneous soil, $\phi = 0$, (a) α^*_{uM} and α^*_{uP} and (b) $\alpha^*_{\theta M}$ and α^*_{uP}. (Reproduced from
Mylonakis & Gazetas (1999), © ASCE.)

in accord with the assumption of this study (Figure 6.18). In that sense, the derived
expressions in Eq. 6.50 are asymptotically "exact" at high frequencies, whilst the empirical expressions in Eq. 6.42 are applicable for low frequencies. Nevertheless, because
the magnitude of the rocking and cross-interaction factors is quite small, both sets of
expressions [i.e., Eqs. 6.42 and 6.45, 6.47] provide acceptable engineering estimates.

6.3.3   Applying interaction factors

An example group of six identical piles in a 2×3 symmetric configuration is shown
in Figure 6.23. Once the stiffness matrix of the individual piles and the interaction
factors between all pile pairs in the group have been established, the response of the
whole pile group can be determined using the superposition method (Poulos 1971b;
Kaynia & Kausel 1982). Firstly, for each pile, i, the flexibility expression describing the

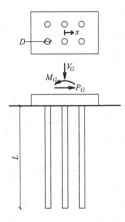

Figure 6.23 Example six pile group configuration.

displacement vector, $\{u\}_i$, as a superposition of the response to external loads applied to each other pile $\{L\}_j$ (including itself) is obtained using the inverse of the stiffness matrix for this pile, $[K]_j^{-1}$, and the interaction factors matrix between these and the pile of interest, $[\alpha]_{ij}$. For a group of N piles, these can be expressed in the form

$$\{u\}_i = \sum_{j=1}^{N} [\alpha]_{ij}[K]_j^{-1}\{L\}_j \tag{6.51a}$$

For loading in one plane, this can be written explicitly as

$$\left\{\begin{array}{c} y \\ \theta \\ w \end{array}\right\}_i = \sum_{j=1}^{N} \begin{bmatrix} \alpha_{up} & \alpha_{uM} & 0 \\ & \alpha_{\theta M} & 0 \\ sym & & \alpha_v \end{bmatrix}_{ij} \begin{bmatrix} K_{11} & K_{12} & 0 \\ & K_{22} & 0 \\ sym & & K_v \end{bmatrix}_j^{-1} \left\{\begin{array}{c} P \\ M \\ V \end{array}\right\}_j \tag{6.51b}$$

where V, w, K_v, and α_v are the vertical load, displacement, stiffness, and interaction factors, respectively, which can be obtained from the corresponding chapter on axial response, and *sym* indicates that the matrices are symmetric. Note that if $i = j$, each non-zero term of the interaction factor matrix $[\alpha]_{ii}$ is equal to 1 (i.e., all the interaction terms between a pile and itself equal 1).

For static loading, these expressions are all real valued, whereas under dynamic loading, all matrices and vectors are generally complex (in the example here, the * indicating complex values is omitted).

This sum can be expressed in the following matrix form:

$$\left\{\begin{array}{c} \{u\}_1 \\ \cdots \\ \cdots \\ \{u\}_N \end{array}\right\} = [\alpha F]\left\{\begin{array}{c} \{L\}_1 \\ \cdots \\ \cdots \\ \{L\}_N \end{array}\right\} = \begin{bmatrix} [\alpha F]_{11} & [\alpha F]_{12} & \cdots & [\alpha F]_{1N} \\ [\alpha F]_{21} & [\alpha F]_{22} & \cdots & \cdots \\ \cdots & \cdots & \cdots & \cdots \\ [\alpha F]_{N1} & \cdots & \cdots & [\alpha F]_{NN} \end{bmatrix} \left\{\begin{array}{c} \{L\}_1 \\ \cdots \\ \cdots \\ \{L\}_N \end{array}\right\}$$

$$\tag{6.52}$$

where $[\alpha F]$ is the overall flexibility ($3N \times 3N$) matrix of the group incorporating all planar degrees of freedom and encompassing interaction between the piles.

For a rigid pile cap, it is simple to relate the displacement components of each pile to the overall displacement components of the pile group, y_G, θ_G, and w_G (the horizontal displacement, rotation, and vertical displacement of the pile cap, respectively):

$$\left\{\begin{array}{c} \{u\}_1 \\ \cdots \\ \cdots \\ \{u\}_N \end{array}\right\} = [T]\left\{\begin{array}{c} y_G \\ \theta_G \\ w_G \end{array}\right\} = \begin{bmatrix} 1 & 0 & 0 \\ 0 & 1 & 0 \\ 0 & -x_1 & 1 \\ \cdots & \cdots & \cdots \\ \cdots & \cdots & \cdots \\ 1 & 0 & 0 \\ 0 & 1 & 0 \\ 0 & -x_N & 1 \end{bmatrix} \left\{\begin{array}{c} y_G \\ \theta_G \\ w_G \end{array}\right\} \tag{6.53}$$

where x_j is the distance along the loading axis of pile j from the origin on loading.

Therefore, the overall response of a rigid pile group under combined lateral (P_G), moment (M_G), and vertical (V_G) loading can be obtained in stiffness form:

$$\left\{\begin{array}{c} P_G \\ M_G \\ V_G \end{array}\right\} = [K_G] \left\{\begin{array}{c} y_G \\ \theta_G \\ w_G \end{array}\right\}$$

(6.54a)

where $[K_G]$ is the overall pile group stiffness matrix obtained as

$$[K_G] = [T]^T [\alpha F]^{-1} [T]$$

(6.54b)

For combined loading along more than one horizontal axis, Eq. 6.51b can be expanded to include six degrees of freedom at the pile head. Alternatively, multiple loads can be considered independently, and the resulting displacements combined using the principle of superposition. For flexible or semi-flexible pile caps, similar formulations can be easily derived using standard matrix structural analysis.

6.3.3.1   Symmetric pile groups

A special case with significant practical implications relates to symmetric pile groups with (i) an even number of piles and (ii) rotational symmetry of order N (i.e., as many planes of symmetry as there are piles in the group, which must include perpendicular planes, e.g., a 2×2 rectangular configuration or any even numbered pile group arranged around the circumference of a circle). In this special case, when the cap is restrained in rotation, all piles carry equal loads (i.e. $P_i = P_G/N$), which greatly simplifies the analysis as it allows reducing the flexibility formulation in Eq. 6.51 to a single equation. For example, for a symmetric group comprising fixed-head piles subjected to a horizontal cap force P_G, one obtains for the arbitrary pile i

$$y_i = y_G = \left(P_G/N\right)/K_{11} \left[1 + \sum_{j=1}^{N,\, i \neq j} \alpha_{ij}\right]$$

(6.55)

where y_i ($= y_G$) is the lateral displacement of the pile, (P_G/N) is the load carried by the pile, and K_{11} is the stiffness of the single pile. The summation stands for the superposition of (fixed head) interaction factors between pile i and all other piles in the group, except for the trivial case $j = i$ that corresponds to $\alpha_{ii} = 1$.

This expression can be easily cast in the following stiffness form:

$$P_G = \left[N\, K_{11} \middle/ \left(1 + \sum_{j=1}^{N,\, i \neq j} \alpha_{ij}\right)\right] y_G$$

(6.56)

in which the term in brackets stands for the stiffness of the group when the pile cap is restrained in rotation, K_G.

In the same vein as in the axial mode, the performance of the group can be expressed in terms of the deflection ratio, R_D:

$$R_D \equiv y_G / (P_G / N \ K_{11}) \tag{6.57}$$

Using Eq. 6.56, the above ratio can be readily determined as

$$R_D = 1 + \sum_{j=1}^{N, \ i \neq j} \alpha_{ij} = 1 + \zeta \sum_{j=1}^{N, \ i \neq j} \psi_{ij} \tag{6.58a}$$

or

$$R_D = 1 + \frac{3}{4} \left(\frac{\lambda^*_{massless}}{\lambda^*} \right)^4 \sum_{j=1}^{N, \ i \neq j} \psi_{ij} \tag{6.58b}$$

which is merely the sum of all interaction factors associated with any of the piles in the group. In the above equation, $\zeta = (3/4) \ (\lambda^*_{massless}/\lambda^*)^4$ stands for the relevant dynamic diffraction factor according to Eq. 6.49. For static loading, the second term of the diffraction factor is unity. It is worth mentioning that according to the Winkler model, the diffraction factor is the same for fixed-head piles and piles whose head is free to rotate (Eq. 6.45b).

A few special cases are considered below:

2 × 1 Group
Considering a horizontal force, P_G, acting on the pile group:

$$y_G = \left[1 + \alpha_{uF}(s, \phi) \right] (P_G / 2) / K_{11} \tag{6.59a}$$

$$R_D = 1 + \alpha_{uF}(s, \phi) = 1 + \frac{3}{4} \left(\frac{\lambda^*_{massless}}{\lambda^*} \right)^4 \psi(s, \phi) \tag{6.59b}$$

where, in this case, s is the spacing between the two piles, and ϕ is the aperture angle between the applied load and this axis.

2 × 2 Square Group
Considering a horizontal force acting within a plane of symmetry parallel to a side of the group, one obtains for any of the four piles

$$y_G = \left[1 + \alpha_{uF}(s, 0) + \alpha_{uF}\left(s, \frac{\pi}{2} \right) + \alpha_{uF}\left(\sqrt{2}s, \frac{\pi}{4} \right) \right] \frac{P_G}{4K_{11}} \tag{6.60a}$$

$$R_D = 1 + \frac{3}{4} \left(\frac{\lambda^*_{massless}}{\lambda^*} \right)^4 \left[\psi(s, 0) + \psi\left(s, \frac{\pi}{2} \right) + \psi\left(\sqrt{2}s, \frac{\pi}{4} \right) \right] \tag{6.60b}$$

where $\psi\left(\sqrt{2}s, \frac{\pi}{4} \right) = \frac{1}{2} \left[\psi\left(\sqrt{2}s, 0 \right) + \psi\left(\sqrt{2}s, \frac{\pi}{2} \right) \right]$.

Therefore,

$$R_D = 1 + \frac{3}{4}\left(\frac{\lambda^*_{massless}}{\lambda^*}\right)^4 \left[\psi(s,0) + \psi\left(s,\frac{\pi}{2}\right) + \frac{1}{2}\left[\psi\left(\sqrt{2}s,0\right) + \psi\left(\sqrt{2}s,\frac{\pi}{2}\right)\right]\right]$$ (6.60c)

For static loading, $\psi(s, 0)/\psi(s, \pi/2) = 2$ (see Eq. 6.43b); the above expression simplifies to

$$R_D = 1 + \frac{9}{16}\left[2\,\psi(s,0) + \psi\left(\sqrt{2}s,0\right)\right]$$ (6.60d)

By splitting any applied load into components parallel and perpendicular to the plane considered, then combining the resulting displacements, it is evident that this stiffness is independent of the direction of the applied load.

6.3.3.2 Illustrative example

Figure 6.24 shows the deflection ratio for 2 × 1 and 2 × 2 pile groups under static loading. The value of R_D decreases as the piles are spaced further apart, approaching a value of 1 (the value for a single pile) as the distance increases. Evidently, the further apart the piles are, the more efficiently they are utilised at the cost of a larger pile cap.

Figure 6.25 shows the deflection ratio for a 2 × 2 pile group under dynamic loading. Note that as the phase of the dynamic interaction factor is non-zero, R_D has both an amplitude and phase. The amplitude is normally of most interest.

The terms in Eq. 6.60 interfere both constructively and destructively depending on different factors, particularly the frequency of the applied loading, spacing of the piles, and shear wave velocity of the soil. Therefore, the amplitude of R_D varies with the frequency of the applied load. If enough of the attenuated pile displacements interfere

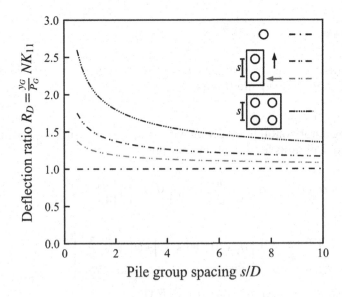

Figure 6.24 Deflection ratio for 2 × 1 and 2 × 2 pile groups under static loading.

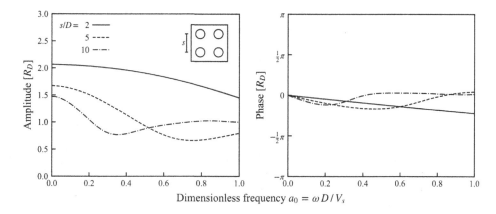

Figure 6.25 Deflection ratio for a 2×2 pile group under dynamic loading $\left(E_p/E_s = 1000, \rho_p/\rho_s = 1.25, \beta_s = 0.05, \beta_p = 0, v_s = 0.4\right)$.

destructively, the amplitude of R_D can be lower than 1, indicating the group is stiffer than the sum of its component piles (not possible under static loading). However, if the attenuated pile displacements interfere constructively, the amplitude of R_D can reach larger values than in the static case.

6.4 SUMMARY AND CONCLUSIONS

Available solutions were reviewed for static and dynamic analysis of laterally loaded single piles, pairs of piles, and pile groups. A number of solutions were derived using the Winker model of soil reaction in conjunction with a virtual work formulation and complex-valued shape functions analogous to those used in spectral FEMs. These solutions can be expressed in closed form and be implemented by means of simple computer programs or spreadsheets for use in routine design. Specifically

1   For single piles under static loading, exact solutions for homogeneous soil are presented in Eqs. 6.4, 6.7, and 6.8 and in Table 6.1. A general approximate energy solution was presented for inhomogeneous soils in Eq. 6.12. Closed-form solutions for linear varying soil stiffness with depth as well as for layered soil are provided in Eqs. 6.17 and 6.19, respectively.

2   It was shown that application of the Winkler model in conjunction with the strength-of-material assumption for pile bending leads to a characteristic length that can be used to normalise the independent variable (depth) in the solution. This parameter can be interpreted as a "mechanical length" as it depends on pile-soil relative stiffness.

3   Some constraints on the powers of monomial expressions for pile stiffness available in the literature (Table 6.2) were discussed. It was shown that the exponent associated with the cross-stiffness term should be equal to the average of the exponents associated with the translational and rocking terms, i.e., $n_{12} = (n_{11} + n_{22})/2$.

4   For single piles under dynamic loading, exact solutions for homogeneous soil are presented in Eq. 6.23 and in Table 6.4. A modified approximate energy solution for inhomogeneous soils based on complex valued shape functions was presented in Eq. 6.28.

5   The physics of the cut-off frequency, associated with the emergence of waves propagating horizontally in the soil medium, was discussed. Some approximate solutions were presented incorporating this effect.

6   Expressions for pile displacement, rotation, bending moment, and shear force as a function of depth are presented in Eqs. 6.30–6.38 and Table 6.5.

7   A reduction in soil shear modulus to account for strain effects is presented in Eq. 6.40.

8   For pairs of piles, exact solutions for interaction factors between piles in homogeneous soil under static loading are presented in Eq. 6.45, and corresponding solutions for dynamic loading are presented in Eq. 6.49.

9   For pile groups, solutions using the superposition method are presented in Eqs. 6.51–6.54. Symmetric pile groups were treated in Eqs. 6.55–6.60.

Satisfactory comparisons against more rigorous continuum solutions of the finite element and boundary element type were presented for both static and dynamic loads.

ACKNOWLEDGEMENTS

The second author would like to thank the Engineering and Physical Sciences Research Council for their support (grant number EP/N509619/1). No new experimental data were collected during this study.

NOTATION

Latin

a_0	Dimensionless frequency
$a_{0,c}$	Dimensionless soil layer cut-off frequency
A_p	Pile cross sectional area
C	Pile head damping (viscosity) matrix
C_{ij}	Pile head damping matrix term
c_1	Strain-displacement proportionality coefficient
$c(z)$	Winkler dashpot viscosity
c_h	Winkler hysteretic dashpot viscosity
c_r	Winkler radiation dashpot viscosity
D	Pile diameter
E_p, E_p^*	Real and complex valued pile Young's modulus, respectively
E_s, E_s^*	Real and complex valued soil Young's modulus, respectively
$E_{s,D}, E_{s,20D}$	Soil Young's modulus at 1 and 20 pile diameters depth, respectively
F	Pile head flexibility matrix
F_{ij}	Pile head flexibility matrix term
G_s, G_s^*	Real and complex soil shear modulus, respectively

H	Depth to underlying rigid layer
h_m	Thickness of soil layer m
I	Pile second moment of area
i	Imaginary unit
K	Pile head stiffness matrix
$K_{floating}$	Pile head stiffness matrix under floating base conditions
K_{fixed}	Pile head stiffness matrix under fixed base conditions
K_{hinged}	Pile head stiffness matrix under hinged base conditions
K_{11}, K_{11}^*	Real and complex valued pile swaying head stiffness, respectively
K_{12}, K_{21}	Real valued pile cross-swaying-rocking head stiffness
K_{12}^*, K_{21}^*	Complex valued pile cross-swaying-rocking head stiffness
K_{22}, K_{22}^*	Real and complex valued pile rocking head stiffness, respectively
K_{free}, K_{free}^*	Real and complex valued pile free head stiffness, respectively
$K_\mu()$	Modified Bessel function of the second kind of order μ
$[K]_j, [K_G]$	Stiffness matrix for pile j and the whole group, respectively
$k(z), k_0, k_m$	Winkler spring stiffness at depth z, the ground surface, and in soil layer m, respectively
k_{sl}	Increase in Winkler spring stiffness per unit depth
$k^*(z)$	Complex Winkler spring stiffness at depth z
L	Pile length
L_a	Pile active length
$\{L\}_j$	Load vector at the head of pile j
$M, M(t)$	Static and dynamic pile head moment, respectively
$M(z)$	Bending moment in the pile at depth z
M_{fix}	Moment required to fix the pile head
M_j, M_G	Moment at the head of pile j and the pile cap, respectively
M_{max}	Maximum bending moment in the pile
N	Number of piles in the group
n_{ij}	Pile-soil stiffness ratio exponent for the corresponding pile head stiff matrix term
n_L	Active length pile-soil stiffness ratio exponent
$P, P(t)$	Static and dynamic pile head load, respectively
P_j, P_G	Horizontal load at the head of pile j and the pile cap, respectively
$P(z)$	Shear force in the pile at depth z
q, q^*	Complex dimensionless frequency
R_D	Deflection ratio
s	Distance between two piles
s, s^*, s_m^*	Complex dimensionless frequency
t	Time
$[T]$	Geometric transformation matrix
$\{u\}_i$	Displacement vector of pile i
V_j, V_G	Vertical load on pile j and the pile cap, respectively
V_c	Soil compression wave velocity
V_{La}	Lysmer's analog wave velocity
$V_s(z) V_s^*$	Real and complex soil shear wave velocity, respectively
$V_{s,20D}$	Soil shear wave velocity at 20 pile diameters depth

w_i, w_G	Vertical displacement of pile i and the pile cap, respectively
$y(z)$, $y^*(z)$	Real and complex valued pile displacement, respectively
y, y_G	Horizontal displacement of pile i and the pile cap, respectively
$y_v(z)$	Virtual pile displacement
y_0	Pile head displacement
$y_{ij}(z)$	Displacement of pile i at depth z due to a load on pile j
x_j	Distance (along axis of loading) of pile j from the pile group coordinate origin
z	Depth below ground surface
z_m	Depth below ground surface of soil layer m
z_{Mmax}	Depth below ground surface of the maximum bending moment in the pile

Calligraphic

$\Im\,[]$	Imaginary component
$\Re\,[]$	Real component

Greek

α	Interaction factor matrix
α_c	Winkler spring stiffness parameter
α_{ij}, α_{ij}^*	Real and complex valued interaction factor, respectively, for degree of freedom i in the receiver pile due to load along degree of freedom j in the source pile
α_v	Vertical interaction factor
$[\alpha]_{ij}$	Interaction factor matrix between pile i and pile j
$[\alpha F]$	Overall flexibility matrix for the whole pile group
β_r	Radiation damping factor
β_p	Pile hysteretic damping factor
β_s	Soil hysteretic damping factor
Δ_{ij}	Response (translation or rotation) atop pile i, due to a load on pile j
γ	Euler's number
$\gamma(z)$	Soil shear strain at depth z
$[\zeta]$	Diffraction factor matrix
ζ_{ij}, ζ_{ij}^*	Real and complex valued diffraction factor, respectively, for degree of freedom i in the receiver pile due to load along degree of freedom j in the source pile
η, η_s, η_u	Dimensionless soil compressibility coefficients
$\theta(z)$	Pile rotation at depth z
θ_i, θ_G	Rotation of pile i and the pile cap, respectively
θ_0	Pile head rotation
λ, λ_m	Load transfer parameter
λ^*	Complex load transfer parameter
$\lambda_{massless}^*$	Complex load transfer parameter assuming a massless pile
μ, μ^*	Real and complex valued load transfer parameter

v_p	Pile Poisson's ratio
v_s	Soil Poisson's ratio
ρ_p	Pile mass density
ρ_s	Soil mass density
ϕ	Aperture angle between the direction of loading and the line connecting pile centres
$\chi_i(z), \chi_j(z)$	Real valued shape function
$\chi_i^*(z), \chi_j^*(z)$	Complex valued shape function
χ_L	Active length multiplier
$\psi(s,\phi)$	Shaft displacement attenuation function
ω	Applied load angular frequency
ω_c	Cut-off soil layer frequency

REFERENCES

Achenbach, J.D. (2004) *Reciprocity in Elastodynamics*, Cambridge University Press, Cambridge, UK.

Agapaki, E., Karatzia, X. & Mylonakis, G. (2018) Higher-order Winkler solutions for laterally loaded piles. In: *Proceedings of the 16th European Conference on Earthquake Engineering*, 18–21 June 2018, Thessalonki, Greece.

Angelides, D.C. & Roesset, J.M. (1981) Nonlinear lateral dynamic stiffness of piles. *Journal of the Geotechnical Engineering Division, ASCE*, 107(11), 1443–1460.

Anoyatis, G. (2013) Contribution to kinematic and inertial analysis of piles by analytical and experimental methods. Ph.D. thesis, University of Patras, Patras, Greece.

Anoyatis, G., Di Laora, R., Mandolini, A. & Mylonakis, G. (2013) Kinematic response of single piles for different boundary conditions: Analytical solutions and normalization schemes. *Soil Dynamics & Earthquake Engineering*, 44, 183–195.

Anoyatis, G. & Lemnitzer, A. (2017) Dynamic pile impedances for laterally–loaded piles using improved Tajimi and Winkler formulations. *Soil Dynamics and Earthquake Engineering*, 92, 279–297.

Anoyatis, G., Mylonakis, G. & Lemnitzer, A. (2016) Soil reaction to lateral harmonic pile motion. *Soil Dynamics and Earthquake Engineering*, 87, 164–170.

API (1982) *Recommended Practice for Planning, Designing, and Constructing Fixed Offshore Platforms*, 13th Edition, American Petroleum Institute, Washington, D.C.

Badoni, D. & Makris, N. (1996). Nonlinear response of single piles under lateral inertial and seismic loads. *Soil Dynamics and Earthquake Engineering*, 15(1), 29–43.

Baguelin, F., Frank, R. & Saïd, Y.H. (1977) Theoretical study of lateral reaction mechanism of piles. *Géotechnique*, 27(3), 405–434.

Banerjee, P.K. (1994) *The Boundary Element Methods in Engineering*, McGraw Hill, London, UK.

Banerjee, P.K. & Davies, T.G. (1978) The behaviour of axially and laterally loaded single piles embedded in nonhomogeneous soils. *Géotechnique*, 28(3), 43–60.

Baranov, V.A. (1967) On the calculation of excited vibrations of an embedded foundation (in Russian). *Voprosy Dinamiki i Prochnoeti*, (14), Polytechnic Institute, Riga, Latvia, 195–209.

Barber, E.S. (1953) Discussion to paper by S.M. Gleser. In: Symposium on Lateral Load Tests on Piles, Presented at the Fifty-*sixth Annual Meeting*, ASTM, Atlantic City, NJ, 1 July 1953. ASTM STP154, 96–99.

Barghouthi, A.F. (1984) Pile response to seismic waves. Ph.D. thesis, University of Wisconsin, Madison, WI.

Basu, D. & Salgado, R. (2008) Analysis of laterally loaded piles with rectangular cross sections embedded in layered soil. *International Journal for Numerical and Analytical Methods in Geomechanics*, 32(7), 721–744.

Basu, D., Salgado, R. & Prezzi, M. (2009) A continuum-based model for analysis of laterally loaded piles in layered soils. *Géotechnique*, 59(2), 127–140.

Berger, E., Mahi, S.A. & Pyke, R. (1977) Simplified methods for evaluating soil-pile-structure interaction effects. In: *Proceedings of the 9th Offshore Technology Conference*, Houston, TX, 2–5 May 1977, Offshore Technology Conference, Houston, TX, USA, OTC paper 2954, 589–598.

Bhattacharya, S. (2019) *Design of Foundations for Offshore Wind Turbines*, Wiley, Hoboken, NJ.

Bhattacharya, S. & Adhikari, S. (2011) Experimental validation of soil–structure interaction of offshore wind turbines. *Soil Dynamics and Earthquake Engineering*, 31(5), 805–816.

Biot, M.A. (1937) Bending of an Infinite Beam on an Elastic Foundation. *Journal of Applied Mechanics ASME*, 4(1), A1–A7.

Blaney, G.W., Kausel, E. & Roesett, J.M. (1976) Dynamic stiffness of piles. In: *Proceedings of the 2nd International Conference on Numerical Methods in Geomechanics*, Blacksburg, VA, 2, 1010–1012.

Boulanger, R.W., Curras, C.J., Kutter, B.L., Wilson, D.W. & Abghari, A. (1999) Seismic soil-pile-structure interaction experiments and analyses. *Journal of Geotechnical and Geoenvironmental Engineering, ASCE*, 125(9), 750–759.

Broms, B.B. (1964a) Lateral resistance of piles in cohesive soils. *Journal of the Soil Mechanics and Foundations Division, ASCE*, 90(2), 27–63.

Broms, B.B. (1964b) Lateral resistance of piles in cohesionless soils. *Journal of the Soil Mechanics and Foundations Division, ASCE*, 90(3), 123–156.

Budhu, M. & Davis, T.G. (1987) Nonlinear analysis of laterally loaded piles in cohesionless soils. *Canadian Geotechnical Journal*, 24(2), 289–296.

Budhu, M. & Davis, T.G. (1988) Analysis of laterally loaded piles in soft clays. *Journal of Geotechnical Engineering, ASCE*, 114(1), 21–39.

Byrne, B.W., Houlsby, G.T., Burd, H.J., Gavin, K.G., Igoe, D.J.P., Jardine, R.J., Martin, C.M., McAdam, R.A., Potts, D.M., Taborda, D.M.G. & Zdravković, L. (2019) PISA design model for monopiles for offshore wind turbines: Application to a stiff glacial clay till. *Géotechnique*, 70(11), 1030–1047.

Cairo, R., Dente, G. & Troncone, A. (1999) Cone model for a pile foundation embedded in a soil layer. In: Brebbia, C.A. & Oliveto, G. (eds) *Earthquake Resistant Structures II*, Transactions on the Built Environment Vol. 41, WIT Press, Southampton, UK, 565–574.

Chidichimo, A., Cairo, R., Dente, G., Taylor, C.A. & Mylonakis, G. (2014) 1-g Experimental investigation of bi-layer response and kinematic pile bending. *Soil Dynamics and Earthquake Geotechnics*, 67, 219–232.

Darendeli, M.B. (2001) Development of a new family of normalized modulus reduction and material damping curves. Ph.D. thesis, University of Texas at Austin, Austin, TX.

Davies, T.G. & Budhu, M. (1986) Non-linear analysis of laterally loaded piles in heavily over-consolidated clays. *Géotechnique*, 36(4), 527–538.

Dezi, F., Carbonari, S. & Leoni, G. (2009) A model for the 3D kinematic interaction analysis of pile groups in layered soils. *Earthquake Engineering & Structural Dynamics*, 38(11), 1281–1305.

Dezi, F., Carbonari, S. & Morici, M. (2016) A numerical model for the dynamic analysis of inclined pile groups. *Earthquake Engineering and Structural Dynamics*, 45(1), 45-68.

Di Laora, R., Mandolini, A. & Mylonakis, G. (2012) Insight on kinematic bending of flexible piles in layered soil. *Soil Dynamics and Earthquake Engineering*, 43, 309–322.

Di Laora, R. & Rovithis, E. (2015) Kinematic bending of fixed-head piles in nonhomogeneous soil. *Journal of Geotechnical and Geoenvironmental Engineering, ASCE*, 141(4), 04014126.

Dobry, R. & Gazetas, G. (1988) Simple method for dynamic stiffness and damping of floating pile groups. *Géotechnique*, 38(4), 557–574.

Dobry, R. & O'Rourke, M.J. (1983) Seismic response of end-bearing piles. *Journal of Geotechnical Engineering Division, ASCE*, 109(5), 778–781.

Dobry, R., Vicente, E., O'Rourke, M.J. & Roesset, J.M. (1982) Horizontal stiffness and damping of single piles. *Journal of Geotechnical Engineering, ASCE*, 108(3), 439–459.

Durante, M.G., Di Sarno, L., Mylonakis, G., Taylor, C.A. & Simonelli, A.L. (2016) Soil-pile-structure interaction: Experimental outcomes from shaking table tests. *Earthquake Engineering and Structural Dynamics*, 45(70), 1041–1061.

El-Marsafawi, H., Han, Y.C. & Novak, M. (1992a) Dynamic experiments on two pile groups. *Journal of Geotechnical Engineering, ASCE*, 118(12), 576–592.

El-Marsafawi, H., Kaynia, A.M. & Novak, M. (1992b) Interaction factors and the superposition method for pile group dynamic analysis. Geotechnical Research Center, Department of Civil Engineering, University of Western Ontario, Canada, Report number GEOT-1-92.

El Naggar, M.H. & Novak, M. (1995) Nonlinear lateral interaction in pile dynamics. *Soil Dynamics and Earthquake Engineering*, 14(2), 141–157.

El Naggar, M.H. & Novak, M. (1996) Nonlinear analysis for dynamic lateral pile response. *Soil Dynamics and Earthquake Engineering*, 15(4), 233–244.

El-Sharnouby, B. & Novak, M. (1985) Flexibility coefficients and interaction factors for pile group analysis. *Canadian Geotechnical Journal*, 23(4), 441–450.

Fan, K., Gazetas, G., Kaynia, A. & Kausel, E. (1991) Kinematic seismic response of single piles and pile groups. *Journal of Geotechnical Engineering, ASCE*, 117(12), 1860–1879.

Finn, W.D.L. (2005) A study of piles during earthquakes: Issues of design and analysis. *Bulletin of Earthquake Engineering*, 3(2), 141–234.

Fiorentino, G., Cengiz, C., De Luca, F., Mylonakis, G., Karamitros, D., Dietz, M., Dihorou, L., Lavorato, D., Briseghella, B., Isakovic, T., Vrettos, C., Topa Gomes, A., Sextos, A. & Nuti, C. (2021) Integral abutment bridges: Investigation of seismic soil-structure interaction effects by shaking table testing. *Earthquake Engineering and Structural Dynamics*, doi:10.1002/eqe.3409 (in press).

Fleming, K., Weltman A., Randolph, M. & Elson, K. (2009). *Piling Engineering*, 3rd edition, Taylor & Francis, Abingdon, UK.

Flores-Berrones, R. & Whitman, R.V. (1982) Seismic response of end-bearing piles. *Journal of Geotechnical Engineering ASCE*, 108(4), 554–569.

Francis, A.J. (1964) Analysis of pile groups with flexural resistance. *Journal of Soil Mechanics and Foundations Division, ASCE*, 90(3), 1–32.

Franklin, J.N. & Scott, R.F. (1979) Beam equation with variable foundation coefficient. *Journal of the Engineering Mechanics Division, ASCE*, 105(5), 811–827.

Gazetas, G. (1984) Seismic response of end-bearing single piles. *International Journal of Soil Dynamics and Earthquake Engineering*, 3(2), 82–93.

Gazetas, G. (1987) Simple physical models for foundation impedances. In: Banerjee, P.K. & Butterfield, R. (eds.) *Dynamic Behaviour of Foundations and Buried Structures (Developments in Soil Mechanics and Foundation Engineering Vol. 3)*, Elsevier Applied Science Publishers Ltd., London, UK, 45–93.

Gazetas, G. (1991) Foundation vibrations. In: Fang, H-Y. (ed.) *Foundation Engineering Handbook, 2nd Edition*, Van Nostrand Reinhold, New York, NY, USA, 553–593.

Gazetas, G. & Dobry, R. (1984a) Horizontal response of piles in layered soil. *Journal of Geotechnical Engineering, ASCE*, 110(1), 20–40.

Gazetas, G. & Dobry, R. (1984b) Simple radiation damping model for piles and footings. *Journal of Geotechnical Engineering, ASCE*, 110(6), 937–956.

Gazetas, G., Fan, K. & Kaynia, A. (1993) Dynamic response of pile groups with different configurations. *Soil Dynamics and Earthquake Engineering*, 12(4), 239–257.

Gazetas, G., Fan, K., Kaynia, A. & Kausel, E. (1991) Dynamic interaction factors for floating pile groups. *Journal of Geotechnical Engineering, ASCE*, 117(10), 1531–1548.

Gazetas, G. & Mylonakis, G. (1998) Seismic soil-structure interaction: New evidence and emerging issues state of the art paper. In: Dakoulas, P., Yegian, M. & Holtz, R.D. (eds.) *Geotechnical Earthquake Engineering and Soil Dynamics III*, 3–6 August 1998, Seattle, WA. ASCE Geotechnical Special Publication, GSP75, Vol. 2, 1119–1174.

Gerber, T.M. & Rollins, K.M. (2008) Cyclic p-y curves for a pile in cohesive soil. In: Zeng, D., Manzari, M.T. & Hiltunen, D.R. (eds.) *Geotechnical Earthquake Engineering and Soil Dynamics IV, Proceedings of the Conference*, 18–22 May 2008, Sacramento, CA. ASCE Geotechnical Special Publication, GSP181.

Gerolymos, N. & Gazetas, G. (2005) Phenomenological model applied to inelastic response of soil-pile interaction systems. *Soils and Foundations*, 45(4), 119–132.

Gohl, W.B. (1991) Response of pile foundations to simulated earthquake loading: Experimental and analytical results. Ph.D. thesis, University of British Columbia, Vancouver, Canada.

Goit, C.S., Saitoh, M. & Mylonakis, G. (2016) Principle of superposition for assessing horizontal dynamic response of pile groups encompassing soil nonlinearity. *Soil Dynamics and Earthquake Engineering*, 82, 73–83.

Guin, J. (1997) Advances in soil-pile-structure interaction and nonlinear pile behaviour. Ph.D. thesis, State University of New York at Buffalo, Buffalo, NY.

Guo, W.D. (2012) *Theory and Practice of Pile Foundations*, CRC Press, Boca Raton, FL.

Guo, W.D. & Lee, F.H. (2001) Load transfer approach for laterally loaded piles. *Numerical and Analytical Methods in Geomechanics*, 25(11), 1101–1129.

Hadjian, A.H., Fallgren, R.B. & Tufenkjian, M.R. (1992) Dynamic soil-pile-structure interaction - The state of practice. In: Prakash, S. (ed.) *Piles Under Dynamic Loads*, 13–17 September 1992, New York, NY. ASCE Geotechnical Special Publication, GSP34, 1–27.

Haiderali, A.E. & Madabhushi, G. (2016) Evaluation of curve fitting techniques in deriving p-y curves for laterally loaded piles. *Geotechnical and Geological Engineering*, 34(5), 1453–1473.

Hetenyi, M. (1946) *Beams on Elastic Foundation*, The University of Michigan Press, Ann Arbor, MI.

Ishihara, K. (1996) *Soil Behaviour in Earthquake Geotechnics*, Clarendon Press, Oxford, UK.

Jia, J. (2018) *Soil Dynamics & Foundation Modelling: Offshore and Earthquake Engineering*, Springer, Cham, Switzerland.

Kagawa, T. & Kraft Jr., L.M. (1981) Lateral pile response during earthquakes. *Journal of the Geotechnical Engineering Division, ASCE*, 107(12), 1713–1731.

Karatzia, X. & Mylonakis, G. (2012) Horizontal response of piles in inhomogeneous soil: Simple analysis. In: Soccodato, C. & Maugeri, M. (eds.) *Proceedings of the 2nd International Conference on Performace-based Design in Earthquake Geotechnical Engineering*, Taormina, Italy, 28–30 May 2012, No. 1117.

Karatzia, X. & Mylonakis, G. (2016) Discussion of "Kinematic bending of fixed-head piles in nonhomogeneous soil" by Raffaele Di Laora and Emmanouil Rovithis. *Journal of Geotechnical and Geoenvironmental Engineering, ASCE*, 142(2), 07015042.

Karatzia, X. & Mylonakis, G. (2017) Horizontal stiffness and damping of piles in inhomogeneous soil. *Journal of Geotechnical and Geoenvironmental Engineering, ASCE*, 143(4), 04016113.

Karatzia, X., Papastylianou, P. & Mylonakis, G. (2014) Horizontal soil reaction of a cylindrical pile segment with a soft zone. *Journal of Engineering Mechanics, ASCE*, 140(10), 04014077.

Karatzia, X., Papastylianou, P. & Mylonakis, G. (2015) Erratum for "Horizontal soil reaction of a cylindrical pile segment with a soft zone" by Xenia Karatzia, Panos Papastylianou and George Mylonakis. *Journal of Engineering Mechanics, ASCE*, 141(11), 08215005.

Kaynia, A.M. (1980) Dynamic stiffness and seismic response of sleeved piles. Department of Civil Engineering, Massachusetts Institute of Technology, Cambridge, MA, Research report no. R80-12.

Kaynia, A.M. (1982) Dynamic stiffness and seismic response of pile groups. Ph.D. thesis, Massachusetts Institute of Technology, Cambridge, MA.

Kaynia, A.M. & Kausel, E. (1982) Dynamic stiffness and seismic response of pile groups. Department of Civil Engineering, Massachusetts Institute of Technology, Cambridge, MA, Research report no. R82-03.

Kaynia, A.M. & Kausel, E. (1991) Dynamics of piles and pile groups in layered soil media. *Soil Dynamics and Earthquake Engineering*, 10(8), 386–401.

Kaynia, A.M. & Mahzooni, S. (1996) Forces in pile foundations under seismic loading. *Journal of Engineering Mechanics, ASCE*, 122(1), 46–53.

Kuhlemeyer, R.L. (1979) Static and dynamic laterally loaded floating piles. *Journal of the Geotechnical Engineering Division, ASCE*, 105(2), 289–304

Lam, I.P. & Martin, G. (1986) Seismic design for highway bridge foundations, US Department of Transportation, Federal Highway Administration Report Nos. FHWA/RD-86/101, FHWA/RD-86/102 and FHWA/RD-86/103.

Madabhushi, G., Knappett, J. & Haigh, S. (2010) *Design of Pile Foundations in Liquefiable Soils*, Imperial College Press, London, UK.

Maheshwari, B.K., Truman, K.Z., El Naggar, M.H. & Gould, P.L. (2004) Three-dimensional nonlinear analysis for seismic soil–pile-structure interaction. *Soil Dynamics and Earthquake Engineering*, 24(4), 343–356.

Makris, N., Badoni, D., Delis, E. & Gazetas, G. (1994) Prediction of observed bridge response with soil-pile-structure interaction. *Journal of Structural Engineering, ASCE*, 120(10), 2992–3011.

Makris, N. & Gazetas, G. (1992) Dynamic pile-soil-pile interaction. Part 2: Lateral and seismic response. *Earthquake Engineering and Structural Dynamics*, 21(2), 145–162.

Makris, N. & Gazetas, G. (1993) Displacement phase differences in a harmonically oscillating pile. *Géotechnique*, 43(1), 135–150.

Mamoon, S.M., Kaynia, A.M. & Banerjee, P.K. (1990) Frequency domain dynamic analysis of piles and pile groups. *Journal of Engineering Mechanics, ASCE*, 116(10), 2237–2257.

Maravas, A., Mylonakis, G. & Karabalis, D. (2007) Dynamic characteristics of structures on piles and footings. In: *4th International Conference on Earthquake Geotechnical Engineering*, 25–28 July 2007, Thessaloniki, Greece. Paper No. 1672.

Maravas, A., Mylonakis, G. & Karabalis, D. (2014) Simplified discrete systems for dynamic analysis of structures on footings and piles. *Soil Dynamics & Earthquake Engineering*, 61–62, 29–39.

Matlock, H. (1970) Correlations for design of laterally loaded piles in soft clay. In: *Proceedings of the 2nd Offshore Technology Conference*, Houston, 22–24 April 1970, Offshore Technology Conference, Houston, TX, OTC paper 1204, 577–588.

Matlock, H. & Reese, L.C. (1960) Generalized solutions for laterally loaded piles. *Journal of the Soil Mechanics and Foundations Division, ASCE*, 86(5), 63–91.

Matsuo, H. & Ohara, S. (1960) Lateral earth pressure and stability of quay walls during earthquakes. In: *Proceeding of the 2nd World Conference on Earthquake Engineering*, 11–18 July 1960, Tokyo, Japan. Vol. 1, 165–181.

McClelland, B. & Focht Jr., J.A. (1956) Soil modulus for laterally loaded piles. *Journal of the Soil Mechanics and Foundations Division, ASCE*, 82(4), paper 1081-1-22.

Meymand, P.J. (1998) Shaking table scale model tests of nonlinear soil-pile superstructure interaction in soft clay. Ph.D. thesis, University of California, Berkeley, CA.

Mylonakis, G. (1995) Contributions to static and seismic analysis of piles and pile-supported bridge piers. Ph.D. thesis, State University of New York at Buffalo, Buffalo, NY.

Mylonakis, G. (2000) A new elastodynamic model for settlement analysis of a single pile in multi-layered soil, Discussion to Paper by K.M. Lee and Z.R. Xiao. *Soils and Foundations*, 40(4), 163–166.

Mylonakis, G. (2001) Elastodynamic model for large-diameter end-bearing shafts. *Soils and Foundations*, 41(3), 31–44.

Mylonakis, G. & Gazetas, G. (1999) Lateral vibration and internal forces of grouped piles in layered soil. *Journal of Geotechnical and Geoenvironmental Engineering, ASCE*, 125(1), 16–25.

Mylonakis, G., Papastamatiou, D., Psycharis, J. & Mahmoud, K. (2001) Simplified modelling of bridge response on soft soil to non-uniform seismic excitation. *Journal of Bridge Engineering, ASCE*, 6(6), 587–597.

Mylonakis, G. & Roumbas, D. (2001) Lateral impedance of single piles in inhomogeneous soil. In: *Proceedings of the 4th International Conference on Recent Advances in Geotechnical Earthquake Engineering and Soil Dynamics*, San Diego, CA, 26–31 March 2001, paper 6.27.

Mylonakis, G., Syngros, C., Gazetas, G. & Tazoh, T. (2006) The role of soil in the collapse of 18 piers of Hanshin Expressway in the Kobe earthquake. *Earthquake Engineering and Structural Dynamics*, 35(5), 547–575.

NEHRP (2012) Soil-structure interaction for building structures. National Institute of Standards and Technology (NIST). Report number: GCR 12-917-21.

Nikolaou, S., Mylonakis, G., Gazetas, G. & Tazoh, T. (2001) Kinematic pile bending during earthquakes: Analysis and field measurements. *Géotechnique*, 51(5), 425–440.

Nogami, T. (1985) Flexural responses of grouped piles under dynamic loading. *Earthquake Engineering and Structural Dynamics*, 13(3), 321–336.

Nogami, T. & Novak, M. (1977) Resistance of soil to a horizontally vibrating pile. *Earthquake Engineering and Structural Dynamics*, 5(3), 249–261.

Nogami, T., Otani, J., Konagai, K. & Chen, H.L. (1992) Nonlinear soil-pile interaction model for dynamic lateral motion. *Journal of Geotechnical Engineering, ASCE*, 118(1), 89–106.

Novak, M. (1974) Dynamic stiffness and damping of piles. *Canadian Geotechnical Journal*, 11(4), 574–598.

Novak, M. (1991) Piles under dynamic loads. In: *Proceedings of the 2nd International Conference on Recent Advances in Geotechnical Earthquake Engineering and Soil Dynamics*, 11–15 March, St. Louis, MO, (2), 2433–2456.

Novak, M., Nogami, T. & Aboul-Ella, F. (1978). Dynamic soil reactions for plane strain case. *Journal of Engineering Mechanics, ASCE*, 104(4), 953–959.

Novak, M. & Sheta, M. (1980) Approximate approach to contact effects of piles. In: O'Neil, M. & Dobry, R. (eds.) *Dynamic Response of Pile Foundations*, ASCE, New York, NY, USA, 53–79.

Novak, M., Sheta, M., El-Hifnawy, L., El-Marsafawi, H. & Ramadan, O. (1993) DYNA4 – A computer program for foundation response to dynamic loads. In: *User's Manual*, Geotechnical Research Centre, University of Western Ontario, London, Canada.

Ohira, A., Tazoh, T., Dewa, K., Shimuzu, K. & Shimada, M. (1984) Observations of earthquake response behaviors of foundation piles for road bridge. In: *Proceeding of the 8th World Conference on Earthquake Engineering*, 21–28 July 1984, San Francisco, CA, Prentice Hall, Vol. 3, 577–584.

O'Neill, M.W., Ghazzaly, O.I. & Ha, H.B. (1977) Analysis of three-dimensional pile groups with nonlinear soil response and pile-soil-pile interaction. In: *Proceedings of the 9th Offshore Technology Conference*, Houston, TX, 2–5 May 1977, Offshore Technology Conference, Houston, TX, OTC paper 2838, 245–256.

O'Rourke, M.J. & Dobry, R. (1978) *Spring and Dash Pot Coefficients for Machine Foundation on Piles*, American Concrete Institute, Detroit, MI. Report number SP-10, 177–198.

Padrón, L.A., Aznárez, J.J. & Maeso, O. (2007a) BEM–FEM coupling model for the dynamic analysis of piles and pile groups. *Engineering Analysis with Boundary Elements*, 31(6), 472–484.

Padrón, L.A., Aznárez, J.J. & Maeso, O. (2007b) Dynamic analysis of piled foundations in stratified soils by a BEM–FEM model. *Soil Dynamics and Earthquake Engineering*, 28(5), 333–346.

Padrón, L.A., Mylonakis, G. & Beskos, D.E. (2009) Importance of footing-soil separation on dynamic stiffness of piled embedded footings. *International Journal of Numerical and Analytical Methods in Geomechanics*, 33(11), 1439–1448.

Padrón, L.A., Mylonakis, G. & Beskos, D.E. (2012) Simple superposition approach for dynamic analysis of piled embedded footings. *International Journal of Numerical and Analytical Methods in Geomechanics*, 36(12), 1523–1534.

Papadopoulou, M.C. & Comodromos, E.M. (2014) Explicit extension of the p-y method to pile groups in sandy soil. *Acta Geotechnica*, 9(3), 485–497.

Parmelee, R.A., Penzien, J., Scheffey, C.F., Seed, H.B. & Thiers, G.R. (1964) *Seismic Effects of Structures Supported on Piles Extending through Deep Sensitive Clays*, Institute of Engineering Research, University of California Berkeley, Berkeley, CA. Report No. 64-2.

Pender, M. (1993) Aseismic pile foundation design analysis. *Bulletin of the New Zealand National Society for Earthquake Engineering*, 26(1), 49–160.

Penzien, J. (1970) Soil-pile foundation interaction. In: Wiegel, R.L. (ed.) *Earthquake Engineering*, Prentice-Hall, Englewood Cliffs, NJ, 349–381.

Penzien, J., Scheffy, C.F. & Parmelee, R.A. (1964) Seismic analysis of bridges on long piles. *Journal of the Engineering Mechanics Division, ASCE*, 90(3), 223–254.

Pipkin, A.C. (1972) *Lectures on Viscoelastic Theory*, Springer, New York, NY.

Poulos, H.G. (1971a) Behavior of laterally loaded piles: I – single piles. *Journal of the Soil Mechanics and Foundations Division, ASCE*, 97(5), 711–731.

Poulos, H.G. (1971b) Behavior of laterally loaded piles: II – pile groups. *Journal of the Soil Mechanics and Foundations Division, ASCE*, 97(5), 733–751.

Poulos, H.G. (2017) *Tall Building Foundation Design*, CRC Press, Boca Raton, FL.

Poulos, H.G. & Davis, E.H. (1980) *Pile Foundation Analysis and Design*, John Wiley & Sons, Inc., New York, NY.

Prakash, S. & Sharma, H.D. (1990) *Pile Foundations in Engineering Practice*, John Wiley & Sons, Inc., New York, NY.

Radhima, J., Kanellopoulos, K. & Gazetas, G. (2021) Static and dynamic lateral non-linear pile-soil-pile interaction. Géotechnique, Ahead of print, https://doi.org/10.1680/jgeot.20.P.250.

Rajapakse, R.K.N.D. & Shah, A.H. (1989) Impedance curves for an elastic pile. *Soil Dynamics and Earthquake Engineering*, 8(3), 145–152.

Randolph, M.F. (1977) A theoretical study of the performance of piles. Ph.D. thesis, University of Cambridge, Cambridge, UK.

Randolph, M.F. (1981) The response of flexible piles to lateral loading. *Géotechnique*, 31(2), 247–259.

Reese, L.C. (1986) Behavior of piles and pile groups under lateral load. Federal Highway Administration. Report FHWA/RD–85/106.

Reese, L.C., Cox, W.R. & Koop, F.D. (1974) Analysis of laterally loaded piles in sand. In: *Proceedings – 5th Annual Offshore Technology Conference*, Houston, TX, 6–8 May 1974, Offshore Technology Conference, Houston, TX, OTC paper 2080, 473–485.

Reese, L.C., Cox, W.R. & Koop, F.D. (1975) Field testing and analysis of laterally loaded piles in stiff clay. In: *Proceedings – 6th Annual Offshore Technology Conference*, Houston, TX, 5–8 May 1975, Offshore Technology Conference, Houston, TX, OTC paper 2312, 473–485.

Reese, L.C., Isenhower, W.M. & Wang, S-T. (2005) *Analysis and Design of Shallow and Deep Foundations*, John Wiley & Sons, Inc., Hoboken, NJ.

Reese, L.C. & Van Impe, W.F. (2011) *Single Piles and Pile Groups Under Lateral Loading*, 2nd Edition, CRC Press, Boca Raton, FL.

Roesset, J.M. (1980) Stiffness and damping coefficients of foundations. In: O'Neil, M. & Dobry, R. (eds.) *Dynamic Response of Pile Foundations*, ASCE, New York, NY, USA, 1–29.

Roesset, J.M. & Angelides, D. (1980) Dynamic stiffness of piles. In: Smith, I.M., George, P. & Rigden, W.J. (eds.) *Numerical Methods in Offshore Piling*, Institution of Civil Engineers, London, UK, 75–81.

Rollins, K.M., Peterson, K.T. & Weaver, T.J. (1998) Lateral load behavior of full-scale pile group in clay. *Journal of Geotechnical and Geoenvironmental Engineering, ASCE*, 124(6), 468–478.

Rovithis, E., Mylonakis, G. & Pitilakis, K. (2013) Dynamic stiffness and kinematic response of single piles in inhomogeneous soil. *Bulletin of Earthquake Engineering*, 11(6), 1949–1972.

Saitoh, M. (2005) Fixed-head pile bending by kinematic interaction and criteria for its minimization at optimal pile radius. *Journal of Geotechnical and Geoenvironmental Engineering, ASCE*, 131(10), 1243–1251.

Saitoh, M. & Watanabe, H. (2004) Effects of flexibility on rocking impedance of deeply embedded foundation. *Journal of Geotechnical and Geoenvironmental Engineering, ASCE*, 130(4), 438–445.

Salgado, R. (2008). *The Engineering of Foundations*, International Edition, McGraw-Hill, New York, NY.

Sanchez-Salinero, I. (1982). Static and dynamic stiffness of single piles. Geotechnical Engineering Report, University of Texas, Austin, TX. Report number: GR82-31.

Sanchez-Salinero, I. (1983). Dynamic stiffness of pile groups: Approximate solutions. Geotechnical Engineering Report, University of Texas, Austin, TX, USA. Report number: GR83-5.

Scott, R.F. (1981) *Foundation Analysis*, Prentice Hall, Englewood Cliffs, NJ.

Scott, R.F., Tsai, C.F., Steussy, D. & Ting, J.M. (1982) Full-scale dynamic lateral pile tests. In: *Proceedings – 14th Annual Offshore Technology Conference*, Houston, TX, 3–6 May 1982, Offshore Technology Conference, Houston, TX, OTC paper 4203, 435–450.

Sextos, A.G., Mylonakis, G.E. & Mylona, E-K.V. (2015) Rotational excitation of bridges supported on pile groups in soft or liquefiable soil deposits. *Computers and Structures*, 155(15), 54–66.

Shadlou, M. & Bhattacharya, S. (2014) Dynamic stiffness of pile in a layered elastic continuum. *Géotechnique*, 64(4), 303–319.

Skempton, A.W. (1951) The bearing capacity of clays. In: *Proceedings of the Building Research Congress*, 11–20 September 1951, London, UK, Vol. 1, 180–189.

Syngros, K. (2004) Seismic response of piles and pile-supported bridge piers evaluated through case histories. Ph.D. thesis, The City College of the City University of New York, New York, NY.

Tabesh, A. & Poulos, H.G. (2000) A simple method for the seismic analysis of piles and its comparison with the results of centrifuge tests. In: *12 WCEE 2000: 12th World Conference on Earthquake Engineering*, 30 January–4 February 2000, Auckland, New Zealand. New Zealand Society for Earthquake Engineering, Upper Hutt, New Zealand, Vol. 5, 55–62.

Tabesh, A. & Poulos, H.G. (2001) Pseudostatic approach for seismic analysis of single piles. *Journal of Geotechnical and Geoenvironmental Engineering, ASCE*, 127(9), 757–765.

Tajimi, H. (1966) Earthquake response of foundation structures (in Japanese). Faculty of Science and Engineering, Nihon University Tokyo, Tokyo, Japan, 1.1–3.5.

Tajimi, H. (1969) Dynamic analysis of a structure embedded in an elastic stratum. In: *Proceeding of the 4th World Conference on Earthquake Engineering*, 13–18 January 1969, Santiago, Chile, Vol. 3, A6–53–70.

Tajimi, H. (1977) Seismic effects on piles. In: *Proceedings of the 9th International Conference on Soil Mechanics and Foundation Engineering*, Tokyo, Japan, Vol. 3, 553.

Tazoh, T., Shimizu, K. & Wakahara, T. (1987) Seismic observations and analysis of grouped piles. In: Nogami, T. (ed.) *Dynamic Response of Pile Foundations: Experiment Analysis and Observation*, 27 April 1987, Atlantic City, NJ. ASCE Geotechnical Special Publication, GSP11, 1–20.

Ting, J.M. (1978) Full scale cyclic dynamic lateral pile response. *Journal of the Geotechnical Engineering Division, ASCE*, 113(1), 30–45.

Trochanis, A.M., Bielak, J. and Christiano, P. (1991) Three-dimensional nonlinear study of piles. *Journal of Geotechnical Engineering, ASCE*, 117(3), 429–447.

Tyson, T.R. & Kausel, E. (1983) Dynamic stiffness of axisymmetric pile groups. Department of Civil Engineering, Massachusetts Institute of Technology, Cambridge, MA, Research report no. R83-07.

Vardanega, P.J. & Bolton, M.D. (2011) Strength mobilization in clays and silts. *Canadian Geotechnical Journal*, 48(10), 1485–1503.

Varun, Assimaki, D. & Gazetas, G. (2009) A simplified model for lateral response of large diameter caisson foundations – linear elastic formulation. *Soil Dynamics and Earthquake Engineering*, 29(2), 268–291.

Velez, A., Gazetas, G. & Krishnan, R. (1983) Lateral dynamic response of constrained-head piles. *Journal of Geotechnical Engineering, ASCE*, 109(8), 1063–1081.

Vesić, A.B. (1961) Beams on elastic subgrade and the Winkler's hypothesis. In: *Proceedings of the 5th International Conference on Soil Mechanics and Foundation Engineering*, Paris 17–22 July 1961, Dunod, Paris, France, Vol. 3, 845–850.

Viggiani, C., Mandolini, A. & Russo, G. (2011) *Piles and Pile Foundations*, CRC Press, Boca Raton, FL.

Vlasov, V.Z. & Leontiev, N.N. (1966) Beams, plates and shells on elastic foundations. Translated from Russian, Israel program for scientific translations, Washington, DC. NIST No. N67-14238.

Vucetic, M. & Dobry, R. (1991) Effect of soil plasticity on cyclic response. *Journal of Geotechnical Engineering, ASCE*, 117(1), 89–107.

Waas, G. & Hartmann, H.G. (1984) Seismic analysis of pile foundations including pile-soil-pile interaction. In: *Proceeding of the 8th World Conference on Earthquake Engineering*, 21–28 July 1984, San Francisco, CA, Prentice Hall, Vol. 5, 55–62.

Weaver, T.J., Rollins, K.M. & Peterson, K.T. (1998) Lateral statnamic load testing and analysis of a pile group. In: Dakoulas, P., Yegian, M. & Holtz, R.D. (eds.) *Geotechnical Earthquake Engineering and Soil Dynamics III*, 3–6 August 1998, Seattle, WA. ASCE Geotechnical Special Publication, GSP75, Vol. 2, 1319–1330.

Wolf, J.P. (1994) *Foundation Vibration Analysis using Simple Physical Models*, Prentice-Hall Inc., Englewood Cliffs, NJ.

Wolf, J.P. & Von Arx, G.A. (1978) Impedance function of a group of vertical piles. In: *Proceedings of the Specialty Conference on Earthquake Engineering and Soil Dynamics*, Pasadena, CA, 19–21 June 1978, ASCE, New York, NY, Vol. 1, 1024–1041.

Wolf, J.P., Von Arx, G.A., De Barros, F.C.P. & Kakubo, M. (1981). Seismic analysis of the pile foundation of the reactor building of the NPP Angra 2. *Nuclear Engineering and Design*, 65(3), 329–341.

Yoshida, I. & Yoshinaka, R. (1972) A method to estimate modulus of horizontal subgrade reaction for a pile. *Soils and Foundations*, 12(3), 1–17.

Chapter 7

Inelastic stiffnesses of floating pile groups

George Gazetas, Joani Radhima, and Evangelia Garini
National Technical University of Athens

CONTENTS

7.1 INTRODUCTION: ELASTIC PILE-TO-PILE INTERACTION

The stiffnesses (axial and lateral) of a group of piles differ from the sum of the stiffnesses of each pile acting alone. This is due to the influence of one pile upon its neighbour, arising from the extent of the displacement field that each pile displacing (vertically or horizontally) generates. Figure 7.1 illustrates the consequences of such influence with a vertically and a horizontally loaded ("source") pile. The affected ("receiver") pile displaces by a fraction (α_V or α_H) the movement of the loaded ("source") pile. Named interaction factors, these fractions, introduced and theoretically computed by Poulos (1968, 1971), have been utilised to obtain the response of a group of piles by superposition of the effects of all pairs of piles (Poulos 1979, 1989; Poulos & Davis, 1980).

For the idealised case of piles perfectly adhering to the surrounding linear elastic homogeneous soil, the interaction factors α (α_V or α_H), for the vertical or horizontal loading, respectively) can be expressed in a dimensionless form as

$$\alpha = f\left(\frac{s}{d}, \frac{E_p}{E_s}, \frac{L}{d}\right) \tag{7.1}$$

in which s = pile spacing (axis to axis); $d = 2r_o$ = pile diameter; E_p, E_s = pile, soil Young's modulus; L = pile length. An example of such a function is shown for α_V in Figure 7.2, adapted from Poulos & Davis (1980). Notice that for the frequently used spacing ratio $s/d = 3$, the vertical interaction factor attains very substantial values, of at least 0.40, for the larger values of E_p/E_s, 500 and 1000. Note that for a given type of pile (and hence of a given E_p), the above large values of the E_p/E_s ratio imply that the

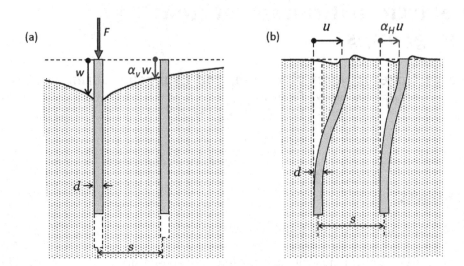

Figure 7.1 Definition of vertical and horizontal pile-to-pile interaction factors. The latter are for zero rotation at their top.

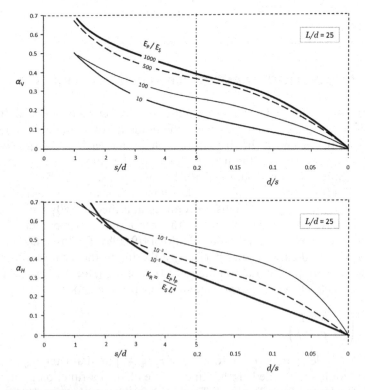

Figure 7.2 Examples of pile-to-pile interaction factors as functions of the spacing ratio, with parameters related to the Young's moduli of pile and soil and pile dimensions. (Adapted from Poulos & Davis (1980).)

soil is relatively soft — a usual case for adopting piles. Also to be noted is that for such soils, α_V is essentially a function of only s, d, and L, as can be judged from the small differences between the 500 and 1000 curves of Figure 7.2. Randolph & Wroth (1979) developed a simple analytical solution for the vertical interaction factor, assumed to be independent of E_p and E_s; for relatively long $(L \geq 15d)$ and stiff $\left(E_p > 500\ E_s\right)$ piles, ignoring the contribution to interaction of the pile-tip settlement, their solution can be expressed approximately (for $v \approx 0.5$) as

$$\alpha_V \approx \frac{\ln\left(1.25\,L/s\right)}{\ln\left(1.25\,L/r_o\right)} \tag{7.2a}$$

This expression is based on the (reasonable) assumption that the shear strain in the soil next to the pile shaft is constant along the length of the pile (see also the early publications of Frank (1974), Baguelin & Frank (1979), and of Randolph & Wroth (1978)). For $L/d = 20$ and $s/d = 3$, Eq. 7.1 gives $\alpha_V \approx 0.54$, slightly higher than the value extrapolated from Figure 7.2a. (Accounting for the pile-tip contribution would reduce this value, anyway.)

Such very high values of α_V are reduced if the soil is inhomogeneous. For Gibson soil (i.e., with shear modulus proportional to depth and $v = 0.5$), we simplify Randolph & Wroth (1979) to write

$$\alpha_V \approx \frac{\ln\left(0.62\,L/s\right)}{\ln\left(0.62\,L/r_o\right)} \tag{7.2b}$$

which gives the slightly reduced interaction factor $\alpha_V \approx 0.45$. A newer more informed approximate expression by Randolph & Poulos (1982) is

$$\alpha_V \approx \frac{0.5\,\ln\left(L/s\right)}{\ln\left(L/d\right)} \tag{7.2c}$$

which gives $\alpha_V \approx 0.32$, a value that implies an even stronger effect of inhomogeneity, among other factors.

For horizontally loaded piles, α_H depends on the angle of departure β of the receiver pile from the loading direction, in addition to the factors influencing the vertical interaction factor. Hence, we will simply write

$$\alpha_H = \alpha_H\left(\beta\right) \tag{7.3}$$

Of particular interest are $\alpha_H(0°)$, if the receiver pile lies along the loading direction in front of the source pile, plotted in Figure 7.2b; $\alpha_H(90°)$, if it lies in the perpendicular direction; and $\alpha_H(180°)$, if it lies along the loading direction but in the back of the source pile. (See Figure 7.3). Results from numerical solutions using the boundary-element or the finite-element method, as well as from Winkler-on-Elastic Foundation method, can be found in the literature (Banerjee & Sen, 1987; Nogami, 1985; Pender, 1993, 1996; Poulos, 1989; Makris & Gazetas, 1992, 1993).

Under dynamic harmonic loading, the interaction factors (α_V and α_H) are functions of frequency, ω, in addition to the parameters shown in Eq. 7.1. Kaynia & Kausel

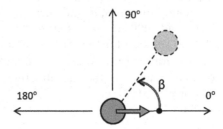

Figure 7.3 Definition of the angle of departure β from the direction of horizontal loading.

(1982) presented plots for these functions using a rigorous semi-analytical method. Early solutions to the dynamic pile-to-pile interaction problem have also been presented by Wolf & Von Arx (1978), Kagawa (1983), Sanchez-Salinero (1983), and Roesset (1984). A simplified analytical wave-propagation–based solution by Dobry & Gazetas (1988), which ignores the influence of pile length and of pile-to-soil stiffness ratio, gave

$$\alpha = \sqrt{\frac{r_0}{s}}\, e^{-\xi\omega s/V} \cdot e^{-i\omega s/V} \tag{7.4a}$$

in which $\alpha = \alpha_V$ or $\alpha_H(0°)$, and correspondingly

$$V = V_s \text{ or } V_{ce} \approx \frac{3.4 V_s}{\pi(1-v)} \tag{7.4b}$$

for the vertical and horizontal (in the loading direction) interaction factors. V_S is the shear wave velocity, and V_{ce} is the apparent compression-extension velocity (introduced by Ricardo Dobry; see Gazetas & Dobry (1984)).

Figure 7.4 plots the real and imaginary parts corresponding to stiffness and damping of α_V for a pile spacing $s = 5d$ and damping ratio $\xi = 0.02$. The important observation

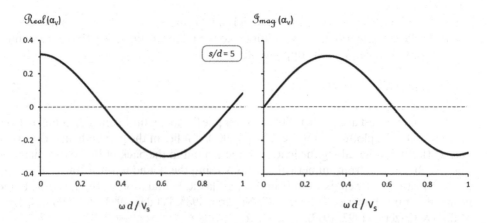

Figure 7.4 The real and imaginary components of the vertical dynamic interaction factor as functions of the dimensionless frequency factor, from Eq. 7.4, for $\beta = 0.02$.

is that $\text{Re}(\alpha_V)$, starting from the static ($\omega = 0$) value, becomes *negative* for a wide range of frequencies — a result of out-of-phase motion of a receiver pile at spacing, s, close to one/half of the wavelength, $\lambda_S/2$. This is a remarkable reversal from the static trend. Similar are the trends for the lateral interaction factor $\alpha_H(\beta)$, $\beta = 0°$, $90°$, or any of the other angles.

7.2   CONSEQUENCES ON ELASTIC PILE-GROUP RESPONSE

The high values of the *static* interaction factors and the strongly oscillatory nature of the *dynamic* factors in linear soil have a significant effect on the response of a group of piles. For a large number of piles in a group, especially, the elastic effects are dramatic. Two examples illustrate this. The first, from Butterfield & Douglas (1981) and Fleming et al. (1991), is portrayed in Figure 7.5. It refers to a group of m rows and n columns (total number: $m \cdot n$ piles) with spacing between two adjacent piles, $s = 3d$. The precise layout of the piles was found by Butterfield and Douglas to have a marginal influence on the group response; a conclusion echoed for the dynamic problem in Chapter 10. We define as *stiffness efficiency of a group* the ratio

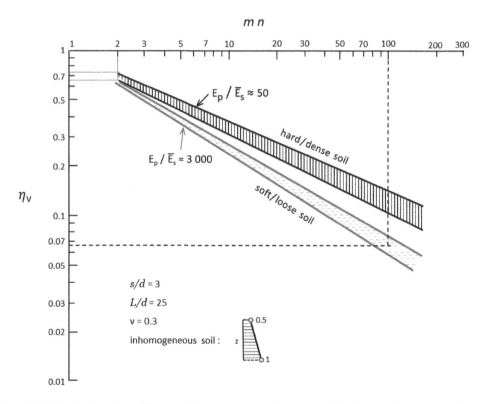

Figure 7.5 Vertical *static* stiffness efficiency of a pile group of m by n piles versus the total number (mn) of piles in the group. (Adapted from Butterfield & Douglas (1981).)

$$\eta = \frac{K^{(mn)}}{mn\,K^{(1)}}$$

(7.5)

i.e., the group stiffness divided by the sum of the individual stiffness of each pile, with η_V and η_H being the efficiency for vertical and horizontal loading, respectively. Figure 7.5 plots η_V as a function of the total number ($m\,n$) of piles. The very small values of efficiency for large groups are remarkable. For example, for a group of 100 piles in relatively soft soil (i.e., of $E_p/\overline{E}_s \approx 2000$), we read: $\eta_V \approx 0.065$. This means that if closely-spaced, 100 piles have the same or smaller static stiffness than a group of *merely* seven widely spaced (and hence not interacting) piles — a dramatic consequence (inefficiency) indeed.

The second example, from the simplified dynamic analysis of Dobry & Gazetas (1988), is portrayed in Figure 7.6. It refers to a 3 × 3 group of fixed-head piles with parametrically variable closest spacing, $s = 2d, 5d, 10d$. Subjected to harmonic steady-state lateral vibration at a circular frequency ω, the dynamic stiffness and damping efficiencies of the group differ substantially from their static values (recovered for $\omega = 0$). With the exception of the very closely spaced group of $s = 2d$, all stiffness and damping curves exhibit strong oscillations, with peaks reaching values of the order of 3 — a complete reversal from the static efficiencies, which are in the range of 0.3–0.5.

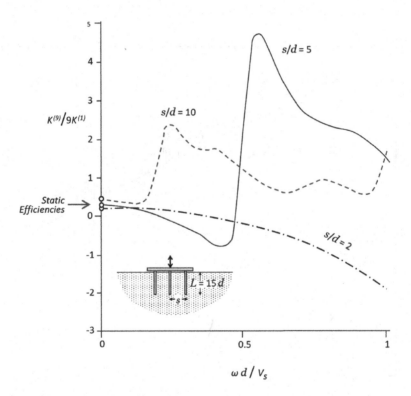

Figure 7.6 Dynamic vertical stiffness efficiency of a 3 × 3 pile group versus the dimensionless frequency, with the spacing ratio as parameter. Elastic homogeneous soil. (Modified from Dobry & Gazetas (1988).)

(Note that even the $s = 2d$ would have exhibited oscillations if we computed the response for higher frequencies.)

It is clear from all the above that the effects of the interaction between piles are very significant if the soil is elastic and homogeneous and the floating piles bonded to the soil.

Soil layering and inhomogeneity tend to tone down these effects (Mylonakis & Gazetas, 1998, 1999; Gazetas & Makris, 1991, 1992; Gazetas et al 1993, Kaynia, 1998; Waas & Hartman, 1984). As an example, Figure 7.7 from Gazetas & Makris (1991) contrasts the dynamic stiffness efficiencies of four different rigidly capped pile groups (1×2; 2×2; 1×4; and 3×3) in homogeneous soil and in soil with linearly increasing modulus with depth, from $G(0)$ at the top to $G(L) = 10\,G(0)$ at the pile tip — almost a

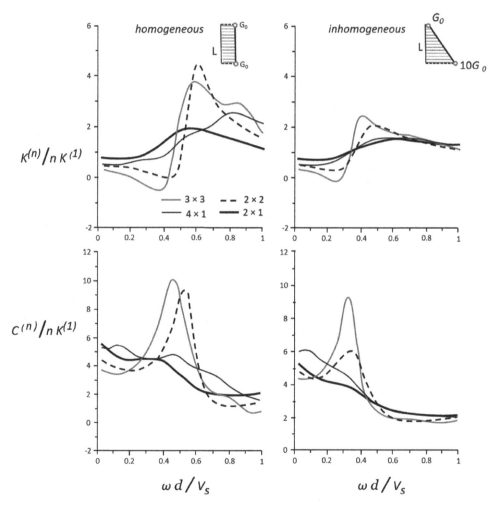

Figure 7.7 Effect of soil-modulus profile and of pile group configuration (2×1; 2×2; 4×1; and 3×3) on the variation of stiffness and damping efficiencies, functions of frequency, for $s/d = 5$. (a) Homogeneous and (b)inhomogeneous soil. (Adapted from Gazetas & Makris (1991).)

"Gibson" soil. In all cases, soil inhomogeneity indeed reduces the peaks observed in the homogeneous efficiency functions.

7.3   PILE–TO–PILE INTERACTION WITH SOIL AND INTERFACE NON-LINEARITIES

The previous results and conclusions were valid for the ideal conditions of soil linearity and fully bonded soil-pile interface. But reality only rarely, at extremely weak loading, justifies such idealisations because

- soil behaviour is inelastic and
- pile–soil contact cannot sustain net tension and has a limited shear capacity (friction).

As a result of the latter, the interface may experience sliding (under vertical loading) and detachment (under lateral loading). Such phenomena have been studied using, among several other methods, finite elements (with Abaqus) by Trochanis et al. (1991) and Kanellopoulos & Gazetas (2020). Results are presented for piles in clay with undrained shear strength $S_u = 50\,kPa$ and parametrically variable low-strain shear modulus $G_o = 200\,s_u \div 1400\,s_u$. Details on the soil constitutive relation and the finite-element model can be found in the cited articles (see also Anastasopoulos et al., 2011). The inelastic soil behaviour is modeled with a Von-Mises failure criterion, kinematic hardening law, and experimentally validated pre-failure non-linearity—the latter being significant for the dynamic problem (Vucetic & Dobry, 1991). The role of interface sliding and detachment is explored below for circular piles of reinforced concrete having $E_p = 25$ GPa, $d = 1$ m, $L = 20$ m.

7.3.1   Vertical loading

Under a large-enough vertical static force, a (single) pile and the ground surface around it will settle as sketched in Figure 7.8a. One part of the settlement will be common for the pile and the attached soil in contact; the other part is additional settlement of the pile alone, as it slides with respect to the soil when the shear traction between them reaches the limiting shear

$$\tau_f = \delta S_u \tag{7.6}$$

in which $\delta \leq 1$ is a decreasing function mainly of the soil shear strength (e.g., Fleming et al., 1992). Figure 7.8b "translates" the above sketch (for the specific pile and soil) into the FE-derived soil settlement as function of distance, $w(s)$, versus the magnitude of the imposed pile head settlement, $w = w(0)$. Notice that up to almost $w \approx 0.7\%$ of the pile diameter, pile and soil-interface are in contact, settling together, and the nearby (at $s = 3d$) pile experiences a more-or-less proportional settlement. Sliding starts immediately thereafter and completes at $w = w_{thr} \approx 0.014d$ (or 1.4% of pile diameter). Thereafter, the soil at the periphery of the pile stops settling, while the pile continues being pushed into the soil. But its effect on the adjacent ($s = 3d$) pile also stops increasing.

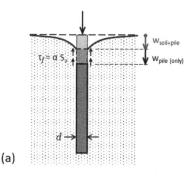

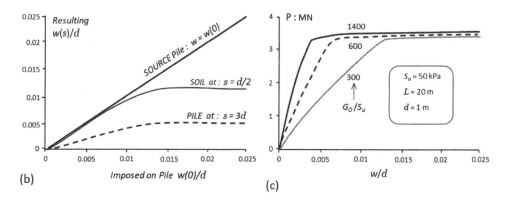

Figure 7.8 (a) Sketch illustrating the effect of sliding at the contact surface of soil and vertically loaded pile; (b) comparison of settlements of the pile, the soil at the periphery of the pile, and a point on the surface at distance $s = 3d$; (c) vertical pile force versus settlement for three different soil rigidity indices.

The *threshold* values w_{thr} of $w(0)$ beyond which slippage takes place depend on the soil rigidity index, G_o/S_u. In Figure 7.8c, slippage initiates at $w_{thr} \approx 0.014d$ when $G_o/S_u = 300$. For the stiffer soils, $G_o/S_u = 600$ and 1400, slippage initiates at $w_{thr} \approx 0.007d$ and $0.003d$, respectively. This is clearly seen with the plots of force vs. settlement for the above three values of the rigidity index. Evidently, with increasing rigidity, the *threshold* settlement w_{thr} decreases.

The effect of sliding and soil non-linearity on the attenuation of vertical displacements away from a forced pile is illustrated in Figure 7.9 in which the displacement contours around a forced-down pile are plotted in two situations: (i) under the idealised linearity and bonding conditions of soil and interface and (ii) under non-linearity conditions imposed by a pile settlement $w = 0.02d$. Evidently, the horizontal range of influence in the second case is substantially diminished compared to the elastic case, thus explaining the much smaller effect on the adjacent pile under non-linear conditions.

The outcome of such reduced effect is reflected on the interaction factor α_V, plotted in Figure 7.10 as a function of settlement $w = w(0)$ for three values of spacing, $s = 3d, 5d$,

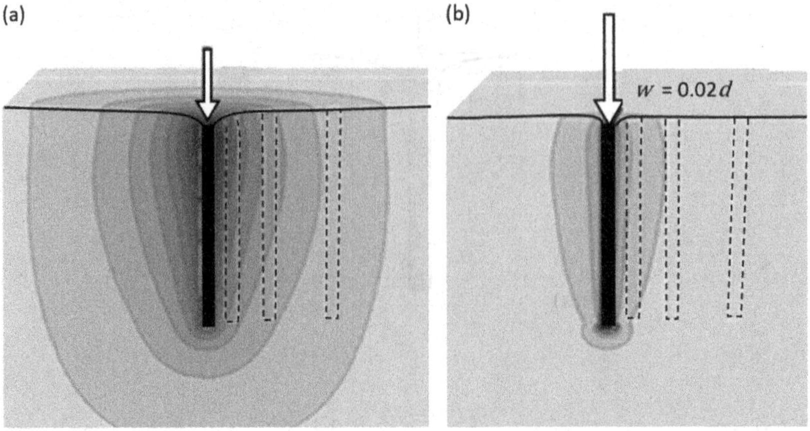

(a)   (b)

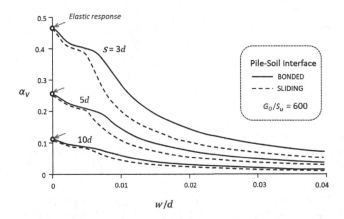

Figure 7.9 Vertical displacement contours from an axially loaded pile under either ideally elastic conditions (a) or fully inelastic soil and interface for an imposed pile settlement 0.02d (b). (Adapted from Kanellopoulos & Gazetas (2020).)

Figure 7.10 Decrease of the vertical *static* interaction factor with increasing soil and interface inelasticity expressed through the imposed pile settlement. (Adapted from Kanellopoulos & Gazetas (2020).)

$10d$. The dramatic decline of α_V from its elastic values is a consequence both of sliding at the interface and inelasticity of the soil.

The consequence of the small values of α_V for large imposed pile-head displacements is the *increase* of pile group efficiency η_V. Figure 7.11 illustrates this for a 2×1 and a 4×4 pile group with $s/d = 3$.

Similar is the effect of non-linearities on the *dynamic* interaction factors. An example is given in Figure 7.12 where, for $s/d = 3$ and $G_o/S_u = 600$, it is seen that the inelastic α_V for $w/d = 1\%$ is about one-half of the elastic one at small frequencies, and more significantly, it just barely becomes negative at higher frequencies.

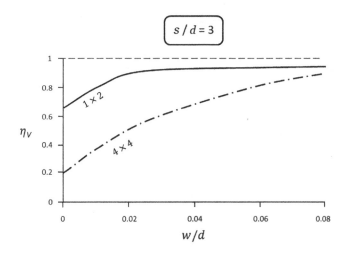

Figure 7.11 Vertical *static* efficiency of two pile groups versus increasing soil and interface inelasticity, as expressed through increasing settlement. Computed with superposition using interaction factors.

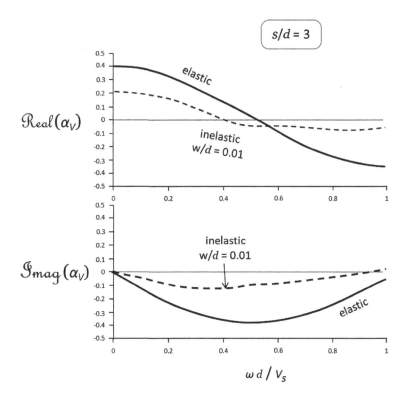

Figure 7.12 Real and imaginary components of the vertical dynamic interaction factor. Comparison between the elastic Eq. 7.4 and the inelastic FE analysis for $w = 001d$.

The implication for the dynamic efficiency, η_V, of a 3×3 pile group is depicted in Figure 7.13, where η_V (stiffness and damping components) is plotted versus the dimensionless frequency $\omega d/V_S$, for $s/d = 5$. The response peaks, arising due to out-of-phase oscillations of the neighbouring piles, reach very high values (≈ 4) under elastic conditions, but they decrease drastically with a mere $w/d = 1\%$. The oscillations are subdued. In fact, for larger displacements, say $w/d > 2\%$, we have found that even ignoring the frequency dependence and choosing the *static* inelastic value, $\alpha_V(\omega) \approx \alpha_V(0)$, may be a reasonable approximation especially for earthquake-induced loading containing a broad range of significant frequencies.

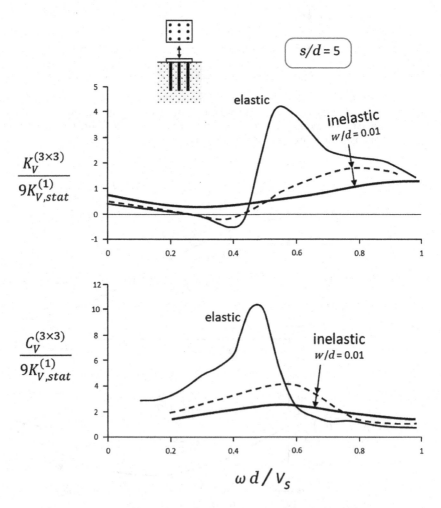

Figure 7.13 Stiffness and damping components of the vertical *dynamic* efficiency of a 3×3 rigidly capped pile group. *Elastic* versus *inelastic* ($w/d = 0.01$) response, the latter computed either with superposition using interaction factors from Eq. 7.4 (dashed line) or the complete FE analysis of the whole group (solid line).

7.3.2   Lateral loading

7.3.2.1   Static

To better appreciate the importance of non-linearities (in the soil *and* the interface) to the interaction between piles, the response of a single fixed-head pile under progressively increasing horizontal force is studied first. The goal is to merely *approach* soil capacity; reaching the *true* capacity is of little use, since to develop a failure mechanism with a "long" (or better, a not-too-"short") pile, plastic hinging of the pile itself must be mobilised. This depends on the structural moment capacity of the pile cross-section (Broms, 1964) and hence will not serve our analysis in which the pile is elastic.

The pile with the smallest E_p/E_{so} ratio, 125, is considered here. Subjected to small top displacement when the soil remains elastic, the pile deforms down to a depth equal to its "active" length. The latter, for a fixed-head horizontally loaded pile, can be estimated either from Randolph's (1981) expression, re-written here for $v = 0.5$ as

$$l_a \approx 1.25d\left(E_p/E_s\right)^{2/7} \tag{7.7a}$$

or from the expression of Gazetas (1991),

$$l_a \approx 1.5d\left(E_p/E_s\right)^{1/4} \tag{7.7b}$$

For soil Young's modulus $E_s = E_{so}$, these two expressions give almost the same value, $l_\alpha \approx 5\,d$. As the top displacement $u = u(0)$ is increased, however, detachment of pile from soil, which obviously initiates at the upper part of the pile, along with soil inelasticity, requires larger soil reactions at greater depths to achieve equilibrium. Hence, the "active" length increases.

We illustrate and verify the above trend, for a number of imposed head displacements increasing from $u = 0.01d$ all the way to $u = 0.20d$. Figure 7.14a plots the distribution with depth of the lateral pile deflections, $u(z)$, normalised by the imposed top displacement, $u = u(0)$. Figure 7.14b plots the resultant soil reaction $p(z)$. A horizontal section (slice) of the pile and soil is shown in Figure 7.14c with the distribution in plan of the normal (σ_r) and shear ($\tau_r\theta$) tractions when $u = 0.10d$. It helps explain the meaning of p. At such large displacements, separation at the back of the pile has occurred, and reactions are limited to almost the front half of the pile. Then $p(z)$ is the integral:

$$p \approx \int_{-\pi/2}^{\pi/2} \left(\sigma_r\cos\beta + \tau_{r\theta}\sin\beta\right) r_o d\theta \tag{7.8}$$

The plots in Figure 7.14b were obtained numerically from Eq. 7.8. The following conclusions emerge:

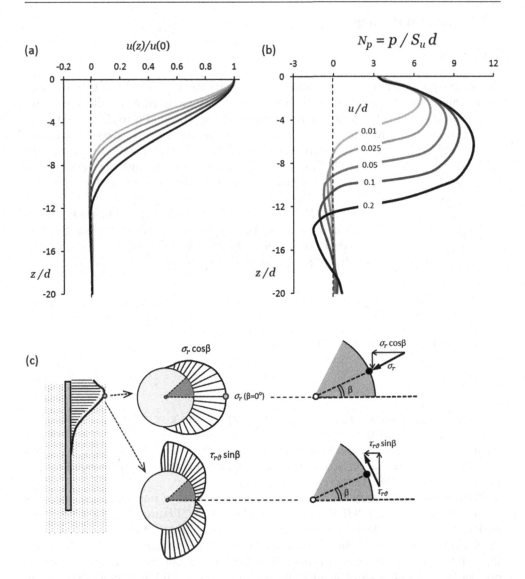

Figure 7.14 Lateral inelastic response of single horizontally loaded pile, at progressively
increasing top displacement: (a) normalised distribution of lateral deflections;
(b) horizontal reaction p, resultant at each depth of the normal and shear
stress components projected in the direction of loading, as illustrated in (c)
and computed according to Eq. 7.8.

- The largest value of p, $p_u \approx 10\, S_u d$, slightly exceeds the conventional semi-
  empirical value, $9\, S_u d$, which prevails in practice (Broms, 1964; Reese & Van
  Impe, 2001), and is about 5/6 of the analytical limit-analysis value of Randolph &
  Houlsby (1984),

$$p_u / S_u d = 4\sqrt{2} + 2\pi \approx 12 \tag{7.9a}$$

for a perfectly rough and bonded interface.

- This peak value develops at a depth $\geq 5d$, only over a limited length of the pile (roughly between 5 and 10 diameters) and only for large imposed displacements (>10% of pile diameter).
- At such large displacements, the (inelastic) "active" length is about 1.7 times the elastic "active" length of Eq. 7.7.
- Near the ground surface, the 3-D *"wedge"* failure mechanism leads to the ultimate value $p_u \approx 3\, S_u\, d$.

Using Figure 7.14b and Eq. 7.8, we could obtain a "reference" ultimate capacity of the pile. The result is

$$P_{u,ref} \approx 6\, d\, \left(1.7\, l_a\right) S_u \tag{7.9b}$$

where $\left(1.7 l_a\right)$ is approximately the inelastic "active" length under a large pile deflection; l_α is the elastic active pile length, equal in this case to about $5d$. Hence, in this particular example, Eq. 7.9b gives

$$P_{u,\,ref} \approx 55 \cdot S_u \cdot d^2 \tag{7.9c}$$

This value is utilised to normalise the applied load P and present the results in terms of

$$F_H = \frac{P_{u,\,ref}}{P} \tag{7.10}$$

in which F_H is just a *surrogate* of the factor of safety against the maximum soil resistance on the deflecting elastic pile.

The effect of a loaded "source" on an adjacent ("receiver") pile depends strongly on the angle of departure from the direction of loading. This dependence stems from the asymmetric displacement field around a loaded pile. Figure 7.15 depicts this field in the form of displacement contours on the ground surface for three pile-head–induced displacements: 1%, 2.5%, and 5% of the pile diameter. Notice that the mentioned asymmetry becomes grossly exaggerated with increasing pile displacement due to soil inelasticity and, especially, due to detachment of the back of the pile from the soil and slippage of its side with respect to the soil.

This behaviour is reflected on the interaction factors. Indeed, $\alpha_H(0°)$, $\alpha_H(180°)$, and $\alpha_H(90°)$ show differences which increase with intensity of loading. Figure 7.16 depicts the variation with $1/F_H$ of the above three factors.

The following trends are worth noticing:

- $\alpha_H(0°) = \alpha_H(180°)$ for elastic soil with fully bonded contact (i.e., when $1/F_H = 0$, or $F_H = \infty$), as there is no difference between compression and tension in pure elasticity.

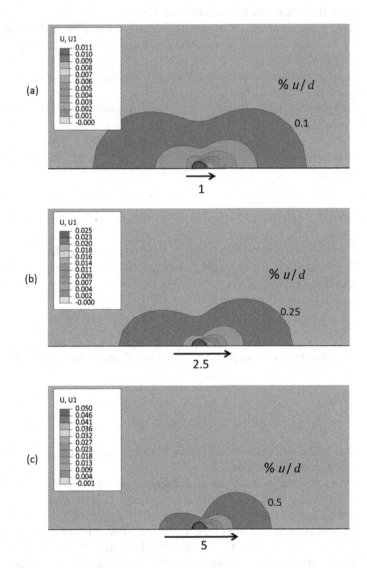

Figure 7.15 Contours of inelastic horizontal displacements on the ground surface around a pile horizontally displacing (a) $u = 0.01d$, (b) $0.025d$, and (c) $0.05d$.

- $\alpha_H(180^\circ) > \alpha_H(90^\circ)$ for all F_H values, a result which stems from the trough in the displacement field observed in Figure 7.15a, even under linear conditions, and the early slippage of the pile side normal to the loading direction.
- For closely spaced piles ($s \leq 3d$), the interaction factor retains a finite value even as F_H approaches 1 (very large displacement). This is understandable since the incompressible (undrained) soil keeps pushing the neighbour, even after the front soil has been fully plasticised.

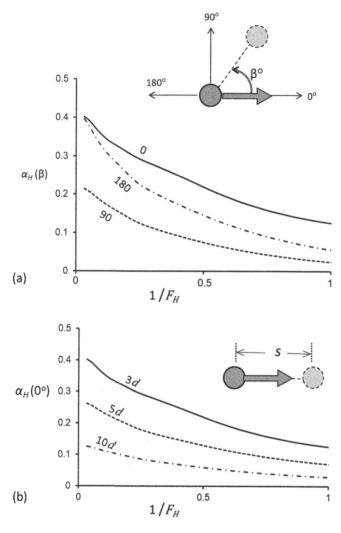

Figure 7.16 Decrease of the horizontal *static* interaction factor with the single-pile lateral factor of safety F_H for (a) three angles of departure β and (b) three pile spacings with $\beta = 0°$.

7.3.2.2 Dynamic

The effect of soil inelasticity on the dynamic interaction factors $\alpha_H(0°)$ and $\alpha_H(90°)$ is portrayed in Figures 7.17 and 7.18, respectively, for $s/d = 5$. Both the real and imaginary parts are shown. The elastic curves were obtained with Eq. 7.4, from the Dobry-Gazetas analytical approximation, while the inelastic curves are from the FE analyses. The observed trends are not different from those noted for the vertical interaction factors: a substantial decrease essentially at all frequencies with increasing amplitude of imposed lateral top deflection and thereby flattening of the curves.

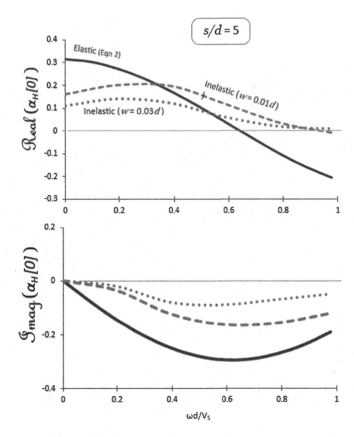

Figure 7.17 Real and imaginary parts of the interaction factor α_H ($\beta = 0\text{o}$): Comparison between the inelastic results for lateral oscillation amplitude of the source pile equal with 0.01d and 0.03d, and the elastic results with the Dobry & Gazetas (1988) approximation. ($G/s_u = 600$).

The implications of the smaller interaction factors are seen in the dynamic efficiency of an amply spaced 3 × 3 pile group, shown in Figure 7.19. The spacing ratio $s/d = 5$. The figure plots the stiffness and damping components of the efficiency. In each sub-figure, there are three curves, the elastic and the two inelastic ones corresponding to two factors of safety against the nominal lateral capacity: $F_H \approx 4$ and $F_H \approx 2$. It is noted that the large constructive-interference-produced peaks seen in the elastic efficiency are being substantially suppressed as non-linearity increases with increasing amplitude of loading.

7.4 CONCLUSION

Under both vertical and lateral loading, the interaction between adjacent piles in a group is of great significance if the soil remains linear-elastic and the pile is fully bonded to the surrounding soil. But once the imposed pile-head displacement amplitude increases,

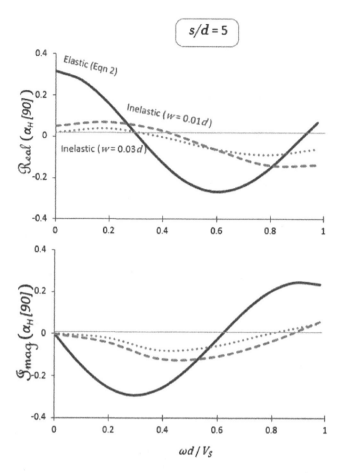

$s/d = 5$

Figure 7.18 Real and imaginary parts of the interaction factor α_H (β = **90**o): Comparison between the inelastic results for lateral oscillation amplitude of the active pile equal with 0.01d and 0.03d, and the elastic results with the Dobry & Gazetas (1988) approximation. (G/s_u = 600).

this interaction starts diminishing. Both inelastic soil behaviour and geometric and material non-linearity at the pile-soil interface are responsible for this phenomenon. As a result, the stiffness efficiency of a group tends to increase. For displacements of the order of 3% of the pile diameter, static and dynamic interaction factors are almost insignificant.

Whether pile-soil-pile interaction could be ignored altogether depends on the actual soil profile and therefore cannot be definitely concluded from our study with the specific idealised soil. However, the numerical results of this chapter may help appreciate important behavioural trends in pile-to-pile interaction and the response of pile groups.

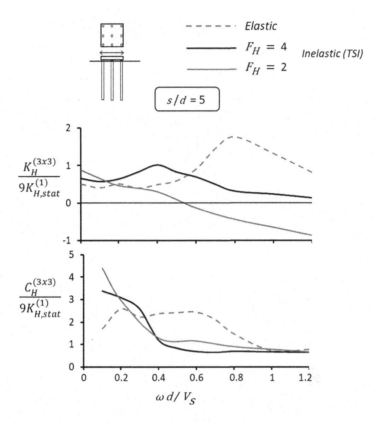

Figure 7.19 Stiffness and damping of the dynamic efficiency factor, *elastic* and *inelastic*, of a 3 × 3 pile group, for *s*/*d* = 5. The inelastic analyses are for two levels of loading corresponding to lateral single-pile factors of safety: 4 (small expected non-linearity) and 2 (moderately large non-linearity).

REFERENCES

Abaqus [Computer software]. ABAQUS, Inc., Dassault Systemes.

Anastasopoulos, I., Gelagoti, E., Kourkoulis, R., & Gazetas, G. (2011). Simplified constitutive model for simulation of cyclic response of shallow foundations: Validation against laboratory tests. *Journal of Geotechnical & Geoenvironmental Engineering*, 137, No. 12, 1154–1168.

Baguelin, F., & Frank, R. (1979). Theoretical studies of piles using the finite element method. *Numerical Methods in Offshore Piling*, 83–89. London: ICE Publishing.

Banerjee, P.K., & Sen, R. (1987). Dynamic behaviour of axially and laterally loaded piles and pile groups. In *Dynamic Behaviour of Foundations and Buried Structures*, P.K. Banerjee & R. Butterfield, eds., London: Elsevier Applied Science, 95–133.

Broms, B.B. (1964). Lateral resistance of piles in cohesive soils. *J. Soil Mech. Foundat. Div.* ASCE 90, No. 2, 27–63.

Butterfield, R. & Douglas, R. A. (1981). *Flexibility coefficients for the design of piles and pile groups.* CIRIA Technical Note 108, London, UK: CIRIA.

Dobry, R., & Gazetas, G. (1988). Simple method for dynamic stiffness and damping of floating pile groups. *Géotechnique*, 38, No. 4, 557–574, https://doi.org/10.1680/geot. 1988.38.4.557.

Fleming, W.G.K., Weltman, A.J., Randolph, M.F., & Elson, W.K. (1992). *Piling Engineering.* Glasgow: Blackie A&P.

Frank, R. A. (1974). *Etude théorique du comportement du pieux sous charge verticale: introduction de la dilatance.* Dr-Ing thesis. Universite Paris VI, France.

Gazetas, G., & Dobry, R. (1984). Horizontal response of piles in layered soils. *Journal of the Geotechnical Engineering,* 110, No. 1, 20–40.

Gazetas, G., Fan, K., & Kaynia, A. (1993). Dynamic response of pile groups with different configurations. *Soil Dynamics and Earthquake Engineering,* 12, No. 4, 239–257.

Gazetas, G., & Makris, N. (1991). Dynamic pile-soil-pile interaction. Part I: Analysis of axial vibration. *Earthquake Engineering and Structural Dynamics,* 20, 115–132.

Gerolymos, N., & Gazetas, G. (2005). Phenomenological model applied to inelastic response of soil—pile interaction systems. *Soils Found,* 45, No. 4, 119–132.

Kagawa, T. (1983). Dynamic lateral pile-group effects. *Journal of the Geotechnical Engineering,* 109, No. 10, 1267–1285.

Kanellopoulos, K. & Gazetas, G. (2020). Vertical static and dynamic pile-to-pile interaction in non-linear soil. *Géotechnique,* 70, No. 5 432–447.

Kaynia, A.M. (1988). Characteristics of the Dynamic Response of Pile Groups in Homogeneous and Nonhomogeneous Media. *Proc. 9th World Conf. Earthquake Engrg,* Tokyo-Kyoto, Japan, 3, 575–580.

Kaynia, A.M., & Kausel, E. (1982). *Dynamic Stiffness and Seismic Response of Pile Groups.* Research Report R82–03. Cambridge, MA: Massachusetts Institute of Technology.

Makris, N., & Gazetas, G. (1992). Dynamic pile-soil-pile interaction. Part II: Lateral and seismic response. *Earthquake Engineering and Structural Dynamics,* 21, 145–162.

Makris, N., & Gazetas, G. (1993). Displacement phase differences in a harmonically oscillating pile. *Géotechnique,* 43, No. 1, 135–150, https://doi.org/10. 1680/geot.1993.43.1.135.

Mylonakis, G., & Gazetas, G. (1998). Settlement and additional internal forces of grouped piles in layered soil. *Géotechnique,* 48, No. 1, 55–72.

Mylonakis, G., & Gazetas, G. (1999). Lateral vibration and internal forces of grouped piles in layered soil. *Journal of Geotechnical and Geoenvironmental Engineering,* 115, No. 1, 16–25.

Nogami, T. (1985). Flexural responses of grouped piles under dynamic loading. *Earthquake Engineering and Structural Dynamics,* 13, 321–336.

Pender, M.J. (1993). Aseismic pile foundation design analysis. *Bulletin of the New Zealand National Society for Earthquake Engineering,* 26, No. 1, 42–160.

Pender, M.J. (1996). Earthquake resistant design of foundations. *Bulletin of the New Zealand National Society for Earthquake Engineering,* 29, No. 3, 155–171.

Poulos, H. G. (1968). Analysis of the settlement of pile groups. *Géotechnique* 18, No. 4, 449–471.

Poulos, H.G. (1971). Behavior of laterally-loaded piles Il: Pile groups. *Journal of the Geotechnical Engineering Division,* 97, No. 5, 733–751.

Poulos, H.G. (1979). Group factors for pile deflection estimation. *Journal of the Geotechnical Engineering Division,* 105, No. 12, 1489–1509.

Poulos, H.G. (1989). Pile behaviour—theory and application. 29th Rankine Lecture. *Géotechnique,* 39, No. 3, 365–471.

Poulos, H.G., & Davis, E.H. (1980). *Pile Foundation Analysis and Design.* New York: Wiley.

Randolph, M.F. (1981). Response of flexible piles to lateral loading. *Géotechnique,* 31, No. 2, 247–259.

Randolph, M.F., & Houlsby (1984). The limiting pressure on a circular pile loaded laterally in cohesive soil. *Géotechnique,* 34, No. 4, 613–623.

Randolph, M.F., & Poulos, H.G. (1982). Estimation of the flexibility of offshore pile groups. Proc. Int. Conf. on Numerical Methods in Offshore Piling, Austin, TX, 313–328.

Randolph, M.F., & Wroth, C.P. (1978). An analysis of deformation of vertically loaded pile. *Journal of the Geotechnical Engineering Division,* 104, No. 12, 1465–1488.

Randolph, M.F., & Wroth, C.P. (1979). An analysis of the vertical loading of pile groups. *Géotechnique*, 29, No. 4, 423–439.

Reese, L., & Van Impe, W. (2001). *Single Piles and Pile Groups under Lateral Loading*. London: CRC Press.

Roesset, J.M. (1984). Dynamic stiffness of pile groups. In *Analysis and Design of Pile Foundations*, J.R. Meyer, ed., New York: American Society of Civil Engineers, 236–286.

Sanchez-Salinero, I. (1983). Dynamic stiffness of pile groups: approximate solutions. Geotech Engng Report GR83-5, Univ. Texas, Austin.

Trochanis, A.M., Bielak, J., & Christiano, P. (1991). Three-dimensional nonlinear study of piles. *Journal of Geotechnical and Geoenvironmental Engineering*, 117, No. 3, 429–447.

Vucetic, M., & Dobry, H. (1991). Effect of soil plasticity on cyclic response. *Journal of Geotechnical and Geoenvironmental Engineering*, 117, No. 1, 89–107.

Waas, G., & Hartmann, H.G. (1984). Seismic analysis of pile foundations including soil–piles–soil interaction. Proc. 8th World Conf. Earthquake Engng 5, 55–62.

Wolf, J.P., & Von Arx, G.A. (1978). Impedance function of a group of vertical piles. Proc. ASCE Specialty Conf. Geotech Earthq. Engng & Soil Dyn, 2, 1024–1041.

Chapter 8

Design of piles under seismic loading

Raffaele Di Laora
Università degli Studi della Campania "Luigi Vanvitelli"

Emmanouil Rovithis
Institute of Engineering Seismology and Earthquake Engineering EPPO-ITSAK

CONTENTS

8.1   INTRODUCTION

When a shallow foundation is not able to guarantee a satisfactory behavior, pile foundations represent the most common design option, having the main function of transferring the structural loads to competent deep soil layers. However, in earthquake-prone regions, piles must be designed to withstand additional loading due

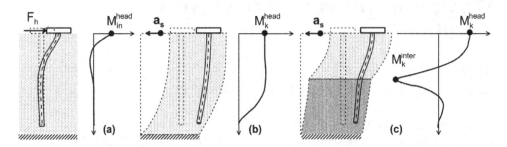

Figure 8.1 Deflected shapes and bending moment (*M*) profiles of a pile due to (a) inertial action reflected in a horizontal force (*F$_h$*) at the pile-head and (b, c) kinematic action of the soil imposing ground surface acceleration (*a$_s$*) in a homogeneous and a two-layer medium.

to (i) the oscillation of the superstructure which imposes forces and moments (the so-called *inertial* actions) at the head of the piles (Figure 8.1a), thus inducing significant internal stresses up to a pile length of typically 10 pile diameters and (ii) the deformation of the surrounding soil during the passage of seismic waves (the so-called *kinematic* actions), which imposes additional pile bending (independently of the presence of the superstructure) that may be significant over the whole pile length (Figure 8.1b). The latter type of bending due to kinematic soil-pile interaction has been documented by post-earthquake observations of failures on piles embedded in non-liquefiable soils (Tazoh et al., 1987; Mizuno, 1987; Nikolaou et al., 2001). These identified pile failures provided field evidence that, besides pile-head, the bending capacity of the pile section may also be exceeded at deeper elevations close to interfaces of soil layers with sharp stiffness contrast (Figure 8.1c), where forces transmitted to the piles from the superstructure are negligible. A concurrent phenomenon which also stems from kinematic soil-pile interaction has been recognized as a potential of piles to reduce the seismic motion transmitted to the superstructure with respect to the ground surface motion at free-field conditions (i.e., in the absence of the structure and its foundation), referred to as the "*filtering effect*".

Therefore, for the pile geometry established from the axial bearing capacity and settlement requirements, the inertial and kinematic bending moments imposed on piles under earthquake loading should be computed to derive pile reinforcement, referring to a structural Ultimate Limit State (ULS) check, and then verify that the ultimate lateral resistance of the pile is not exceeded under the inertial load, referring to a geotechnical ULS check (e.g. Broms, 1964).

Within the above design framework, this chapter deals with bending moments and possible filtering effects in the design of piles under seismic loading, while topics related to the above-mentioned geotechnical ULS checks are not covered. The soil is considered as non-liquefiable, and therefore, pile loading or failure mechanism due to liquefaction phenomena is also not addressed.

It is convenient from a conceptual standpoint to assume that the seismic response of the piles is the superposition of a kinematic and an inertial part in the realm of the substructure approach (Kausel et al., 1978; Makris et al., 1996; Gazetas & Mylonakis, 1998). The latter offers the advantage of obtaining piles' seismic response by means of the following consecutive steps: (a) a ground response analysis in free-field conditions to derive the acceleration and deformation field demanded by the surrounding soil, (b) a kinematic interaction analysis to compute (i) the kinematic-induced bending along the pile and (ii) the pile-head motion taking into account the piles' "filtering effect", (c) an inertial interaction analysis where the superstructure is supported on springs and dashpots, representing the dynamic stiffness of the soil-foundation system, and it is excited by the pile-head motion computed in step (b), to derive the forces and deformations transmitted to the foundation, and (d) computation of the inertia-induced pile bending. The above decomposition method is rigorous under the basic assumption of equivalent linear behavior of the materials, which is employed herein.

This chapter provides guidance on the seismic design of piles with reference to steps (a), (b), and (d) of the substructure approach mentioned above by first rationalizing the phenomenology of the kinematic interaction mechanism, following the derivation of the free-field ground response, and then providing simple formulae, which allow an easy assessment of the kinematically induced pile bending and filtering effect; simple formulae to derive the inertial-induced bending moments are also reported. Relevant issues such as pile group effects, the combination of inertial and kinematic forces, and the role of pile diameter are discussed. Finally, a flowchart of the seismic design procedure is provided, followed by a numerical example of a pile embedded in a real subsoil by which a foundation designer can address the issue of pile bending based either on ground response analysis results or on relevant seismic code provisions.

8.2 KINEMATIC INTERACTION

8.2.1 Phenomenology – physical mechanism

Consider a pile embedded in a layered soil. If the pile were infinitely flexible (i.e., with zero flexural stiffness), it would follow exactly soil movement, and thereby, the pile curvature would be equal to the soil curvature at any depth. Note that such curvature is infinite at the interface between two consecutive layers with different stiffness, but bending moment is everywhere zero. If the pile were rigid with infinite flexural stiffness, with the rotation at its head being restrained, it would displace uniformly and thereby its curvature would be zero at any depth, with the bending moment being always finite. In reality, piles' stiffness is finite, and thereby, pile behavior lies between these two extreme scenarios. It is thus understandable that the larger the flexural stiffness of the pile, the larger is its resistance against the deformation demanded by the surrounding soil; therefore, the pile will experience lower curvatures paying the price of adsorbing larger bending moments. If we consider that failure occurs because of an excessive deformation of the pile cross section, it is reasonable to consider flexible piles more sensitive to the problem of resisting kinematic actions compared to piles having higher flexural stiffness. However, this might not be true for reasons that will become clear later in this chapter.

Conceptually, it is possible to interpret the kinematic bending as the outcome of two distinct, yet simultaneous phenomena: (i) the displacement profile that instant by instant, the soil is trying to impose on the pile and (ii) the reluctance of the pile to deform according to the deflected shape imposed by the surrounding soil, owing to its flexural stiffness. The first phenomenon is a ground response problem in free-field conditions, while the second one is a pile-soil interaction problem. It is evident that the soil displacement profile under seismic loading defines the demand of curvature and bending on piles; therefore, the first step is to perform a ground response analysis.

8.2.2 Assessment of ground response in free-field conditions

From a physical point of view, the problem of ground response is to assess the effect of local soil conditions on the characteristics of a seismic motion specified at the bedrock level and predict the seismic motion at the free surface or at any depth of a soil stratum. Mathematically, the problem is one of wave propagation in a continuous medium. In practice, analytical solutions and numerical analyses of ground response refer often to the simple case of shear waves propagating vertically. This is justified by the fact that when earthquake body waves travel away from the source in all directions, they tend to propagate more vertically as they approach the ground surface due to the gradual decrease of soil stiffness. For the most common case of level or smoothly sloping sites, with horizontal boundaries between the soil layers, one-dimensional ground response analysis, where the soil and bedrock are assumed to extend infinitely in the horizontal direction, has been proved to be sufficiently accurate to predict the response of soil under vertically propagating shear waves.

The simplest approach one could follow to predict the response at ground surface is to merely modify the bedrock motion parameters, derived by means of a probabilistic hazard analysis, by simple coefficients which are also provided by codes. In this regard, the effect of local soil conditions on ground surface motion is taken into account in an approximate manner.

Toward a more detailed analysis, pertinent analytical solutions, which relate bedrock motion to surface response, or specialized geotechnical software are available, which allow equivalent-linear or nonlinear modelling of the soil behavior. These options are described in the ensuing, following the definition of the main parameters that are involved in ground response analysis.

8.2.2.1 Main definitions

Figure 8.2 presents the main definitions used for ground response analysis in order to assess site effects on seismic motion. More specifically, ground response analysis aims at evaluating the change in amplitude, duration, and frequency content that the reference motion, which is generally defined at the bedrock level (R), undergoes when travelling through the overlying media until reaching ground surface (S). However, an alternative evaluation of ground response, which assumes the motion at the outcrop (O) as reference seismic motion is often more interesting from an engineering viewpoint, given that generally earthquake recordings are available at this location. Note that the motions in (R) and (O) coincide only in the hypothesis of infinite stiffness for the rock formation.

The seismic motion may be represented either in time or frequency domain. In the former case, the motion is generally described by a single parameter, that is the peak acceleration. In the latter case, the time history is decomposed in a series of harmonics, each of them having different amplitudes, through a relevant mathematical operation

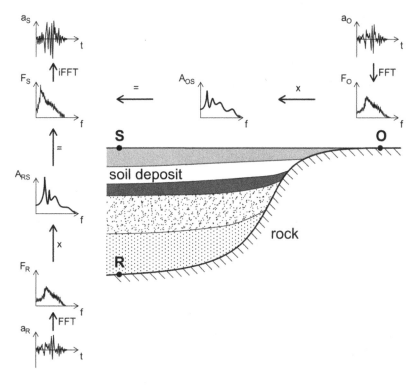

Figure 8.2 Key definitions of 1D ground response analysis.

named Fourier transform. Therefore, the quantitative assessment of ground response for time domain analysis may be performed by means of the ratio between maximum acceleration at surface and bedrock (or outcrop). In frequency domain, a more de-tailed assessment may be performed through an amplification function which relates the change in amplitude of the signal over the whole frequency range.

8.2.2.2 Analytical solutions for the amplification function

If soil behaves as a linear elastic material, the dynamic equilibrium of a soil element in 1D conditions simply states that, at any depth, the variation in shear stress is equal to the inertia force. This leads to a partial differential equation that can be easily solved by standard mathematical methods. Upon imposing the boundary condition of zero stress at ground surface, the solution leads to the bedrock-to-surface amplification function of a soil stratum with shear wave propagation velocity V_s and thickness H as

$$A_{RS}(\omega) = \frac{a_S(\omega)}{a_R(\omega)} = \frac{1}{\cos(kH)} \tag{8.1}$$

where k (= ω/V_s) is the soil wavenumber, ω (= $2\pi f$) is the circular excitation frequency, and f is the excitation frequency. This also coincides with the amplification function from the outcrop to surface under the assumption of rigid bedrock. In the more real-istic hypothesis of elastic bedrock, the waves travelling downward toward the bedrock are not completely reflected back in the soil layer, but they are partially transmitted in the rock. Thus, part of the energy is also dissipated by radiation of waves in the stiffer layer. In this case, while the amplification between bedrock and surface remains still the same, the amplification between outcrop and surface is lower, and can be expressed as

$$A_{OS}(\omega) = \frac{a_S(\omega)}{a_O(\omega)} = \frac{1}{\sqrt{\cos^2(kH) + \left[\dfrac{\rho_s V_s}{\rho_r V_r}\sin(kH)\right]^2}} \tag{8.2}$$

where ρ_s is soil mass density, while V_r and ρ_r are shear wave velocity and mass density of the rock.

However, soil possesses an inherent capability of dissipating energy by hysteresis. This can be expressed though a damping ratio ξ. It can be shown that the above re-lations are still valid for a damped soil, if the soil shear wave propagation velocity is expressed in a complex form $\left[V_s^* = V_s(1+i\xi)\right]$. A phase lag between bedrock and sur-face acceleration is then generated, and accordingly, the amplification of the seismic motion between bedrock (or outcrop) and ground surface is a complex function of the excitation frequency, referred to as *"transfer function"* (TF).

Figure 8.3 shows the amplification as affected by the soil-to-bedrock impedance contrast (= $\rho_s V_s/\rho_r V_r$) (Figure 8.3a) and the hysteretic damping of soil (Figure 8.3b) by means of Eqs. (8.2) and (8.1), referring to the amplitude of the TF. It is apparent that soil damping is a key parameter which governs ground response and must be therefore accurately estimated. Likewise, the assumption of infinite stiffness for the bedrock is overconservative for determining the surface motion.

For layered soils consisting of more than one horizontal layer, the solution to the 1D wave equation is repeated for each layer and the TF can be derived by applying the compatibility of displacements and continuity of shear stresses at the interfaces between successive layers.

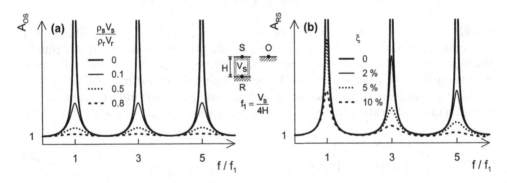

Figure 8.3 Effect of (a) elastic bedrock and (b) soil hysteretic damping on amplification function.

In the case of more complex geotechnical conditions involving continuously inhomogeneous soils, a number of closed-form solutions have been reported (Dobry et al., 1976; Gazetas, 1982; Towhata, 1996; Semblat & Pecker, 2009; Rovithis et al., 2011; Vrettos, 2013; Mylonakis et al., 2013 among others) for different variations of soil stiffness with depth.

8.2.2.3 Practical computation of ground response

The practical importance of the TF stems from the fact that it allows an easy computation of the ground surface motion in the time domain for a specified time history of bedrock or outcrop (input) motion by means of fast Fourier transform (FFT) and inverse fast Fourier transform (iFFT). It should be mentioned that unless an earthquake motion recorded at a certain depth from a borehole instrument is available to be employed as the input motion in ground response analysis, the rock outcrop conditions should be adopted. For a given input motion at rock outcrop level, $a_O(t)$, or at bedrock, $a_R(t)$, the ground surface motion, $a_S(t)$, can be derived as (Figure 8.2)

$$a_S(t) = iFFT\left[F_S(\omega)\right] = \begin{cases} iFFT\left[TF_{OS}(\omega) \cdot FFT\left(a_O(t)\right)\right] \\ iFFT\left[TF_{RS}(\omega) \cdot FFT\left(a_R(t)\right)\right] \end{cases} \tag{8.3}$$

Thus, TF determines how each frequency in the outcrop or bedrock (input) motion is amplified (or de-amplified under certain conditions) by the soil deposit. Generally speaking, a TF relating the motion between two arbitrary depths can be built. In this manner, any other response parameter such as shear stress, shear strain, and displacement can be easily calculated at any depth.

8.2.2.4 Equivalent linear approach

The assumption of linear soil response is valid only for very low levels of shear strain (γ) induced by the seismic motion, which typically range below $10^{-3}\%$ (Ishihara, 1996). Thus, the shear modulus and hysteretic damping of soil introduced in the above linear solutions refer to the low-strain values of shear modulus G_0 ($= V_{s0}^2 \rho_s$) and damping ratio ξ_0, respectively. With increasing shear strain, the behavior of soils under cyclic loading is represented by a nonlinear shear stress (τ)-shear strain (γ) hysteresis loop (Figure 8.4a) depending on the type and stress state of the soil. The slope of the line that connects the end points of the hysteresis loop represents an "average" shear stiffness of the soil (the "secant" shear modulus $G = \tau_c/\gamma_c$, where τ_c and γ_c are the shear stress and shear strain amplitudes), which decreases with increasing amplitude of strain. The relationship between shear modulus and strain amplitude is typically characterized by a normalized modulus reduction curve $G/G_0-\gamma$ (%) (Figure 8.4b). On the other hand, the breadth of the loop is related to the area A_{loop}, which is a measure of the energy dissipated during a loading cycle, and it can be described by the material damping ratio ξ ($= A_{loop}/2\pi G\gamma_c^2$). The energy dissipation and, thus, ξ increase with increasing shear strain as reflected in a typical ξ (%)-γ (%) curve (Figure 8.4b).

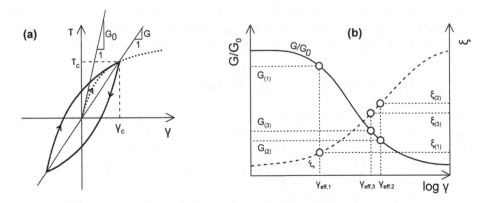

Figure 8.4 (a) Nonlinear shear stress (τ)-strain (γ) behavior of soils under shear cyclic loading; (b) typical trends of $G/G_0-\gamma$ (%)$-\xi$ (%) curves employed for an EL ground response analysis.

A variety of experimentally derived $G/G_0-\gamma$ (%)$-\xi$ (%) curves and relevant analytical expressions to fit the experimental data have been reported in the literature for different types of soils (Hardin & Drnevich, 1972; Seed & Idriss, 1970; Vucetic and Dobry, 1991; Ishibashi & Zhang, 1993; Darendeli, 2001; Kallioglou, 2003 among others). The above data are necessary to perform an *equivalent linear* (EL) ground response analysis where the actual nonlinear behavior of the soil is approximated by modifying the low-strain shear modulus and hysteretic damping according to the shear strain which is exerted by the earthquake motion. More specifically, with reference to a layered soil, the EL approach involves the following iterative procedure: (i) an initial ground response analysis under a prescribed input motion at the base of the soil profile, where the low-strain stiffness and damping properties are assigned to each soil layer, (ii) an effective shear strain (γ_{eff}) is defined from the time history of the generated shear strain within each soil layer as a fraction of the corresponding maximum shear strain (γ_{max}) (i.e. $\gamma_{eff} = R_\gamma \gamma_{max}$), where R_γ depends on the magnitude (M) of the earthquake and can be estimated from [(M−1)/10] (note that the usually employed value of $R_\gamma = 0.65$ refers to a $M = 7.5$), (iii) relevant $G/G_0-\gamma$ (%)$-\xi$ (%) curves are employed to update G and ξ values corresponding to γ_{eff} for the next iteration, and (iv) the computational steps (ii) and (iii) are repeated until the difference between G and ξ values derived in two successive iterations is below a predefined threshold of convergence. The above procedure has been coded in computer software like SHAKE2000 (Schnabel et al., 1972; Gustavo, 2012), STRATA (Kottke & Rathje, 2009), and DEEPSOIL (Hashash et al., 2016), which allow derivation of ground response under EL considerations. Such an approach may be employed for levels of the shear strain up to approximately 0.2% (Kaklamanos et al., 2013).

8.2.2.5   Nonlinear models

Various nonlinear constitutive laws have been developed to model the stress-strain relationship of soils under loading and unloading-reloading conditions, as affected by a series of parameters including strain amplitude, confining pressure, number of loading cycles, plasticity index, overconsolidation ratio, and frequency of loading. A comprehensive review of the subject is reported in Hashash et al. (2010). With reference to pile analysis and design, such constitutive models should be employed when the mobilized strains are too high (say larger than 0.2%) to employ an equivalent linear model for soil behavior. It follows that for very large shear strains, the design of piles under EL considerations (which allow the assessment of the mobilized soil stiffness and the superposition of kinematic and inertial effects) is not applicable anymore and, therefore, more sophisticated finite element analyses involving piles, soil, and structure with proper constitutive models should be performed. Of course, along with large computational effort, high expertise and a large amount of quality data for characterizing materials' behavior are required for this task.

8.2.3 Simplified formulae for pile kinematic bending

The issue of pile kinematic bending has received considerable attention by the geotechnical research community, and a large number of simplified formulae are available to derive kinematic pile bending at the head of the pile (de Sanctis et al., 2010; Di Laora et al., 2013; Di Laora & Rovithis, 2015) or at the elevation of an interface between successive soil layers (Dobry & O'Rourke, 1983; Mylonakis, 2001; Nikolaou et al., 2001; Maiorano et al., 2009; Dezi et al., 2010; Sica et al., 2011; Di Laora et al., 2012). The role of salient features such as the diameter of the pile and the frequency of excitation on the relevant importance of inertial and kinematic bending moments has been highlighted (Kaynia & Mahzooni, 1996; Mylonakis, 2001; Saitoh, 2005; Di Laora et al., 2017a). The issue has been recognized in contemporary seismic design codes, which mandate the consideration of kinematic bending moments in pile design when combined requirements for the seismicity of the area, the importance of the pile-supported structure, and the ground profile are met. Yet, no means for deriving this type of pile bending have been accommodated by the codes.

8.2.3.1 Bending moments at pile head

The simplest approach to assess kinematic bending is to neglect pile-soil interaction altogether, so that pile curvature equals soil curvature at any depth. This assumption was first formulated by Margason in 1975 and seems to oversimplify the problem. Further, this approach cannot be applied at the interface of two layers with different stiffness, as soil curvature is infinite, and returns a finite curvature at pile head, which cannot be true for free-head piles. However, more recent numerical works (de Sanctis et al., 2010; Di Laora et al., 2013) show that such a simple approach is able to provide bending moments at the top of fixed-head piles embedded in homogeneous soil or in a two-layer soil when the interface between the layers is far enough from the pile head. Di Laora et al. (2013) showed that if the thickness (h_1) of the first layer is larger than approximately 6d, where d is the pile diameter, pile-head kinematic bending may be reasonably predicted by

$$M_k^{head} = E_p I_p \frac{a_s \rho}{G_1} \tag{8.4}$$

where M_k^{head} is the kinematic moment, E_p and I_p are Young's modulus and cross-sectional moment of inertia of the pile, while ρ and G_1 are mass density and shear modulus of the first soil layer. When the interface is shallow (i.e., $h_1 < 6d$), the moment given by the above formula may overestimate the actual bending at the pile head. In this case, a better approximation may be achieved if the bending moment computed for a deep interface (Eq. 8.4) is multiplied by a reduction factor which is a function of the interface depth, the pile diameter, and the stiffness contrast between the two layers. After manipulating the expression proposed by Stacul & Squeglia (2020), one may get the simple formula:

$$M_k^{head} = E_p I_p \frac{a_s \rho}{G_1} \left[\frac{h_1}{6d} \left(\frac{G_2}{G_1} \right)^{\frac{1}{20}\left(\frac{h_1}{d} - 6 \right)} \right] \tag{8.5}$$

The above expression may be considered valid only if $h_1 < 6d$ and as long as it does not yield a value of pile-head moment lower than

$$M_k^{head} = \frac{E_p I_p a_s \rho}{G_2} \tag{8.6}$$

which refers to a homogeneous soil with properties those of the second layer.

In many circumstances, the stiffness profile of a real subsoil may be well idealized through a continuous variation with depth. Di Laora & Rovithis (2015) analyzed the kinematic bending atop a fixed-head pile embedded in a subsoil with shear stiffness varying with depth according to the law

$$G_s(z) = G_{sd} \left[a + (1-a) \frac{z}{d} \right]^n \tag{8.7}$$

where $G_s(z)$ is the depth-varying soil shear modulus, G_{sd} is the shear modulus at the depth of one pile diameter (i.e., at $z = d$), while a and n are dimensionless factors which control the stiffness profile. It is easy to verify that for $n = 0$ or $a = 1$, a uniform stiffness is obtained, while $n = 1$ corresponds to the case of a linear distribution of shear modulus with depth. The above authors found out by numerical analyses that the bending moment at the top of the pile, where it is maximum, may be accurately calculated by

$$M_k^{head} = E_p I_p \frac{a_s \rho}{G_s(z = L_a/2)} \tag{8.8}$$

In the above expression, L_a is the active length of the pile, which defines the portion of the pile interacting strongly with the surrounding soil, and can be calculated through the following expression (Di Laora & Rovithis, 2015; Karatzia & Mylonakis, 2016):

$$L_a = d \frac{1}{1-a} \left\{ \left[a^{\frac{n+4}{4}} + \frac{5}{16}(n+4)(1-a) \left(\frac{\pi}{2} \frac{E_p}{E_{sd}} \right)^{\frac{1}{4}} \right]^{\frac{4}{4+n}} - a \right\} \tag{8.9}$$

where E_{sd} is the soil Young's modulus at the depth of one pile diameter. Equation (8.8) denotes that the kinematic bending moment experienced by the pile in an inhomogeneous soil is equal to that occurring in an equivalent homogeneous soil of shear modulus equal to the one at the depth of one half of the active length of the pile. As a corollary of the above, the soil stiffness which rules the interaction phenomenon is the one at very shallow depths (usually 3–4 pile diameters). It is also evident that if the subsoil is formed by a surficial inhomogeneous layer with thickness larger than the pile's active length, the presence of an underlying layer does not affect kinematic bending at the pile-head, and Eq. (8.8) is still applicable.

8.2.3.2 Bending moments at soil layers' interface

Since piles are often designed to transfer loads from the structure to a competent soil stratum, they often penetrate soil layers with better mechanical properties as compared to the shallower soil strata. At the interface between two consecutive layers of different shear moduli, the shear strain is discontinuous. It is therefore not physically sound to assume that pile follows soil, as an unbroken pile will always experience a finite curvature (i.e., no jump in sectional rotation) and thereby a finite bending moment. To overcome this problem, a number of simplified procedures have been developed for estimating kinematic bending moments in layered soils. A simple model to account for pile-soil interaction in a two-layer soil was proposed by Dobry & O'Rourke (1983), under the assumptions that (i) both soil layers are infinitely thick and (ii) each layer is subjected to a uniform static shear stress field, τ, which generates constant shear strains $\gamma_1 = \tau/G_1$ and $\gamma_2 = \tau/G_2$, where G_1 and G_2 refer to the shear modulus of the first and the second layer, respectively. The above authors proposed the following explicit expression for the kinematic pile bending moment at the interface as

$$M_k^{inter} = 1.86 \left(E_p I_p \right)^{0.75} G_1^{0.25} \gamma_1 F \tag{8.10}$$

F is a dimensionless function of the shear modulus ratio (G_2/G_1) between the two layers, given by

$$F = \frac{\left(1 - c^{-4}\right)\left(1 + c^3\right)}{\left(1 + c\right)\left(c^{-1} + 1 + c + c^2\right)} \tag{8.11}$$

where $c = (G_2/G_1)^{1/4}$. The peak soil shear strain (γ_1) in the upper layer at the interface depth (i.e., $z = h_1$) may be computed from a free-field site response analysis or by the approximate expression suggested by Seed & Idriss (1982):

$$\gamma_1 = \frac{\rho_s a_s h_1}{G_1} \left(1 - 0.015 h_1\right) \tag{8.12}$$

where a_s stands for the acceleration at the soil surface.

Later, Mylonakis (2001) proposed an enhanced model where both layers are assumed to be thick (though not to an infinite extent), and the seismic excitation is truly dynamic. To quantify the interaction phenomenon, the above study introduced a "strain transmissibility" function relating peak pile bending strain, $\varepsilon_p \left(= M_k^{inter} d / 2 E_p I_p\right)$, and γ_1. In the static limit $(\omega \to 0)$ this function is given by

$$\left(\frac{\varepsilon_p}{\gamma_1}\right)_{st} = \frac{1}{2} c^{-4} \left(c^2 - c + 1\right) \left\{\left[3\left(\frac{k_1}{E_p}\right)^{1/4} \left(\frac{h_1}{d}\right) - 1\right] c(c-1) - 1\right\} \left(\frac{h_1}{d}\right)^{-1} \tag{8.13}$$

where k_1 ($= \delta E_1$) represents the Winkler modulus of the first layer and is primarily related to its Young's modulus E_1 through a coefficient δ (Roesset, 1980; Dobry et al., 1982; Gazetas & Dobry, 1984). Following the early work of Kavvadas & Gazetas (1993), a simplified expression of δ [$= 6(E_p/E_{s1})^{-1/8}$] is suggested by Mylonakis (2001), which yields a value around 2.5 for typical ratios of E_p/E_s. The static value of transmissibility is then multiplied by a coefficient Φ ranging between 1 and 1.5 to take into account dynamic effects. Thus, the complete expression of the kinematic interface moment in the dynamic regime can be formulated as

$$M_k^{inter} = \left(E_p I_p\right) \left(\frac{\varepsilon_p}{\gamma_1}\right)_{st} \gamma_1 \Phi \frac{d}{2} \tag{8.14}$$

Based on a parametric study carried out by means of a beam on dynamic Winkler foundation (BDWF) model of a pile embedded in a two-layer soil subjected to harmonic steady-state excitation, Nikolaou et al. (2001) derived a simple approximate formula for the maximum bending moment at the interface under resonant conditions:

$$M_{k,res}^{inter} = \beta \tau_{char} d^3 \left(\frac{L}{d}\right)^{0.3} \left(\frac{E_p}{E_1}\right)^{0.65} \left(\frac{V_{s2}}{V_{s1}}\right)^{0.5} \tag{8.15}$$

where $\beta = 0.042$, L/d is the pile slenderness, E_p/E_1 is the ratio of the Young's modulus between pile and soil, V_{s2}/V_{s1} is the ratio of the shear-wave velocities of the two layers, and τ_{char} is a characteristic shear stress being proportional to the actual shear stress that is likely to develop at the interface:

$$\tau_{char} = a_s \rho_1 h_1 \tag{8.16}$$

with ρ_1 the mass density of the first layer. It is noted that the above formula predicts infinite moments for both very slender piles ($L/d \to \infty$) and soils with high layer stiffness contrasts ($V_{s2}/V_{s1} \to \infty$), which is a counterintuitive trait, while overpredicts pile bending under transient seismic excitation. In this regard, Nikolaou et al. (2001) suggest that the steady-state amplitudes of the interface moment given by Eq. (8.15) should be reduced by a factor η to obtain the peak values of transient bending moments:

$$M_k^{inter} = \eta M_{k,res}^{inter} \tag{8.17}$$

where n attains values between 0.2 and 0.3, depending on the relationship between the fundamental period of the soil deposit and the dominant period range of the earthquake motion. A simple approximate expression for η was proposed by Cairo et al. (2009) by means of BEM analyses as

$$\eta = 1.31 - 0.2\frac{T_1}{T_m} \tag{8.18}$$

where T_1 refers to the fundamental period of the deposit and may be evaluated, as a first approximation, by $4h_1/V_{s1}$, and T_m stands for the mean period of the excitation, which may be derived by suitable equations in the literature (Rathje et al., 1998) or by pertinent signal processing tools (e.g., SeismoSignal). On the basis of dynamic FE analyses, Maiorano et al. (2009) suggested to employ in the solutions of Mylonakis (2001) and Nikolaou et al. (2001) the dynamic transient value $\gamma_{1,dyn}$ obtained from a ground response analysis and to adopt $\Phi = 1.32$ and $\beta = 0.071$. Following the same approach, Sica et al. (2011) derived Φ and β at 1.15 and 0.053, respectively, by means of regression analysis of results obtained from a BDWF model.

Di Laora et al. (2012) highlighted that bending moment at a layers' interface is made up of two distinct contributions of opposite sign, as shown in Figure 8.5. Upon considering low-frequency excitation and taking the transmissibility developing for $G_2/G_1 = 1$ (i.e., in a homogeneous soil) as conventionally negative, an increase in the stiffness of the second layer furnishes a higher restrain for pile rotation, resulting, accordingly, in an increasingly positive contribution to bending. Therefore, contrary to the common perception – also implied by code prescriptions – that a layered soil always generates higher kinematic bending as compared to homogeneous soil conditions, the overall interface moment in absolute value is lower than that developing in a homogeneous soil with the properties of the first layer, except for very large stiffness contrasts between layers ($G_2/G_1 \geq 10$). The above authors proposed the following fitting formula based on transient dynamic FE analyses:

$$\varepsilon_p = \chi\,\gamma_1\left[\left(\frac{E_p}{E_1}\right)^{-0.25}(c-1)^{0.5} - \frac{1}{2}\left(\frac{h_1}{d}\right)^{-1}\right] \tag{8.19}$$

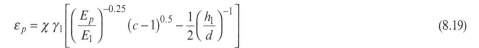

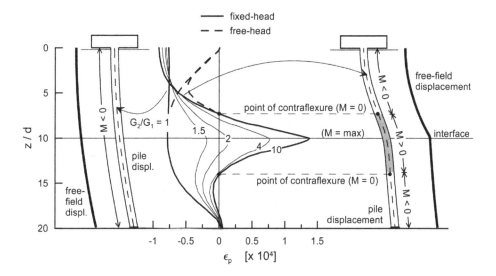

Figure 8.5 Development of kinematic pile bending at the interface of soil layers for low-frequency excitation (Di Laora et al., 2012).

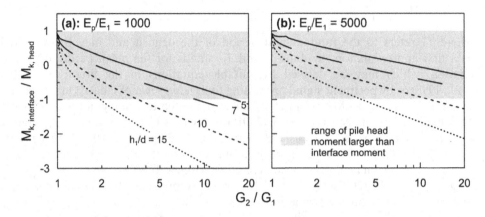

Figure 8.6 Interface to pile-head kinematic bending moment against G_2/G_1 for E_p/E_1 at 1000 (a) and 5000 (b).

where $\chi = 0.93$ takes also into account dynamic effects. An average value of $\Phi = 1.25$, independent of the earthquake record, was derived from back analyses of the results, in agreement with the recommendations of Mylonakis (2001) and close to the values suggested by Maiorano et al. (2009) and Sica et al. (2011).

Upon implementing Eqs. (8.4), (8.5), (8.12), and (8.19), the ratio of interface to pile-head kinematic moment is plotted in Figure 8.6 for different soil-pile configurations expressed by the dimensionless parameters h_1/d, G_2/G_1, and E_p/E_1.

8.2.3.3 Practical use of simplified formulae

Numerical works employing advanced constitutive models of soil behavior (Martinelli et al., 2016; Mucciacciaro & Sica, 2018) and experimental studies referring to 1-g shaking table (Chidichimo et al., 2014) or centrifuge tests (Garala et al., 2020) have shown that the simplified methods detailed in the previous section provide accurate predictions of kinematic pile bending.

All the formulae are proportional to a parameter describing free-field ground response – the demand of kinematic bending – through various expressions which thereby express pile-soil interaction. For pile-head bending, the value of a_s to be considered is the maximum free-field acceleration at the depth of pile head. As far as γ_1 is concerned, the maximum shear strain during the whole time history should be considered, while the mobilized stiffness of soil should be employed for the pile-soil interaction counterpart (e.g., in E_p/E_1 in Eq. 8.19). Both these parameters can be obtained by performing EL 1D analyses via freely available software (e.g., STRATA, DEEPSOIL). Relevant code provisions which suggest soil amplification factors and reduction coefficients to derive ground surface acceleration and mobilized soil stiffness, respectively, may alternatively be employed toward a more simplified procedure without the need to perform a ground response analysis. In the same spirit, γ_1 can be roughly estimated by simplified expressions if ground surface acceleration is known. For the computation of the interface bending moment, the mobilized shear modulus of soil at 3 diameters above and below the interface is a reasonable assumption. The reduction of pile Young's modulus can be also accounted for in a similar manner if a moment-curvature diagram of the pile section is available. A quick iterative procedure may be followed, assuming the initial value of E_p, calculating pile curvature and then deriving the new E_p based on the moment-curvature relationship and iterating until convergence is reached.

8.2.4   The *filtering* mechanism of piles: physical interpretation and simplified formulae

Earthquake recordings from instrumented pile-supported structures have provided field evidence that piles filter the high-frequency components of the ground surface motion at free-field conditions (Kawamura et al., 1977; Ohta et al., 1980; Gazetas, 1984). Such a behavior stemming from the kinematic interaction between soil and piles is referred to as *"filtering effect"* and indicates a beneficial mechanism of fixed-head piles to reduce the seismic motion which is transmitted to the superstructure. The above phenomenon is commonly quantified in the dynamic regime by the ratio (I_u) between pile response at the pile-head in terms of horizontal displacement, $u_p(0)$, or acceleration, $a_p(0)$, over free-field horizontal displacement at ground surface $u_{ff}(0)$ or acceleration $a_{ff}(0)$:

$$I_u = \frac{u_p(0)}{u_{ff}(0)} = \frac{a_p(0)}{a_{ff}(0)} \tag{8.20}$$

Theoretical advancements on this topic since the original contribution of Flores-Berrones & Whitman (1982), where the first closed-form analytical solution was reported for piles restrained against head rotation by a rigid cap, have been published during the last three decades, which allow assessment of the filtering effect for a wide range of soil-pile configurations (Gazetas, 1984; Kaynia & Kausel, 1991; Fan et al., 1991; Nikolaou et al., 2001; Anoyatis et al., 2013; Di Laora & de Sanctis, 2013; Rovithis et al., 2015; Di Laora et al., 2017a; Rovithis et al., 2017; Iovino et al., 2019).

As an example, for a fixed-head, infinitely long pile in a homogeneous half-space with shear wave propagation velocity V_s under harmonic SH waves of circular frequency ω, I_u is given by (Flores-Berrones & Whitman, 1982; Nikolaou et al., 2001; Anoyatis et al., 2013)

$$I_u = \left[1 + \frac{1}{4}\left(\frac{q}{\lambda}\right)^4\right]^{-1} \tag{8.21}$$

where q (= ω/V_s) is the wavenumber of the harmonic SH wave, and λ is the Winkler parameter:

$$\lambda = \left(\frac{k}{4E_pI_p}\right)^{\frac{1}{4}} \tag{8.22}$$

where k is defined in the same spirit as k_1 in Eq. (8.13). From a physical point of view, the filtering effect has been attributed to the inability of piles, due to their stiffness, to follow the wavelength of the soil motion which becomes progressively shorter with increasing frequency of excitation (Eq. 8.21). An alternative holistic interpretation of the physical mechanism governing filtering effect is attempted herein, in line with the interpretation of kinematic pile bending reported above.

In the theoretical case of a fixed-head rigid pile, the latter will displace uniformly along its whole length, thus averaging the nonuniform profile of displacement imposed by the surrounding soil. However, as real piles possess finite flexural stiffness, the hypothesis of infinite pile stiffness and the associated averaging behavior may be assumed along an effective depth, which may be defined as a fraction b of the pile's active length. It follows that pile's lateral displacement u_p may be written as (Iovino et al., 2019)

$$
u_p = \frac{1}{bL_a} \int_0^{bL_a} u_{ff}(z)\,dz
\tag{8.23}
$$

where $u_{ff}(z)$ is the free-field displacement profile with depth. Thus, I_u may be expressed by

$$
I_u = \frac{u_p}{u_{ff}(0)} = \frac{\dfrac{1}{bL_a} \displaystyle\int_0^{bL_a} u_{ff}(z)\,dz}{u_{ff}(0)}
\tag{8.24}
$$

which for a homogeneous soil takes the form

$$
I_u = \frac{\displaystyle\int_0^{bL_a} u_{ff}(0)\cos\!\left(\frac{\omega z}{V_s}\right)dz}{bL_a u_{ff}(0)} = \frac{\sin\!\left(\dfrac{\omega b L_a}{V_s}\right)}{\dfrac{\omega b L_a}{V_s}}
\tag{8.25}
$$

This ratio is equal to unity when the dimensionless ratio $\omega b L_a / V_s$ tends to zero, indicating that when the wavelength in the soil, $V_s/f = 2\pi V_s/\omega$, is very large compared to the portion of pile which is capable of averaging soil motion (i.e bL_a), the displacement profile that the soil is trying to impose is uniform, and therefore, pile displacement equals soil displacement. With decreasing wavelength in the soil as compared to bL_a, the average soil displacement reduces and eventually becomes zero when the wavelength of the soil motion is equal to $2bL_a$ or, equivalently, when $\omega b L_a / V_s = \pi$, since in this case, there is a perfect balance between positive and negative contribution of soil displacements on pile response (Figure 8.7). It follows that for a given excitation frequency, I_u decreases with increasing pile diameter and Young's modulus and with decreasing soil stiffness, in agreement with the physical perception mentioned above that stiffer piles are more reluctant to follow soil motion, but this resistance from the pile gets attenuated when the soil that is dictating the motion becomes stiffer. This simple model allows understanding of a further concept, harder to catch at first sight,

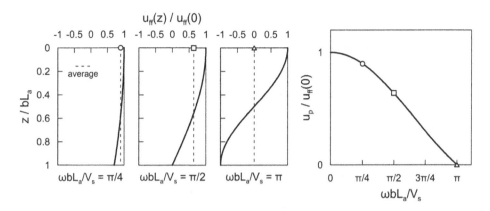

Figure 8.7 *Figure 8.7* Physical interpretation of the filtering mechanism of piles averaging free-field displacement of the surrounding soil.

involving the role of excitation frequency, the increase of which leads to a reduction of I_u for a given pile-soil configuration. The above model was conceptualized under the assumptions of constant stiffness and zero damping for the soil. Upon considering soil hysteretic damping, the amplitude of I_u will always attain values larger than zero. However, I_u is minimized at the same frequency $V_s/2bL_a$.

In the case of a more general profile of soil stiffness such as that described by Eq. (8.7), reference can be made to an average shear wave velocity $V_{s,\,av}$ within $L_a/2$ (Di Laora & Rovithis, 2015):

$$V_{S,av} = V_{Sd}\frac{\left(L_a/4d\right)(a-1)(n-2)}{\left[a+(1-a)\dfrac{L_a}{2d}\right]^{1-n/2} - a^{1-n/2}}$$ (8.26)

where V_{sd} is the shear wave velocity at one pile diameter depth. In this case, Iovino et al. (2019) introduced a general dimensionless filtering factor $a_{eff,\,La}$:

$$a_{eff,La} = \frac{\omega L_a}{V_{s,av}}$$ (8.27)

and found the following simplified expression for I_u which is applicable for all pile-soil configurations:

$$I_u = \begin{cases} \left[1+0.02\,a^3_{eff,La}\right]^{-1} & \text{for } a_{eff,La} < 5 \\ 0.29 & \text{for } a_{eff,La} \geq 5 \end{cases}$$ (8.28)

The practical usefulness of I_u is that it can be readily employed to obtain pile-head response $a_p(t)$ in time domain. For a given free-field motion at ground surface $a_s(t)$, which may be derived by means of Eq. (8.3), $a_p(t)$ can be computed from the following formula involving FFT and iFFT:

$$a_p(t) = iFFT\left[I_u(\omega) * FFT(a_s(t))\right] \tag{8.29}$$

The simplified procedure formed by Eqs. (8.28) and (8.29) was followed by Iovino et al. (2019) to derive a mean spectral acceleration ratio ξ, referring to the average response spectrum of the pile-head motion ($S_{a, p, av}$) over the average response spectrum of the free-field motion ($S_{a, s, av}$). To this end, different earthquake recordings were specified as input motions at the base of the soil profile. Results are shown in Figure 8.8 for different soil-pile configurations. Spectral ratios derived from rigorous FE analyses are also plotted in Figure 8.8, highlighting the efficiency of the simplified I_u formula (Eq. 8.28). A larger reduction of structural response is observed with increasing pile diameter and decreasing soil average stiffness and degree of inhomogeneity. It is also noteworthy that while for the homogeneous case, the role of pile diameter is critical, it becomes less important for the proportional soil profile, denoting that the effect of

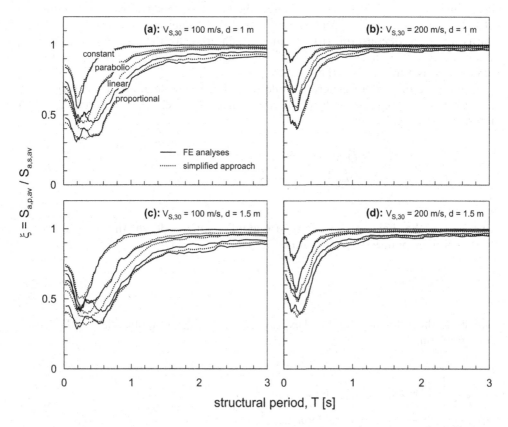

Figure 8.8 Comparison of mean spectral acceleration ratios ξ between the simplified procedure (Eqs. 8.27 and 8.28) and rigorous FE analyses for different soil-pile configurations. In all plots, E_p = 30 GPa and hysteretic damping of soil at 10% (Iovino et al., 2019).

pile diameter on the filtering action of piles is diminishing with increasing degree of soil inhomogeneity. This should be associated with the proportionality of $a_{eff, La}$ to pile diameter. Indeed, it is easy to verify that $a_{eff, La}$ is proportional to d for the homogeneous case, while for the proportional profile, $a_{eff, La}$ is proportional to $d^{2/5}$, indicating that $a_{eff, La}$ is progressively insensitive to pile diameter with increasing levels of inhomogeneity. Furthermore, it is observed that long-period structures are insensitive to pile filtering as their response is not affected by the high-frequency components of the seismic motion which in turn are filtered by piles. Simplified analytical expressions to derive characteristic points (T, ξ) of spectral ratios are also reported in the above study, which allow a quick preliminary assessment of the importance of the filtering action of piles by means of hand calculations.

8.3   INERTIAL INTERACTION

Given the dynamic nature of the forces applied by the superstructure on the pile foundation, a harmonic treatment of the inertial interaction problem should be provided, in the same spirit as for the kinematic interaction. However, while soil-foundation-structure interaction (SFSI) may play a major role in the magnitude of seismic forces transmitted to the foundation due to structural oscillation, these can be applied statically at the pile head as pile inertia does not modify substantially the profile of bending moment. In this section, we start from the knowledge of the inertial horizontal force on the single pile, taking for granted that this force has been derived by considering pile-soil-pile interaction and SFSI in the structural analysis if such effects are deemed to be relevant.

For an elastic pile embedded in an elastic soil, the horizontal displacement u and the rotation θ under pile head loading with a horizontal force H and a moment M are given in matrix form as (Pender, 1993)

$$
\begin{bmatrix} u \\ \theta \end{bmatrix} = \begin{bmatrix} f_{uH} & f_{uM} \\ f_{\theta H} & f_{\theta M} \end{bmatrix} \begin{bmatrix} H \\ M \end{bmatrix}
\tag{8.30}
$$

where f_{uH}, f_{uM}, $f_{\theta H}$, and $f_{\theta M}$ refer to flexibility coefficients. In the case of fixed-head piles (i.e., $\theta = 0$), the above equation yields the fixing moment M_F at the pile head:

$$
M_F = -\frac{f_{\theta H}}{f_{\theta M}} H
\tag{8.31}
$$

The inversion of the pile's flexibility matrix provides the pile-head stiffness matrix which allows Eq. 8.31 to be rewritten in stiffness terms as

$$
M_F = \frac{K_{HM}}{K_{HH}} H
\tag{8.32}
$$

where the indices "HH" and "HM" denote stiffness in swaying and cross swaying-rocking pile motion, respectively. If an infinitely long pile embedded in a homogeneous soil is idealized as a Winkler beam, Eq. 8.32 simplifies to (Hetenyi, 1946)

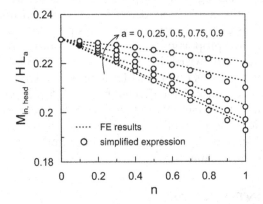

Figure 8.9 Dimensionless inertial moment against inhomogeneity factors a and n.

$$\frac{M_F}{H} = \frac{1}{2\lambda} \tag{8.33}$$

where λ is given in Eq. 8.22. Under inertial loading, δ may be considered equal to 1.2 (Roesset, 1980). Randolph (1981) provides a FE-based expression which however provides very similar results.

For layered or continuously inhomogeneous soils, the Winkler method may be applied in conjunction with a layer transfer matrix approach known as the Haskell-Thompson technique to derive pile stiffness (Thompson, 1950; Mylonakis, 1995; Rovithis et al., 2013).

Upon considering the generalized law of soil stiffness given in Eq. 8.7, the inertial moment $\left(M_{in}^{head} \right)$ at the pile head may be considered proportional to the product of the horizontal force H acting on the pile-head times the active length L_a of the pile in the same spirit as the relevant expression in Randolph (1981). Based on a series of rigorous FE analyses, the following simplified expression may be employed to derive M_{in}^{head}:

$$\frac{M_{in}^{head}}{HL_a} = 0.23 - 0.035n\left(1 - 0.9a^2\right) \tag{8.34}$$

which approximates FE results with a maximum error of 1% (Figure 8.9).

8.4 KINEMATIC AND INERTIAL INTERACTION

8.4.1 Combination of kinematic and inertial effects

In general, time histories of kinematic and inertial actions are out of phase. As kinematic and inertial effects are estimated herein with reference to their maxima, a problem arises: how should we combine them? Tokimatsu et al. (2005) suggest that if the fundamental period of the structure is lower than that of the soil deposit, kinematic and inertial effects are in phase, otherwise they are out of phase showing a phase lag of 180°.

While in the former case it is reasonable to sum their maximum values in order to derive the overall bending moment at the pile head, in the latter case, an engineering judgment should be used to decide whether to combine the maximum values by means of a SRSS procedure (square root of the squares of maxima) or, less conservatively, to consider the maximum between inertial and kinematic maxima.

8.4.2 The role of pile size

Given the different nature of kinematic and inertial loading, pile size, and in particular pile diameter, exerts a different effect on the development of bending moment at pile head. It is straightforward to recognize that a long pile embedded in a subsoil with constant stiffness almost follows soil deformations under kinematic loading so that if pile remained sufficiently long, kinematic bending at pile head would be essentially proportional to pile cross-section moment of inertia, i.e., to d^4. On the contrary, for the inertial action, increasing pile diameter will lead to a proportional increase of inertial bending under the same horizontal force. This is because, despite the rotation of a free-head pile decreases in inverse proportion to d^2, a progressively larger moment, proportional to d^3, has to be exerted by the restraint, represented by the structural elements connecting piles' heads, to bring the rotation to zero.

With reference to more general geotechnical scenarios, it is easy to verify by inspecting Eqs. (8.8) and (8.9) that for depth-proportional stiffness ($a = 0$, $n = 1$), the pile head kinematic moment is proportional to pile diameter raised to a lower exponent ($d^{3.2}$ as opposed to d^4 for constant stiffness). All cases with a and n between 0 and 1 are bounded by these two extreme scenarios.

Identifying the effect of pile diameter on the inertial moment requires further considerations. Consider a design problem where very often pile diameter and length may vary within a certain range to guarantee a certain safety factor against bearing capacity failure under an axial load P_p, so that one could select a larger diameter with a lower length and vice versa under constant values of the above parameters. This way, it is easy to derive from Eqs. (8.9) and (8.34) that inertial moment is proportional to d for constant stiffness ($n = 0$) and to $d^{1.8}$ in the case of soil stiffness varying proportionally with depth ($a = 0$, $n = 1$). Note that consideration of tip resistance does not alter the exponent, as it merely reduces length for a given safety factor.

As far as section capacity M_y is concerned, it is possible to show that (Di Laora et al., 2017b, 2019) this is proportional to pile diameter raised to an exponent between 1 and 3, depending on the relative contribution of steel and concrete to the overall section capacity, and on whether we assume a constant reinforcement area A_s or we consider an increase in A_s with increasing pile diameter. If the reinforcement area is increased proportionally to the section area of the pile, M_y is roughly proportional to d^3.

Figure 8.10 depicts a typical trend of demand and capacity with pile diameter, under the hypothesis of zero phase lag between kinematic and inertial moments. It is easy to recognize that capacity increases with pile diameter more than inertial moment and less than the kinematic demand.

Di Laora (2020) analyzed the case of two subsoils, namely a sandy soil and a normally consolidated clay, where the profile of acceleration and mobilized soil stiffness was determined by a ground response analysis. Figure 8.11 depicts inertial and

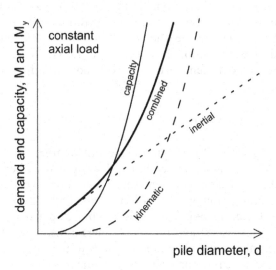

Figure 8.10 Schematic trend of kinematic, inertial, and combined demand and section capacity as a function of pile diameter.

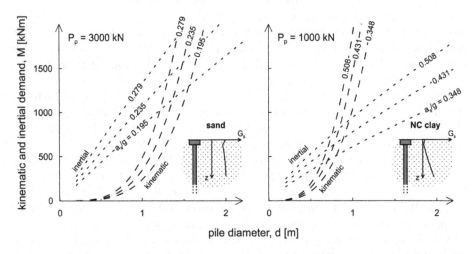

Figure 8.11 Kinematic and inertial demand as function of pile diameter for different acceleration levels and for the two subsoil under investigation (Di Laora, 2020).

kinematic moment demand for the two subsoils as a function of pile diameter, for different values of the ground surface acceleration. Results are obtained with reference to a linear stiffness profile interpolating mobilized stiffness below pile head. It is worth noticing that

a   both kinematic moment and inertial moment increase with acceleration;
b   kinematic bending is much more severe for the pile in clay; this is due to both the higher accelerations occurring at surface and the lower mobilized stiffness;

c   kinematic moments tend to dominate over the inertial ones for large diameter of
the pile embedded in clay;

d   despite the large difference in the working load, the inertial bending moment of
the pile in the clay is just slightly lower than the corresponding value derived for
the sand; this is due to a partial compensation of the higher load by the higher
acceleration and lower mobilized stiffness, which increase bending.

Let us now think of designing a pile for seismic actions, so that we select diameter,
length, and reinforcement area which provide adequate capacity to resist seismic demand. Suppose to increase pile diameter, with length and reinforcement area set at a
constant value. This way, safety against a bearing capacity failure increases, as well as
cost. Figure 8.12 illustrates this scenario. The selected values of reinforcement areas
correspond to 1%, 2%, and 4% of a 0.8 m diameter pile. It is clear that increasing pile
diameter, despite the increase in cost and safety against bearing capacity failure, may
lead to a tremendous decrease in safety against pile structural collapse. This is particularly pronounced for the pile in clay, where despite the large amount of reinforcement
for a pile of 0.8 m diameter, increasing size above this value leads rapidly to failure.
This occurs because the remarkable increase of kinematic moment (proportional to
diameter raised to an exponent between 3.2 and 4) is not balanced by the increase in
section capacity (roughly proportional to $d^{1.4}$). As a side comment, note that the increase in diameter under constant load also results in a decrease of the normal stress
in the section which is detrimental for the flexural capacity. This means that large
diameters are safer only if accompanied by a substantial increase in reinforcement to
withstand the increasing kinematic demand. To shed further light on the last concept,
one may think of a pile designed only for inertial action, for example, ensuring that the
inertial moment is half of the section capacity. Figure 8.13 shows the seismic demand
normalized by the moment capacity for different acceleration levels as a function of
pile diameter. It is evident that increasing pile diameter leads quickly to failure. Now

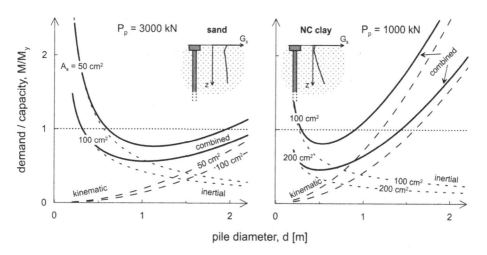

Figure 8.12 Kinematic, inertial, and combined demand over capacity as a function of pile
diameter for different values of A_s and for the two subsoil under investigation
(Di Laora, 2020).

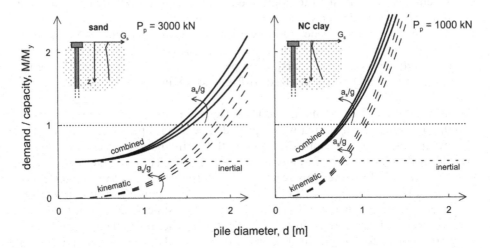

Figure 8.13 Demand over capacity for a pile designed only for inertial action, as a function of pile diameter for different acceleration levels and for the two subsoils under investigation (Di Laora, 2020).

the assertion mentioned in Section 8.2.1 that increasing flexural stiffness $E_p I_p$ of the pile is not always beneficial becomes clear: while increasing pile Young's modulus also corresponds to an increase in concrete strength and is thereby beneficial, increasing pile cross sectional moment of inertia may be severely detrimental.

8.5   GROUP EFFECTS AND GENERAL REMARKS FOR THE SEISMIC DESIGN OF PILES

In contrast to pile group response under inertial loading, where pile-to-pile interaction plays a major role, numerous studies in the literature have proven that kinematic bending for a single pile can be reasonably extended to all the piles in a group. This denotes that group effects may be ignored for kinematic bending (Kaynia & Mahzooni, 1996) which is understandable from a physical viewpoint for a group of piles which are not closely spaced (say a piles' spacing not less than 3 pile diameters). In this case, the soil between the piles deforms similarly to the one in free-field conditions, and thereby all piles are forced to withstand the same displacement demand. For a smaller spacing between piles, it has been demonstrated that referring to the bending profile of the single pile is a conservative assumption (Dezi & Poulos, 2017).

The results reported in this chapter in terms of pile kinematic bending and filtering action introduce an issue regarding the role of pile diameter. On one hand, increasing pile diameter may result in large kinematic bending moments (a detrimental effect) but on the other hand may reduce seismic actions on low-period structures (generally a beneficial effect). In other words, large diameter piles in soft soils have the potential to reduce the seismic action transmitted to a low-period structure at the expense of attracting bending moments on themselves. A possible exception to the above beneficial effect may be generated by the additional rotational component of a capped pile group kinematic motion with no counterpart in free-field conditions. Di Laora et al. (2017b) provided a simple method to derive the kinematic interaction factors for a pile group and showed that (i) the rotational component may be significant only for slender structures founded on small pile groups, and (ii) the translational component of the group is essentially the one associated with the single fixed-head pile.

Along with the technical issues discussed so far, the choices in design cannot avoid cost considerations. It is possible to show that generally the choice of a lower diameter and a larger length will also lead to lower foundation cost. For a preliminary assessment of the pile design geometry, it is therefore suggested to choose a pile diameter slightly larger than the minimum diameter which is required to withstand inertial action while deriving the length from geotechnical ULS considerations under the static axial load. A summary of all the computational steps required to design piles under seismic action is reported in Figure 8.14 as a flowchart which encompasses all the issues covered in this chapter.

8.6 NUMERICAL EXAMPLE

Consider a fixed-head pile of 1 m diameter and 20 m length in a two-layered soil consisting of an upper layer of fully saturated normally consolidated clay and a deeper layer of overconsolidated clay. Both layers have a thickness of 15 m, and below 30 m, a stiff rock is encountered. The pile carries an axial load (N_p) equal to 1200 kN, and the horizontal load (H) due to the structure oscillations under earthquake is considered 0.5 times the product of surface acceleration (a_s) times the mass carried by the pile:

$$H = 0.5 a_s \frac{N_p}{g} \tag{8.35}$$

The design peak rock acceleration at the outcrop level is 0.15 g. The low-strain shear modulus in the first layer is described through the relation

$$G_0(z) = 12 + 3z \tag{8.36}$$

where G_0 is in MPa and z in meters. The second layer has a constant shear modulus G_0 at 120 MPa. The shear wave velocity of the underlying the rock is considered at 800 m/s. The saturated mass density is 1.8, 2.0, and 2.2 Mg/m^3 for the three materials, respectively.

Bending moments due to kinematic and inertial interaction will be derived following two different procedures with respect to free-field ground response by (i) performing a 1D equivalent-linear analysis which is suggested and (ii) adopting simplifying code suggestions.

8.6.1 Prediction of pile bending based on ground response analysis

Figure 8.15 shows the results of a ground response analysis performed by considering seven earthquake motions at the outcrop level being compatible with the design spectrum, all scaled at the design peak rock acceleration of 0.15 g. For the example presented herein, the modulus reduction and damping curves proposed in Vucetic &

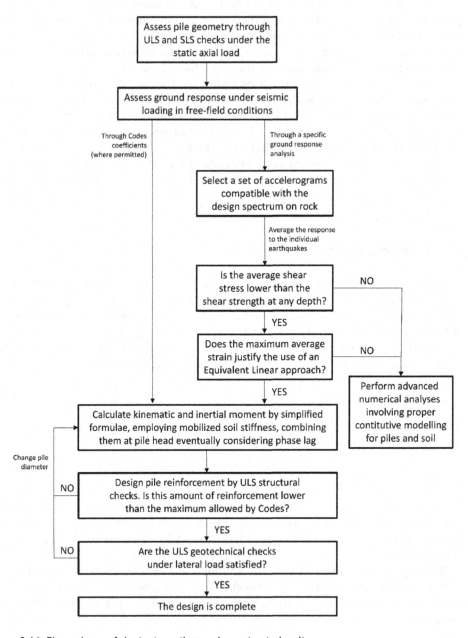

Figure 8.14 Flow chart of designing piles under seismic loading.

Dobry (1991) for PI = 50% are assumed for both clay layers, among the various models that are available in the relevant literature. The profile of the mobilized stiffness in the first layer may be approximated through the relation

$$G(z) = 8.8 + 2.2z \tag{8.37}$$

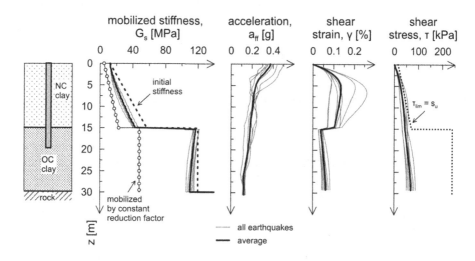

Figure 8.15 Ground response analysis results.

which results in $G_{sd} = 11\,\text{MPa}$, $a = 0.8$, $n = 1$, $L_a = 7.0\,\text{m}$, and $G_s(L_a/2) = 16.5\,\text{MPa}$. The average peak acceleration at surface a_s calculated from the analysis is equal to $3.67\,\text{m/s}^2$. Assuming a pile Young's modulus of $30\,\text{GPa}$, the kinematic moment at the pile head is then calculated through Eq. 8.8 as

$$M_k^{head} = E_p I_p \frac{a_s \rho}{G_s\left(L_a/2\right)} = 30{,}000{,}000 \cdot \frac{\pi \cdot 1^4}{64} \cdot \frac{3.67 \cdot 1.8}{16{,}500} = 590\,kNm \tag{8.38}$$

The above calculation could also be performed using the average shear strain following Di Laora & Rovithis (2015).

For the interface moment, it is reasonable to assume as the constant stiffness of the first layer the one at 3 pile diameters above the interface. The corresponding value of the mobilized shear modulus is thus equal to

$$G_1\left(h_1 - 3d\right) = 8.8 + 2.2(15 - 3) = 35.2\,MPa \tag{8.39}$$

and accordingly, $E_1(h_1-3d) = 3G_1(h_1-3d) = 105.6\,\text{MPa}$. The second layer is considered to have a constant mobilized shear modulus of $117\,\text{MPa}$ derived from the analysis. The average shear strain immediately above the interface is found from the analysis to be 0.0791%. The interface moment may be calculated through Eq. 8.19:

$$M_k^{int} = \frac{E_p I_p}{d/2}\varepsilon_p = \frac{E_p I_p}{d/2}0.93\,\gamma_1\left[-\frac{1}{2}\left(\frac{h_1}{d}\right)^{-1} + \left(\frac{E_p}{E_1}\right)^{-0.25}(c-1)^{0.5}\right]$$

$$= \frac{30{,}000{,}000 \cdot \dfrac{\pi \cdot 1^4}{64}}{1/2} \cdot 0.93 \cdot 0.000791 \cdot \tag{8.40}$$

$$\left\{-\frac{1}{2}\cdot\frac{1}{15} + \left[\frac{30{,}000{,}000}{105{,}600}\right]^{-0.25} \cdot \left[\left(\frac{117{,}000}{35{,}200}\right)^{0.25} - 1\right]^{0.5}\right\} = 240\,kNm$$

The bending moment at the pile head coming from inertial interaction is calculated by means of Eq. 8.34:

$$M_{in}^{head} = \chi H L_a = \left[0.23 - 0.035 \cdot 1 \cdot \left(1 - 0.9 \cdot 0.8^2\right) \right] \cdot 0.5 \cdot 3.67 \cdot \frac{1200}{9.81} \cdot 7.0 = 338 \, kNm \quad (8.41)$$

8.6.2 Prediction of pile bending based on simplified code suggestions

We will now perform the calculations following simple code suggestions instead of ground response analysis. The surface acceleration may be taken as 1.35 times peak rock acceleration following EC8 (CEN, 2004) for Seismic action Type 1 and Ground Type D, resulting in $a_s = 0.2025\,g$. EC8 also suggests, for the acceleration under consideration, to consider a mobilized shear modulus equal to 0.4 times the low-strain value, which underestimates (in the order of 50%) the mobilized stiffness in both layers (Figure 8.15). The active length of the pile L_a is now 8.0 m, and thus kinematic moment at head is

$$M_k^{head} = E_p I_p \frac{a_s \rho}{G_s (L_a/2)} = 30{,}000{,}000 \cdot \frac{\pi \cdot 1^4}{64} \cdot \frac{0.2025 \cdot 9.81 \cdot 1.8}{4800 + 1200 \cdot \dfrac{8.0}{2}} = 548 \, kNm \quad (8.42)$$

It follows that the simplified approach slightly underestimates the kinematic moment at head. This close agreement should be attributed to the fact that the underestimation of the mobilized stiffness is counterbalanced by a similar underestimation of the surface acceleration.

To calculate interface moment, we first need to estimate the shear strain at the interface by means of Eq. 8.12:

$$\gamma_1 = \frac{a_s \rho h_1}{G(h_1)} (1 - 0.015 h_1) \; [\text{with } h_1 \text{ in m}] = 0.182\% \quad (8.43)$$

which is much larger than the one previously calculated, despite the lower surface acceleration. Kinematic moment at the interface is then equal to

$$M_k^{int} = \frac{30{,}000{,}000 \cdot \dfrac{\pi \cdot 1^4}{64}}{1/2} \cdot 0.93 \cdot 0.00182 \cdot$$

$$\left\{ -\frac{1}{2} \cdot \frac{1}{15} + \left[\frac{30{,}000{,}000}{14{,}400 + 3600 \cdot (15 - 3)} \right]^{-0.25} \cdot \left[\left(\frac{48{,}000}{4800 + 1200 \cdot (15 - 3)} \right)^{0.25} - 1 \right]^{0.5} \right\}$$

$$= 364 \, kNm$$

$$(8.44)$$

This time, the simplified approach provided a conservative result over more accurate calculations.

For the inertial-driven moment at the pile-head, only the value of surface acceleration changes in the simplified formula (Eq. 8.34) resulting in

$$M_{in}^{head} = \chi HL_a = \left[0.23 - 0.035 \cdot 1 \cdot \left(1 - 0.9 \cdot 0.8^2 \right) \right] \cdot 0.5 \cdot 0.2025 \cdot 1200 \cdot 8.0 = 210 \, kNm$$

(8.45)

It is observed that both approaches yielded a pile-head inertial moment which is lower
than the kinematic one, confirming that for large-diameter piles in soft soils, kinematic effects may be more important than inertial ones.

ACKNOWLEDGEMENTS

The authors would like to deeply thank Professors Alessandro Mandolini, Stefano
Aversa, Luca de Sanctis, Nikolaos Klimis, Kyriazis Pitilakis and Emilios Comodromos, who kindly provided fruitful comments and suggestions to improve the quality
of this Chapter. Thanks are also due to Dr. Chiara Iodice and Dr. Maria Iovino for
proofreading the original manuscript. Their contribution is gratefully acknowledged.

REFERENCES

Anoyatis, G., Di Laora, R., Mandolini, A., and Mylonakis, G. (2013) Kinematic response of sin-
gle piles for different boundary conditions: Analytical solutions and normalization scheme.
Soil Dynamics and Earthquake Engineering, 44, 183–195.
Broms, B.B. (1964) Lateral resistance of piles in cohesive soils. *Journal of Soil Mechanics and
Foundations Division*, 90(2), 27–64.
Cairo, R., Conte E., Dente G., Sica, S., and Simonelli, A.L. (2009) Soil-pile kinematic interaction: New perspectives for EC8 improvement. In E. Cosenza (ed.), *Eurocode 8 Perspectives
from the Italian Standpoint Workshop*, Doppiavoce, Napoli, Italy, 263–275.
Chidichimo, A., Cairo, R., Dente, G., Taylor, C.A., and Mylonakis, G. (2014) 1-g experimental
investigation of bi-layer soil response and kinematic pile bending. *Soil Dynamics and Earthquake Engineering*, 67, 219–232.
Darendeli, M.B. (2001) Development of a new family of normalized modulus reduction and
material damping curves. PhD Thesis, University of Texas at Austin.
de Sanctis, L., Maiorano, R.M.S., and Aversa, S. (2010) A method for assessing kinematic
bending moments at the pile head. *Earthquake Engineering and Structural Dynamics*, 39(4),
1133–1154.
Dezi, F., Carbonari, S., and Leoni, G. (2010) Kinematic bending moments in pile foundations.
Soil Dynamics and Earthquake Engineering, 30(3), 119–132.
Dezi, F., and Poulos, H. (2017) Kinematic bending moments in square pile groups. *International
Journal of Geomechanics*, 17(3), 04016066.
Di Laora, R. (2020) Pile design in seismic areas: Small or large diameter? *Geotechnical Engi-
neering Journal of the SEAGS & AGSSEA*, 51(2), ISSN 0046-5828.
Di Laora, R., and de Sanctis, L. (2013) Piles-induced filtering effect on the foundation input
motion. *Soil Dynamics and Earthquake Engineering*, 46, 52–63.
Di Laora, R., Galasso, C., Mylonakis, G., and Cosenza, E. (2019) A simple method for N-M
interaction diagrams of circular reinforced concrete cross sections. *Structural Concrete*, 21(1),
48–55.

Di Laora, R., Grossi, Y., de Sanctis, L., and Viggiani, G.M.B. (2017b) An analytical solution for the rotational component of the Foundation Input Motion induced by a pile group. *Soil Dynamics and Earthquake Engineering*, 97, 424–438.

Di Laora, R., Mandolini, A. and Mylonakis, G. (2012) Insight on kinematic bending of flexible piles in layered soil. *Soil Dynamics and Earthquake Engineering*, 43, 309–322.

Di Laora, R., Mylonakis, G., and Mandolini, A. (2013) Pile-head kinematic bending in layered soil. *Earthquake Engineering & Structural Dynamics*, 42(3), 319–337.

Di Laora, R., Mylonakis, G., and Mandolini, A. (2017a) Size limitations for piles in seismic regions. *Earthquake Spectra*, 33(2), 729–756.

Di Laora, R. and Rovithis, E. (2015) Kinematic bending of fixed-head piles in nonhomogeneous soil. *Journal of Geotechnical and Geoenvironmental Engineering*, 141(4), 04014126.

Dobry, R., and O'Rourke, M.J. (1983) Discussion on 'Seismic response of end-bearing piles' by Flores-Berrones R and Whitman R.V. *Journal of the Geotechnical Engineering Division*, 109(5), 778.

Dobry, R., Oweis, I., and Urzua, A. (1976) Simplified procedures for estimating the fundamental period of a soil profile. *Bulletin of the Seismological Society of America*, 66(4), 1293–1321.

Dobry, R., Vicente, E., O'Rourke, M., and Roesset, M. (1982) Horizontal stiffness and damping of single piles. *Journal of the Geotechnical Engineering Division*, 108, 439–459.

EN 1998-5 (2004), Eurocode 8: Design of structures for earthquake resistance. Part 5: Foundations, retaining structures, geotechnical aspects. CEN, Brussels.

Fan, K., Gazetas, G., Kaynia, A., Kausel, E., and Ahmad, S. (1991) Kinematic seismic response analysis of single piles and pile groups. *Journal of the Geotechnical Engineering Division*, 117(12), 1860–1879.

Flores-Berrones, R., and Whitman, R.V. (1982) Seismic response of end bearing piles. *Journal of the Geotechnical Engineering Division ASCE*, 108(4), 554–569.

Garala, T.K., Madabhushi, S.P.G., and Di Laora, R. Experimental investigation of kinematic pile bending in layered soils using dynamic centrifuge modelling. Géotechnique (accepted).

Gazetas, G. (1982) Vibrational characteristics of soil deposits with variable wave velocity. *International Journal for Numerical and Analytical Methods in Geomechanics*, 6, 1–20.

Gazetas, G. (1984) Seismic response of end-bearing single piles. *Soil Dynamics and Earthquake Engineering*, 3(2), 82–93.

Gazetas, G., and Dobry, R. (1984) Horizontal response of piles in layered soils. *Journal of Geotechnical Engineering*, 110, 20–40.

Gazetas, G., and Mylonakis, G. (1998) Seismic soil-structure interaction: New evidence and emerging issues. *Geotechnical Earthquake Engineering & Soil Dynamics III*, 2, 1119–1174.

Gustavo, A. (2012) *SHAKE2000–A Computer Program for the 1-D Analysis of Geotechnical Earthquake Engineering Problems*. GeoMotions, LLC: Lacey, Washington.

Hardin, B., and Drnevich, V. (1972) Shear modulus and damping in soils: Design equations and curves. *Journal of the Soil Mechanics and Foundations Division*, 95, 667–692.

Hashash, Y.M.A., Musgrove, M.I., Harmon, J.A., Groholski, D.R., Philips C.A., and Park, D. (2016) DEEPSOIL 6.1 user manual.

Hashash, Y.M.A., Phillips, C., and Groholski, D.R. (2010) Recent advances in non-linear soil response analysis. In *Proceedings of the 5th International Conference on Recent Advances in Geotechnical Earthquake Engineering and Soil Dynamics*, 24–29 May 2010, San Diego, CA.

Hetenyi, M. (1946) *Beams on Elastic Foundation*. University of Michigan Press: Ann Arbor, MI.

Iovino, M., Di Laora R., Rovithis, E., and de Sanctis, L. (2019) The beneficial role of piles on the seismic loading of structures. *Earthquake Spectra*, 35(3), 1141–1162.

Ishibashi, I., and Zhang, X. (1993) Unified dynamic shear moduli and damping ratios of sand and clay. *Soils and Foundations*, 33(1), 182–191.

Ishihara, K. (1996) *Soil Behaviour in Earthquake Geotechnics*. Oxford Science Publications, Clarendon Press: Oxford.

Kaklamanos, J., Bradley, B.A., Thompson, E.M., and Baise, L.G. (2013) Critical parameters affecting bias and variability in site-response analyses using KiK-net downhole array data. *Bulletin of the Seismological Society of America*, 103(3), 1733–1749.

Kallioglou, P. (2003) The study of dynamic properties of soils on resonant-column apparatus. PhD Thesis, Aristotle University of Thessaloniki.

Karatzia, X., and Mylonakis, G. (2016) Discussion of kinematic bending of fixed-head piles in nonhomogeneous soil by Raffaele Di Laora and Emmanouil Rovithis. *Journal of Geotechnical and Geoenvironmental Engineering*, 142(2), 07015042.

Kausel, E., Whitman, R.V., Morray, J.P., and Elsabee, F. (1978) The spring method for embedded foundations. *Nuclear Engineering and Design*, 48, 377–392.

Kavvadas, M., and Gazetas, G. (1993) Kinematic seismic response and bending of free-head piles in layered soil. *Géotechnique*, 43(2), 207–222.

Kawamura, S., Umemura, H., and Osawa, Y. (1977) Earthquake motion measurement of a pile-supported building on reclaimed ground. In *Proceedings of 6th World Conference on Earthquake Engineering*, 10–14 January 1977, New Delhi, India.

Kaynia, A.M., and Kausel, E. (1991) Dynamics of piles and pile groups in layered soil media. *Soil Dynamics and Earthquake Engineering*, 10(8), 386–401.

Kaynia, A.M., and Mahzooni, S. (1996) Forces in pile foundations under earthquake loading. *Journal of Engineering Mechanics*, 122(1), 46–53.

Kottke, A., and Rathje, E. (2009) Technical manual for Strata. Pacific Earthquake Engineering Research, University of California, Berkeley, Report number: 2008/10.

Maiorano, R.M.S., de Sanctis, L., Aversa, S., and Mandolini, A. (2009) Kinematic response analysis of piled foundations under seismic excitations. *Canadian Geotechnical Journal*, 46(5), 571–584.

Makris, N., Gazetas, G., and Delis, G. (1996) Dynamic soil-pile-foundation-structure interaction: Records and predictions. *Géotechnique*, 46(1), 33–50.

Margason, E. (1975) Pile bending during earthquakes. [Lecture] ASCE/UC-Berkeley seminar on design construction & performance of deep foundations.

Martinelli, M., Burghignoli, A., and Callisto, L. (2016) Dynamic response of a pile embedded into a layered soil. *Soil Dynamics and Earthquake Engineering*, 87, 16–28.

Mizuno, H. (1987) Pile damage during earthquake in Japan. In: Nogami, T. (ed.) *Dynamic Response of Pile Foundations*, ASCE Special Publication.

Mucciacciaro, M., and Sica, S. (2018) Nonlinear soil and pile behaviour on kinematic bending response of flexible piles. *Soil Dynamics and Earthquake Engineering*, 107, 195–213.

Mylonakis, G. (1995) Contributions to static and seismic analysis of piles and pile-supported bridge piers. PhD Thesis, State University of New York at Buffalo.

Mylonakis, G. (2001) Simplified model for seismic pile bending at soil layer interfaces. *Soils and Foundations*, 41(4), 47–58.

Mylonakis, G.E., Rovithis, E., and Paraschakis, H. (2013) 1D harmonic response of layered inhomogeneous soil: Exact and approximate analytical solutions. In: Papadrakakis, M., Fragiadakis, M., and Plevris, V. (eds.) *Computational Methods in Earthquake Engineering*. Springer: Dordrecht, Vol. 2, pp. 1–32.

Nikolaou, A., Mylonakis, G., Gazetas, G., and Tazoh, T. (2001) Kinematic pile bending during earthquakes: Analysis and field measurements. *Géotechnique*, 51(5), 425–440.

Ohta, T., Uchiyama, S., Niwa, M., and Ueno, K. (1980) Earthquake response characteristics of structure with pile foundation on soft subsoil layer and its simulation analysis. In *Proceedings of the 7th World Conference on Earthquake Engineering*, 8–13 September 1980, Istanbul, Turkey.

Pender, M. (1993) Aseismic pile foundation design. *Bulletin of the New Zealand National Society for Earthquake Engineering*, 26(1), 49–160.

Randolph, M.F. (1981) The response of flexible piles to lateral loading. *Géotechnique*, 31(2), 247–259.

Rathje, E.M., Abrahamson, N.A., and Bray, J.D. (1998) Simplified frequency content estimates of earthquake ground motions. *Journal of Geotechnical and Geoenvironmental Engineering*, 124(2), 150–159.

Roesset, J.M. (1980) Stiffness and damping coefficients of foundations. In: O'Neil, M., and Dobry, R. (eds.) *Dynamic Response of Pile Foundations*. ASCE, pp. 1–30.

Rovithis, Emm., Di Laora, R., and de Sanctis, L. (2015) Foundation motion filtered by piles: effect of soil inhomogeneity. In Winter, M.G., Smith, M.D., Eldred, P.J.L., and Toll, D.G. (eds.) *Geotechnical Engineering for Infrastructure and Development: Proceedings of the XVI European Conference on Soil Mechanics and Geotechnical Engineering*, 13–17 September 2015, Edinburgh, UK.

Rovithis, Emm., Di Laora, R., Iovino, M., and de Sanctis, L. (2017) Reduction of seismic loading on structures induced by piles in inhomogeneous soil. In Papadrakakis, M., and Fragiadakis, M. (eds.) *Compdyn 2017: Proceedings of the 6th ECCOMAS Thematic Conference on Computational Methods in Structural Dynamics and Earthquake Engineering, COMPDYN2017*, 15–17 June 2017, Rhodes Island, Greece. Paper No. 18333.

Rovithis, E., Mylonakis, G., and Pitilakis, K. (2013) Dynamic stiffness and kinematic response of piles in in homogeneous soil. *Bulletin of Earthquake Engineering*, 11(6), 1949–1972.

Rovithis, E., Parashakis, Ch., and Mylonakis, G. (2011) 1D harmonic response of layered inhomogeneous soil: Analytical investigation. *Soil Dynamics and Earthquake Engineering*, 31(7), 879–890.

Saitoh, M. (2005) Fixed-head pile bending by kinematic interaction and criteria for its minimization at optimal pile radius. *Journal of Geotechnical and Geoenvironmental Engineering*, 131(10), 1243–1251.

Schnabel, P.B., Lysmer, J., and Seed, H.B. (1972) SHAKE: A computer program for earthquake response analysis of horizontally layered sites. University of California, Berkeley, Report No. UCB/EERC–72/12.

Seed, H.B., and Idriss, I.M. (1970) Soil moduli and damping factors for dynamic response analysis. Earthquake Engineering Research Center, Berkeley, California Report No. EERC70–10.

Seed, H.B., and Idriss, I.M. (1982) Ground motions and soil liquefaction during earthquakes. Earthquake Engineering Research Institute Monograph, Oakland, CA.

Semblat, J.F., and Pecker, A. (2009) *Waves and Vibrations in Soils: Earthquakes, Traffic, Shocks, Construction Works*. IUSS Press: Pavia.

Sica, S., Mylonakis, G., and Simonelli, A.L. (2011) Transient kinematic pile bending in two-layer soil. *Soil Dynamics and Earthquake Engineering*, 31(7), 891–905.

Stacul, S., and Squeglia, N. (2020) Simplified assessment of pile-head kinematic demand in layered soil. *Soil Dynamics and Earthquake Engineering*, 130, 105975.

Tazoh, T., Shimizu, K., and Wakahara, T. (1987) Seismic observations and analysis of grouped piles. In: ASCE Geotechnical Special Publication No. 11, Dynamic response of pile foundations, pp. 1–20.

Thompson, W.P. (1950) Transmission of elastic waves through a stratified soil stratum. *Journal of Applied Physics*, 21, 89–93.

Tokimatsu, K., Suzuki, H., and Sato, M. (2005) Effects of inertial and kinematic interaction on seismic behavior of pile with embedded foundation. *Soil Dynamics and Earthquake Engineering*, 25, 753–762.

Towhata, I. (1996) Seismic wave propagation in elastic soil with continuous variation of shear modulus in the vertical direction. *Soils and Foundations*, 36(1), 61–72.

Vrettos, Ch. (2013) Dynamic response of soil deposits to vertical SH waves for different rigidity depth-gradients. *Soil Dynamics and Earthquake Engineering*, 47, 41–50.

Vucetic, M., and Dobry, R. (1991) Effect of soil plasticity on cyclic response. *Journal of Geotechnical Engineering*, 117(1), 89–107.

Chapter 9

Pile foundations in liquefiable soils

S.P. Gopal Madabhushi
University of Cambridge

CONTENTS

9.1   INTRODUCTION

Pile foundations are extensively used in civil engineering to transfer heavy superstructure loads onto competent soil strata. This is particularly important when the surface layers are quite soft, for example, as in the case of offshore sites or other marginal ground. In such cases, the pile foundations are used to carry the axial loads to stiffer sands or clays that exist deeper below, which are identified by a site investigation. Piles are predominantly designed to carry axial loads although there will be applications where the lateral load-carrying capacity of piles is required. One such example is the earthquake loading occurring at a site. The inertial loads from the superstructure during a seismic event can impose significant lateral loads on piles. Similarly, wind loading on the superstructure can also induce lateral loads on piles.

Piles can be classified in different ways. Based on the pile material, they can be steel, tubular piles or reinforced concrete piles. Based on the method of installation, they may be classified as follows:

- Driven piles
- Bored or cast in-situ piles

The way the pile foundation responds to axial and lateral loading can change depending on both the pile material and the method of installation. The earth pressure coefficient K_s mobilised around the pile depends both on the type of pile material and the installation method (driven or cast in-situ piles). The load-carrying mechanisms can be very different for these two types of piles. Driven piles can generate large horizontal stresses during the driving process which leads to a large shaft friction. Large stresses are also generated at the tip of the pile during the driving process, particularly if the pile end is solid or when it forms a soil plug. These stresses can remain high, even after the pile driving has stopped and the pile has been driven to the required depth. This gives these piles a large end bearing capacity. Bored piles on the other hand are created by auguring a vertical hole in the ground and placing the reinforcing steel cage and concrete. This construction process means that the horizontal stresses around the pile shaft are not as large and therefore the initial shaft friction is not as high. Similarly, the end bearing capacity can be small soon after casting the pile in-situ. However, on the application of first load, the pile can start to settle and mobilise both shaft friction and end bearing.

In this chapter, we shall focus on the pile foundations under earthquake loading. In particular, we shall focus on the pile foundations passing through liquefiable soils, i.e., loose, saturated sands, or silts. It is therefore necessary to understand the term liquefaction used in this context and how it is assessed in the field.

9.2 SOIL LIQUEFACTION

Soil liquefaction occurs when loose, saturated sands or silts are subjected to cyclic shear strains induced by the earthquake loading. Loose sands have a tendency to contract in volume upon shearing. When they are fully saturated, this tendency to suffer volumetric contraction is manifested as an increase in the pore water pressure. As these pore water pressures are over and above the hydrostatic water pressure at any given depth, they are termed as excess pore pressures. The excess pore pressures that are generated due to earthquake loading can be so large that they can be nearly equal to the total stress in the soil. When this occurs, it is said that full liquefaction has occurred. Research at Cambridge has shown that even when the excess pore pressures do not reach the magnitude of total stress, i.e., when partial liquefaction occurs, the strength and stiffness of the soil can degrade significantly, Madabhushi & Schofield (1993). Under these circumstances, the natural frequency of the degraded soil-structure system can come close to one of the earthquake's driving frequencies resulting in resonance that can lead to excessive vibrations and damage.

In the field, liquefaction assessment is normally carried out by conducting in-situ tests such as SPT or CPT tests. Historically, SPT tests are quite common especially in the United States due to their perceived advantage of soil sample recovery and visual inspection of the recovered samples from the soil strata. CPT testing is quite popular in Europe due to the continuous measurement of soil strength with depth. Liquefaction susceptibility of soil strata can be assessed based on SPT or CPT profiles obtained from in-situ testing. Low SPT blow count (N) in SPT tests or low cone penetration resistance (q_c) is indicative of high liquefaction potential of the soil strata. In addition, sometimes the soil profiling is carried out using SASW testing and establishing the shear wave velocity (V_s) profile with depth. There are many researchers who have tried to link the in-situ measurements of soil strength to the liquefaction potential of the soil layers. In their seminal work in the early 1970s, Seed & Idriss (1971) have linked the cyclic stress ratio (CSR) to the SPT blow count. The CSR is calculated based on the earthquake strength, relative density of the soil strata, fines content, etc. and is indicative of the driving forces that cause the cyclic shear stresses in the soil strata. These original charts have been updated over the years by many researchers, for example, see Boulanger & Idriss (2014). Also, efforts have been made to link the CSR or the cyclic resistance ratio (CRR) to the cone tip resistance (q_c) for use with CPT testing, for example, see Robertson & Wride (1998). In Eurocode 8, a version of the liquefaction curves is recommended for magnitude $M_w = 7.5$ earthquakes (Eurocode 8, Part 5, 2003). Boulanger & Idriss (2014) have suggested that for a reference earthquake magnitude of $M_w = 7.5$ and effective overburden pressure of $\sigma'_v = p_a$ (100 kPa), the CRR can be calculated using following equation:

$$\text{CRR}_{M_W=7.5, \, \sigma_v'=p_a} = \exp\left(\frac{q_{c1N,cs}}{113} + \left(\frac{q_{c1N,cs}}{1000} \right)^2 - \left(\frac{q_{c1N,cs}}{140} \right)^3 + \left(\frac{q_{c1N,cs}}{137} \right)^4 - 2.80 \right) \quad (9.1)$$

where $q_{c1N, \, cs}$ is the corrected CPT tip resistance normalised for overburden and fines content. The CRR should be multiplied by the earthquake magnitude scaling factor (MSF) given in Eq. 9.2 and by the effective overburden stress factor ($K\sigma$) given in Eq. 9.4.

$$\text{MSF} = 1 + \left(\text{MSF}_{\text{max}} - 1 \right) \left(8.64 \, \exp\left(-\frac{M_w}{4} \right) - 1.325 \right) \quad (9.2)$$

where M_w is the moment magnitude, and MSF_{max} is given by Eq. 9.3.

$$\text{MSF}_{\text{max}} = 1.09 + \left(\frac{q_{c1N,cs}}{180} \right)^3 \leq 2.2 \quad (9.3)$$

$$K_\sigma = 1 - \frac{1}{37.3 - 8.27 \left(q_{c1N,cs} \right)^{0.264}} \ln\left(\frac{\sigma_v'}{p_a} \right) \leq 1.1 \quad (9.4)$$

Equation 9.1 can be plotted graphically as shown in Figure 9.1 for an earthquake with a moment magnitude $M_w = 7.5$. The curved lines in this figure demarcate the 'liquefaction zone' and the 'no liquefaction zone', i.e., if the corrected $q_{c1N, \, cs}$ value of a soil layer plots below the line, the soil layer is deemed to be non-liquefiable, and if it plots above the line, it is considered to be 'liquefiable'. Using Eqs. 9.2 and 9.3, plots are created for different moment magnitude (Mw) earthquakes. Lower magnitude earthquake ($Mw \leq 7.5$) shifts the curve to the left, whilst stronger earthquakes ($Mw > 7.5$) will shift the curve to the right.

A word of caution is necessary whilst using liquefaction susceptibility curves such as the ones shown in Figure 9.1. As explained above, these curves are used in practice as a binary distinguisher of liquefaction, i.e., either liquefaction occurs, or it does not. However, recent research at Cambridge has shown that significant pore pressures can be generated when a particular soil plots close to but below the line.

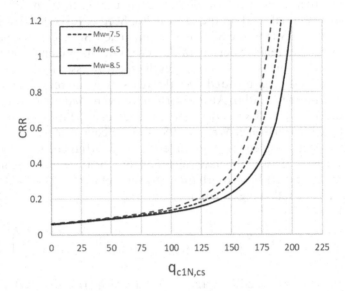

Figure 9.1 Liquefaction triggering chart (Boulanger & Idriss, 2014).

Recently, an in-flight, miniature CPT testing device was developed and used at the Schofield Centre, University of Cambridge as part of the Liquefaction Experiment and Analysis Project (LEAP, Kutter et al., 2018). Using this device, it was possible to carry out in-flight CPT testing on level beds of saturated Hostun F-65 sand in a centrifuge test before and after earthquake loading. This is a clean sand with very little fine content. These tests revealed the increase in corrected $q_{c\text{-}1N}$ resistance post earthquake as the soil densifies following a liquefaction event and subsequent reconsolidation. In addition, it was also possible to measure the excess pore pressures using miniature pore pressure transducers during the actual liquefaction event. The excess pore pressure ratio (r_u) can therefore be obtained independently and directly from the experimental measurements. The cyclic stress ratio (CSR) was calculated using the acceleration measurements within the soil model during the earthquake events. In Figure 9.2, the Robertson & Wride (1998) curve for liquefaction potential is replotted and extended towards low $q_{c\text{-}1N}$ values. Using the centrifuge experimental data, it was possible to plot at various depths within the soil model where both the accelerations and excess pore pressures were measured. This allows for a direct comparison of the experimental data to the liquefaction potential curve identified by Robertson & Wride (1998). In Figure 9.2, it can be seen that although this curve broadly identifies the liquefiable and non-liquefiable strata, there are quite a lot of data points on either side (Dobrisan, 2018). For example, there are several points below the curve at which full liquefaction has occurred. Similarly, there are points above but close to the curve where liquefaction did not occur. Further, at some of the points, the excess pore pressure ratios are also shown. It can be seen that these r_u values can be quite high on either side of the curve. It is therefore argued that care must be exercised when using such curves, and geotechnical engineers must expect that significant excess pore pressures may be generated in the soils that fall close to the curve, even on the non-liquefiable side.

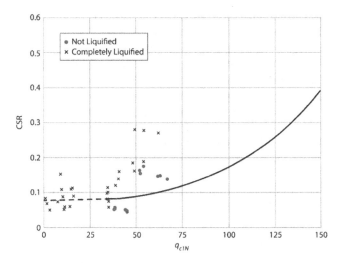

Figure 9.2 Recent centrifuge data compared to a liquefaction potential curve (Dobrisan 2018).

In this chapter, we shall focus on the pile foundations that pass through liquefiable ground. In many cases, we would expect that such loose sand layers will be underlain by either dense sand, stiff clay, rock, or other competent bearing strata. The piles would rarely be terminated in such a loose sand layer, as the axial loads being applied to the piles need to be transferred to a competent bearing stratum. The main concern for pile foundations is the additional loading that may arise due to soil liquefaction that needs to be borne by the piles or the stability of the pile foundation itself. Typically, 10–20 m of saturated, loose sand or silt layer may be present at the surface that may be susceptible from a liquefaction point of view. Below that depth, the pile foundations would enter non-liquefiable, competent strata.

9.3   PILES IN LIQUEFIABLE SOILS

Pile foundations are often used to support buildings, bridge piers, and other types of structures. When loose, saturated sands or clay layers are present at the ground surface, the pile foundations transfer the loads from the superstructure into more competent strata below these weak layers. During liquefaction events, the piles can be subjected to several additional loads. Piles in a group can be subjected to additional dynamic push-pull loads due to the moments imposed on them by the rocking tendency of the superstructure. In addition, they are subjected to lateral loads due to earthquake loading itself. These lateral loads are normally classified as inertial and kinematic loads. The inertial loads arise due to the inertia of the superstructure relative to the pile foundations that move with the ground and are dominant in the upper regions of the piles and pile-pile cap interface. Kinematic loads arise due to the differences in the lateral deformations between the pile and the surrounding soil during earthquake loading. These kinematic loads can be large where there are large differences in lateral deformations between the pile and the soil such as the pile heads or where the piles pass through soil strata with large stiffness contrast, i.e., at the interface between soft and stiff soil layers. More specific details on the inertial and kinematic loads were recently investigated by Garala et al. (2020) and Garala & Madabhushi (2018).

When piles pass through sloping ground, additional lateral loads are imposed on them when liquefaction-induced lateral spreading occurs. There are many examples of lateral spreading causing damage to civil engineering infrastructure. When liquefaction occurs, lateral spreading of the banks of a river can lead to the sloping ground on either side to move towards the river channel. This lateral spreading can lead to the rotation and tilting of the piers. This is demonstrated in Figure 9.3 in which the piers of a bridge near Nantou, Taiwan have suffered tilting following the 921 Chi-Chi earthquake.

Even when the bridge piers are substantial and do not suffer visible damage, the lateral spreading can impose significant lateral loads. This can be seen in Figure 9.4a and b where the lateral spreading has caused significant ground deformation at the base of the concrete pier of a new bridge being constructed near Nantou in Taiwan.

The lateral loading due to slope movements along the edge of a river can be so large as to cause actual shearing of a concrete pile. This was observed at a bridge in Christchurch during the New Zealand earthquakes of 2011/12. The piles were supporting the bridge abutment, and the lateral spreading applied sufficient loading that resulted in the shearing of the RC square pile as seen in Figure 9.5. The bridge abutment itself suffered rotation due to the lateral spreading of the ground towards the Avon river. There were several other bridges that suffered damage during the Christchurch earthquakes, and in many of these cases, the pile foundations supporting the abutments saw damage. More details of these mechanisms can be found in Haskell et al. (2013).

It is clear from the above examples that pile foundations are vulnerable from soil liquefaction and its consequences such as the lateral spreading of sloping ground. In the next sections of this chapter, we shall look at the mechanisms by which pile foundations can suffer damage or even failure. In this chapter, we shall focus on the behaviour of piles in level ground only. However, more details on the behaviour of piles in sloping ground that suffers lateral spreading can be found in Madabhushi et al. (2009).

Figure 9.3 Tilting of bridge piers due to lateral spreading following 921 Chi-Chi earthquake in Taiwan.

(a) Lateral loading due to spreading soil   (b) Lateral displacement of soil

Figure 9.4 Lateral spreading next to concrete bridge pier.

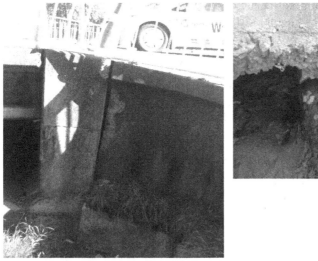

Figure 9.5 Rotation of a bridge abutment and shearing of the RC square, abutment piles.

9.4   PILES AND PILE GROUPS IN LEVEL GROUND

From a structural engineering point of view, piles are long, slender columns that carry large axial loads. In level ground, they carry these axial loads from the superstructure and transfer them through marginal ground to a more competent soil stratum or bed rock below. If one considers typical sizes of piles, their L/D ratio will be often greater than 50 or slenderness ratio greater than 100. For example, a 30 m long pile (L) that has 500 mm diameter (D) will have a L/D ratio of 60. Whilst no structural engineer would design a column that long and with such high L/D or slenderness ratios without the use of struts or bracings to prevent buckling, geotechnical engineers do not normally consider the slenderness ratio of piles. The reason behind this is that we inherently assume that the pile is supported laterally by the ground at all levels, and therefore, buckling of piles is not an issue unless in very soft soils such as marine clays, etc. Pile buckling in sandy soils is therefore never considered. However, when the ground suffers liquefaction due to earthquake loading, the lateral support offered by the ground to the pile diminishes whilst it is in the state of liquefaction. The axial loads carried by the pile remain and can even be enhanced due to the inertial loading coming from the rocking of the superstructure. The long, slender piles can therefore suffer buckling under these conditions.

In addition to the loss of lateral support offered by the ground during liquefaction, another condition can play an important role in determining whether the pile buckling occurs or not. From a structural engineering point of view, for buckling to occur, the ends of the column should be restrained from having any axial displacements. The fixity of the ends of the column will determine the mode of buckling and the Eulerian buckling loads. With the pile foundations, this end fixity can occur if the pile is driven into bedrock and rock-socketed. If the tip of the pile is resting in dense sand or stiff clay, it may be easier for the pile to suffer additional settlement rather than buckle. Clearly, there will be a limit on the length of the pile that is in the dense sand or stiff clay layer which determines whether settlement or buckling would occur. One can imagine a pile driven deep into such layers, where the axial load is unable to cause additional settlement of the pile, i.e., the zero-displacement condition is achieved due to the long length of the pile in the competent strata. In this case, it may be easier for the pile to suffer buckling. On the other hand, if the pile only penetrates $10D$ or so (where D is the diameter of the pile) into the competent strata, then settlement of the pile may be easier than buckling. This, of course, depends on the level of liquefaction and the consequent excess pore pressure ratio generated in the area surrounding the tip of the pile as will be seen in Section 9.6. Equally, the slenderness ratio can play a role here. Long, slender piles can suffer buckling more easily. On the other hand, stout piles with large diameters may suffer settlement more easily rather than buckle.

9.5 PILE BUCKLING

Whilst pile buckling was considered previously in soft clays, very little research existed for piles in liquefiable soils. Using dynamic centrifuge modelling, liquefaction-induced buckling of piles has been investigated at Cambridge by Bhattacharya (2003) and Bhattacharya et al. (2004, 2005) who considered single piles. Although this research established the plausibility of pile buckling in liquefied soil, very few cases in the field exist where a single pile is used. One example of this is the Showa bridge that failed during Niigata earthquake of 1964 in Japan (Bhattacharya et al. 2005), which consisted of piers that were supported on a single row of piles. As such cases seldom occur in practice, it was necessary to see if pile groups can also suffer liquefaction-induced instability. This work was carried out by Knappett (2006), who investigated the more commonly used 2×2 pile groups. In both these test series, the pile tips were modelled as rock-socketed, thereby avoiding the possibility of pile settlements such that instability could be studied in isolation.

If we consider the simplest idealisation of liquefied soil as a material having zero strength and stiffness, a pile passing through a fully liquefied stratum of soil is expected to behave similar to an unsupported column. Under these conditions, the critical axial load, sometimes referred to as the Euler load at which buckling occurs, is given by Euler's equation:

$$P_{cr} = \frac{\pi^2 EI}{(\beta L_p)^2} \tag{9.5}$$

where E is the Young's modulus of the pile material, I is the second moment area of the pile section, L_p is the pile length, and β is a factor accounting for the fixity at both ends of the pile. Values of β for common fixity conditions are shown in Figure 9.6. Essentially, $\beta \times L_p$ is the equivalent length of a column with pinned-pinned ends, which has the same critical load as a column with the given fixity conditions.

Using Figure 9.6 along with the analogy of piles surrounded by liquefiable soil as unsupported structural columns, for single cantilever piles (fully fixed at the base by the rock socket), $\beta = 2$, whilst for piles which are part of a pile group (i.e., fixed against rotation at the top by the pile cap, but whole pile group is free to sway), $\beta = 1$. Sometimes, βL_p can be termed the effective length of the pile and can be normalised by the radius of gyration r_g of the pile section to give the slenderness ratio, $\beta L_p/r_g$. At low values of slenderness ratio, the critical loads are much higher than the plastic squashing loads such that the latter effect controls failure; such piles are considered 'stocky'. Piles which fail by buckling rather than squashing are 'slender' and have higher values of slenderness ratio. The stocky piles that will reach failure in compression can be separated from slender piles that reach failure due to buckling instability defined by Eq. 9.1. In Figure 9.7, the horizontal line indicates the failure due to axial stresses reaching the yield stress for stocky piles, whilst the curved line indicates the failure due to buckling instability. The demarcation between these two falls at a slenderness ratio of about 50. Figure 9.7 also shows a summary of the results of dynamic centrifuge tests on both single piles and piles in groups carrying axial load in fully liquefied soil which clearly shows that such piles will buckle if P_{cr} is exceeded. A poor performance relates to the pile(s) suffering sudden catastrophic failure.

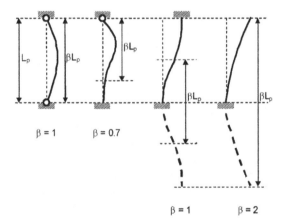

Figure 9.6 Mode shapes for buckling with various β values.

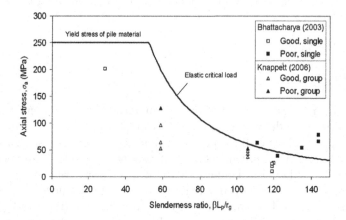

Figure 9.7 Summary of dynamic centrifuge test results on rock-socketed piles in fully liquefied soil.

It is also apparent from Figure 9.7, however, that piles can become unstable at loads below critical. The reason for this is that piles in level ground will often have small imperfections from vertical. This may be due to the lateral displacement at the pile head which may occur cyclically due to earthquake shaking. In the field, the imperfections may also occur for driven piles, as the pile tip can 'wander' away from the vertical alignment during the pile driving process. The presence of axial load will cause amplification of these lateral displacements. If the amplified displacement creates sufficient curvature in the pile to cause yield, the pile will collapse at an axial load lower than critical. Continuing the analogy of a pile being unsupported column, i.e., liquefied soil will not offer any lateral support to the pile from earlier in this section, the effect of imperfections on axial failure loads can be expressed in terms of the Perry-Robertson equation:

$$(\sigma_a)^2 - \left[\sigma_y + \left(1 + \frac{\Delta_0 D_0}{2r_g^2}\right)\frac{P_{cr}}{A_p}\right]\sigma_a + \sigma_y \frac{P_{cr}}{A_p} = 0 \tag{9.6}$$

where σ_y is the column yield stress, Δ_0 is the size of the imperfection, D_0 is the pile diameter, A_p is the cross-sectional area of the pile, and r_g and P_{cr} are as defined earlier in this chapter. Eq. 9.2 plots as shown in Figure 9.8 as a continuous curve on axes of axial stress in the pile σ_a versus slenderness ratio (i.e., as in Figure 9.7).

For increasing imperfection size, the maximum stress which can be sustained before collapse occurs reduces for a given slenderness ratio. In the centrifuge tests of Knappett (2006), the size of imperfections in the pile groups before liquefaction occurred was measured, and the data collected have been compared to Eq. 9.2 for a typical value of Δ_0 measured during the tests. This is shown in Figure 9.8, which shows close agreement between the experimental data and Eq. 9.2.

As well as reducing the axial collapse load, initial imperfections may be amplified (increased) due to the axial loads present in the piles. Amplification of lateral displacements has additionally been studied for unsupported columns by Timoshenko & Gere (1961) in which the amplification factor is related to the fraction of critical load carried according to

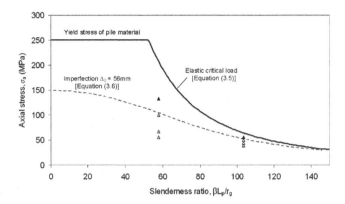

Figure 9.8 Comparisons of computed behaviour for perfect and imperfect piles with centrifuge test data.

$$\text{Amplification} = \frac{1}{1 - \dfrac{P}{P_{cr}}} \tag{9.7}$$

In the dynamic centrifuge tests reported by Knappett (2006), lateral deflections were measured at the pile heads before, during, and after liquefaction. Comparison between these results and Eq. 9.3 are shown in Figure 9.9. It can be seen that the match between observed and predicted behaviour becomes worse as the piles become more slender. For stockier piles, the pile flexibility will tend to dominate compared to that of the soil, whilst for more slender piles, the soil will have a much greater influence on the overall response. This would suggest that the differences between the two sets of results are due to the stiffness

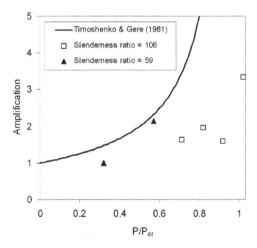

Figure 9.9 Comparison of measured amplifications from dynamic centrifuge tests against predictions using a zero strength and stiffness soil model.

of the liquefied soil being severely underestimated by considering it to be zero. Improved estimates of soil stiffness would therefore be expected to improve the match at high slenderness ratio whilst having a smaller effect on the better match at lower slenderness.

In the case of both single piles and pile groups, the onset of instability leads to collapse due to the formation of plastic hinges in the piles, i.e., the plastic moment capacity of the pile section being exceeded. For soil liquefied over its full depth, the locations of maximum bending moment (and therefore also plastic hinges) would be expected to occur at the bottom of the piles socketed into underlying firm ground (e.g., bedrock) and immediately beneath the cap (in the case of a pile group). However, observations of collapsed piles and pile groups from centrifuge model tests revealed that whilst hinges did form beneath the pile cap, the lower set of hinges occurred at an intermediate depth in the piles. Bhattacharya (2003) proposed that this was due to resistance developed within the liquefied soil as the pile shears the soil. Simplifying the soil reaction as a constant value with depth which is independent of relative soil-pile displacement gives the following equation for the depth below pile head level at which the pile hinge would form in a cantilever (free-head) pile:

$$z_h = 1.165\sqrt{\frac{EI}{P}} \tag{9.8}$$

where P = axial pile load and EI = pile bending stiffness. Eq. 9.4 was found to predict hinges at an intermediate depth within a single free-head pile and gave a reasonable match to observed hinge locations. In Figure 9.10, the extracted pile following a

Figure 9.10 Plastic hinge forms approximately at 2/3rd length in a single pile.

centrifuge test is shown. In this figure, it can be clearly seen that the plastic hinge occurs *not* at the base of the pile but above the base at about 2/3rd the length of the pile.

Experimental measurements of the response of liquefied soil to large monotonic relative soil-pile displacement by Takahashi et al. (2002) and Towhata et al. (1999) have revealed that the lateral soil-pile reaction is highly non-linear with lateral displacement. At low lateral soil-pile displacement, the resistance of the soil is at a low residual value. After a certain displacement, however, the local excess pore pressures around the pile start to drop sharply, leading to a consequent increase in soil-pile resistance. Takahashi et al. (2002) termed this displacement (when normalised by pile diameter) to be the reference strain of transformation (γ_L) as it qualitatively indicates a phase transformation point at which the soil switches from contractile to dilative behaviour. More details of this were discussed by Madabhushi et al. (2009).

Whilst it is easy to understand the buckling of a single pile in liquefied soil, as explained earlier, very few cases in the field exist where single piles are used. Most common use of piles will be in a group (2×2, 3×3 etc.) with a pile cap. Knappett & Madabhushi (2009a, b) and Knappett (2006) have investigated the buckling of a 2×2 pile group in liquefied soil. An example of the buckled 2×2 pile group is shown in Figure 9.11. This pile group has been extracted from the centrifuge model after being subjected to an earthquake-induced liquefaction event, during which it has buckled forming plastic hinges at locations shown in this figure.

The location of the plastic hinges in this pile in the 2×2 pile group shown in Figure 9.11 also indicates that the liquefied soil offers some resistance to the lateral movement of pile during the buckling process. In modelling the response of pile

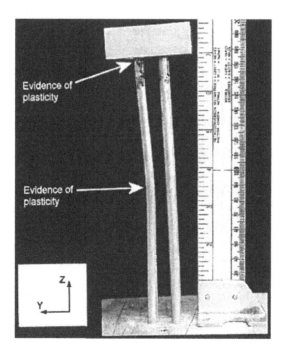

Figure 9.11 Buckling of 2×2 pile group.

groups, Knappett & Madabhushi (2009b) undertook numerical modelling using a beam on non-linear Winkler foundation (BNWF) model such that non-linear lateral soil-pile behaviour due to the resistance offered by the liquefied soil could be modelled. These models were analysed using a Riks algorithm (Riks, 1972, 1979) which can model the full post-buckling response (plastic collapse) of pile groups as the axial critical load is approached. Prior to collapse, the amplification of lateral deflections can also be examined within the same model. Figure 9.12 shows a comparison between the collapsed mode shape and pile stresses from the numerical model with the corresponding mode shape and observed locations of plasticity from the corresponding centrifuge model shown in Figure 9.11. The locations of plastic hinges in Figure 9.12a were identified by looking at the Von Mises stresses in the pile material and identifying the locations where they reach the yield stress of the pile material.

The numerical model is also able to give improved predictions of amplification as shown in Figure 9.13. In this figure, the ratio of the predicted and measured amplification is plotted on the y axis, and therefore, the data points close to 1 (drawn as a solid line in Figure 9.13) indicate a good match. Clearly, the numerical model described above is able to capture the amplification much better compared to the Timoshenko & Gere (1961) approximation given in Eq. 9.3. This also confirms that it is the liquefied soil response which is responsible for the mismatch between predicted and measured amplification shown in Figure 9.9.

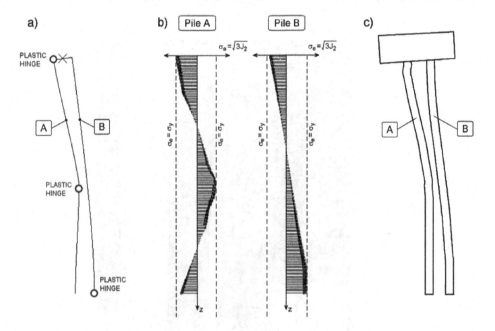

Figure 9.12 Numerical post-buckling simulations using Riks algorithm: (a) collapse mechanism from simulation; (b) Von-Mises stresses within piles showing yield locations; (c) permanent (plastic) deformations of centrifuge model measured post-test (displacements at exaggerated horizontal scale).

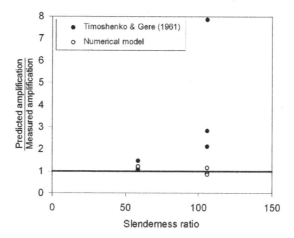

Figure 9.13 Amplification of lateral displacements due to axial load acting on 2 × 2 pile groups in fully liquefied soil, marking the improvement in prediction using the soil-pile interaction behaviour.

9.6 PILE SETTLEMENTS

In the above section, we have seen that when piles cannot settle during an earthquake-induced liquefaction event, they can suffer instability failures, i.e., suffer buckling. The inability of piles to settle may occur if the pile tips are rock-socketed or the piles are driven to large depths into the competent non-liquefiable soil layers that underlay the upper, liquefiable layers. However, in many sites around the world, the piles or pile groups are only driven some $10D$ (D is the outer diameter of the pile) into the underlying layers due to reasons of pile refusal during driving or when these competent strata are quite deep. In such cases, the settlement of piles or pile groups becomes very important, particularly when the surface layers are liable to liquefaction. Madabhushi et al. (2009) describe a procedure to demarcate the occurrence of buckling versus the settlement of piles. In this section, the settlement of piles following a liquefaction event will be considered.

Piles or pile groups can suffer settlement during a liquefaction event caused by an earthquake due to a number of reasons. The shaft friction offered by the loose sand layer prior to liquefaction may reduce significantly during liquefaction, thereby increasing the axial loads in the lower regions of the piles. This increase in axial loads can cause additional settlements. The competent stratum such as a dense sand layer itself may lose some of its strength due to excess pore pressure generation and consequent drop in effective stress. This can lead to a reduction in base capacity of the pile leading to further settlements.

Whilst a significant number of case histories exist of pile foundations suffering failure or excessive settlements for example, Berrill et al. (2001), Haskell et al. (2013), it is difficult to get all the relevant data from case histories to establish the mechanisms of failure that lead to pile settlements.

Dynamic centrifuge tests offer a way forward to investigate the settlements of piles in liquefiable soils. Knappett (2006) and Stringer (2012) have both investigated settlements of piles that are passing through an upper liquefiable layer and penetrating into a dense sand layer below. Knappett's experiments focussed on the measurement of pile loads at the base and top of the pile, whilst Stringer's experiments measured the axial loads in the piles along the length of the pile. In this section, we shall focus on the settlements that were recorded during these two sets of experiments. When the pile cap of the pile foundation is clear off the ground surface as shown in Figure 9.14 or if the surface layers are quite soft, then the pile foundations and the pile cap can suffer excessive settlement. On the other hand, if the pile cap is resting on the ground surface that is reasonably competent, then the settlements will be relatively smaller. Knappett & Madabhushi (2008) tested a 2 × 2 pile group in the centrifuge with this type of configuration. The pile diameter was about 500 mm, and the pile load was about 1.25 MN per pile with a static factor of safety (SSF) of 2.2. A model earthquake of about 0.2 g peak acceleration was applied at the bedrock that caused liquefaction of the loose sand layer. Significant pore pressures were also observed in the dense sand layer below.

The settlement suffered by the pile group that is resting on the ground surface (see Figure 9.14) is plotted in Figure 9.15. The seismic loading that occurs over a period of about 60 s has caused a settlement of about 500 mm. It is also interesting to note that most of the settlement occurs during the seismic loading, and only a relatively small settlement of about 25 mm occurs in the post-seismic period, mostly due to the consolidation of the soil layers in this period. Nevertheless, the total settlements are quite substantial.

Stringer & Madabhushi (2012) have conducted similar centrifuge tests on 2 × 2 pile group in a similar soil profile of a loose sand overlying a dense sand layer. The pile cap for these piles was clear off the ground surface as seen in Figure 9.14. The piles were much more heavily loaded axially in these centrifuge tests. A sequence of earthquakes was considered allowing sufficient time to dissipate the excess pore pressures between

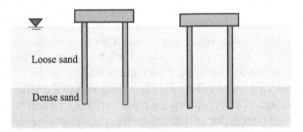

Figure 9.14 Schematic of the cross-section of a pile group.

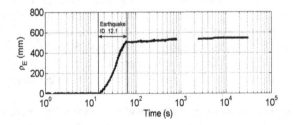

Figure 9.15 The settlement suffered by the pile group.

the earthquakes. This was similar to the Christchurch earthquakes of 2010 and 2011 that occurred with an interval of few months, but both caused liquefaction events to occur near the city of Christchurch. In Figure 9.16, the settlements suffered by the pile group with its pile cap clear off the ground surface (see Figure 9.14) are plotted for two earthquakes E1 and E2, both of which caused liquefaction of the shallow, loose layer. In this figure, it can be seen that the first earthquake causes excessive settlement of nearly 1.1 m, whilst the second earthquake E2 caused a settlement of about 0.2 m. The reason for this may be the increased density of the lower layer following the first earthquake and also the locked in stresses after the first earthquake that can beget strong dilative response from the dense sand during the second earthquake, thereby reducing the settlements.

What is interesting about these settlements is that the settlements between the two earthquakes seem to be continuous. In order to visualise this better, the cumulative settlements for both the earthquakes are plotted in Figure 9.17. In this figure, it can clearly be seen that the settlement of the pile group that occurred during E1 earthquake seems to continue exactly from where it stopped at the end of E1 event and continue smoothly during the E2 earthquake. The slope, which is the rate of settlement, is continuous between the two earthquakes, although in terms of time of occurrence, the two earthquakes are well-separated, allowing for sufficient time for complete dissipation of the excess pore pressures between the two events. This suggests that the soil has a good memory of the stress history and stiffness that were mobilised during the settlement of the pile group in previous earthquake events.

Knappett & Madabhushi (2009c) also tested pile groups with different SSFs under vertical loading. These pile groups had the pile cap resting on the ground surface (see Figure 9.14). Clearly, a lightly loaded pile group will have a higher SSF as is likely to settle less, whilst a heavily loaded pile group will have a lower SSF and is likely to settle more. In Figure 9.18, the normalised settlements with respect to the outer diameter of

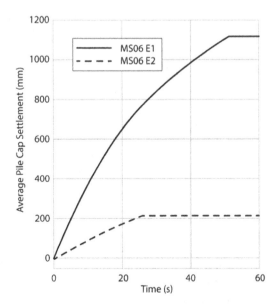

Figure 9.16 Settlement of a pile group during two earthquakes.

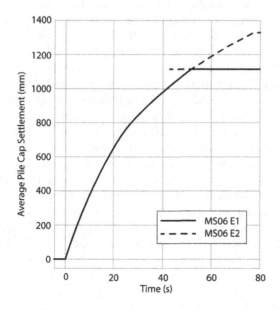

Figure 9.17 Cumulative settlement plot for E1 and E2 earthquakes.

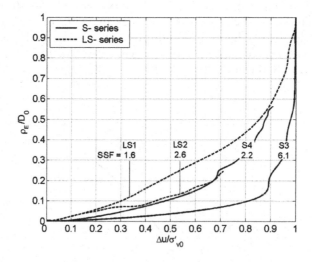

Figure 9.18 Mobilisation of normalised settlement with excess pore pressure ratio.

the pile (D_0) are plotted against the excess pore pressure ratios generated by the earthquake loading. It must be noted that as the earthquake loading progresses with more cycles, the excess pore pressure ratio increases. In Figure 9.18, it can be seen that pile groups with relatively low SSF values require only a small excess pore pressure ratio to start to suffer settlements. On the other hand, for a pile group with a high SSF of 6.1 (i.e., lightly loaded with a load of 0.45 MN per pile), a significant excess pore pressure ratio of 0.9 or higher is required before the settlements start to increase rapidly.

The above observation can be of importance in assessing the vulnerability of an existing pile foundation against liquefaction-induced settlements. Piles that are lightly loaded with high SSFs can tolerate higher excess pore pressures ratios before suffering significant settlements. Such piles may perform better during small to medium strength earthquakes where only a moderate excess pore pressure is expected especially near the area around the tip of the pile.

9.7 LOAD-CARRYING MECHANISMS DURING LIQUEFACTION

In this section, we will look at the load-carrying mechanisms in the pile groups once there has been an onset of liquefaction. We shall focus on the two cases of the pile foundations shown in Figure 9.14 with the case of the pile cap free off the ground surface and the case where the pile cap rests on the ground surface. For the latter, the vertical load from the superstructure is normally in equilibrium with the base and shaft resistances. Accordingly, for a pile group with n piles, we can write

$$V = n(Q_b + Q_s) + Q_{cap} \tag{9.9}$$

where V is the axial load on the pile group and Q_b, Q_s, and Q_{cap} are the base capacity, shaft friction, and the pile cap bearing capacities, respectively. Where the pile group is free off the ground surface, Q_{cap} will be zero. In normal pile design, the contribution of the Q_{cap} is usually neglected as any pile cap bearing capacity adds to the factor of safety against bearing failure. Without any liquefaction, Eq. 9.9 above would be true from a vertical equilibrium point of view.

During liquefaction, we would like to understand the contribution made by each of the load-carrying components identified in Eq. 9.9. If we first consider the case of a free-standing pile, the axial load V/n acting on the pile needs to be supported by a combination of shaft friction and base capacity of each pile in the group. Stringer & Madabhushi (2013) have investigated a 2×2 pile group carrying large axial loads as explained in the previous section. In these centrifuge tests, the loads along the pile shaft were monitored using miniature strain gauges during the earthquake loading. In Figure 9.19, the axial loads recorded at the top of the pile (SG-E) and close to the pile tip (SG-A) are shown along with the acceleration recorded on the pile cap. In this figure, it can be seen that prior to the earthquake loading, the axial loads of 450 and 480 kN were recorded by SG-E and SG-A, with an axial load drop of 30 kN due to the shaft friction between top of the pile and the pile tip.

Similar measurements were also made for the case of the pile cap resting on the ground surface (see Figure 9.14). In Figure 9.20, the axial loads recorded at the top of the pile (SG-E) and the pile tip (SG-A) are shown along with the input accelerations at the bedrock level. In this case, the axial loads recorded before the earthquake were 410 and 380 kN at the pile top and tip, respectively, with a shaft friction of about 30 kN. However, with the application of earthquake loading and the onset of liquefaction, the pile loads both at the top of the pile and pile tip reduce to near-zero values as seen in Figure 9.20. They start to recover post-earthquake, as the excess pore pressures dissipate with time, and the liquefied soil starts to reconsolidate and regain strength.

Similar observations were made at intermediate locations using strain gauges along the piles for both pile group cases shown in Figure 9.14. These are presented for the two

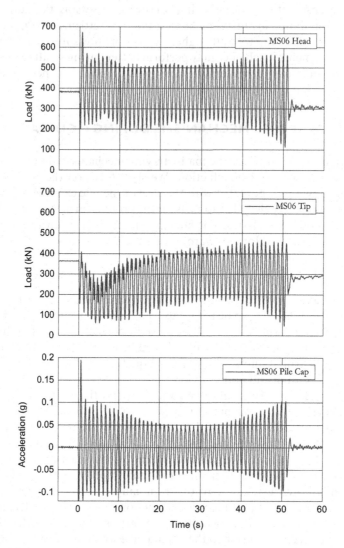

Figure 9.19 Axial loads in a free-standing pile.

cases in Figure 9.21. The axial loads are a measure of the skin friction along the pile (i.e., difference in axial loads between any two elevations will yield the shaft friction mobilised between the elevations). In Figure 9.21, it can be seen that for the case of a free-standing pile group, the initial distribution of axial loads along a pile (marked as 1) starts to reduce with the onset of liquefaction (2). As the pore pressures start to dissipate, the axial loads start to recover (3, 4, and 5). At the end of the dissipation of all the excess pore pressures, the axial load distribution (5) is similar to but slightly lower than the initial load distribution (1). The additional load (difference between 1 and 5) that is no longer supported by the shaft resistance has to be carried by the pile base in the post-earthquake scenario. Another interesting feature in this experimental data is that

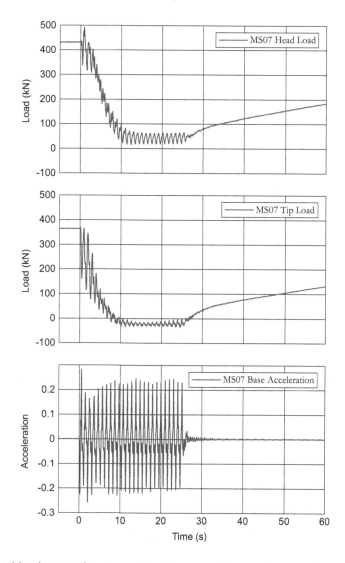

Figure 9.20 Axial loads in a pile group with pile cap resting on the ground surface.

even with full liquefaction (2), the axial load along the pile is about 200 kN on average. It does not reduce below this value. This is because under full liquefaction condition, the end bearing capacity is fully mobilised (as the pile group is suffering settlement). In order to have an instantaneous equilibrium of forces, the axial load (V/n) has to add up to the end bearing capacity and the axial loads in the pile. Thus, a 200 kN force is registered along the pile.

Now let us consider the case of the pile group with the pile cap resting on the ground surface (see Figure 9.14). In this case, the axial loads in the pile in Figure 9.21 (pile cap contact case) start at very similar values to the free-standing case before the earthquake and measured as approximately 400 kN on average (marked as 1 on the

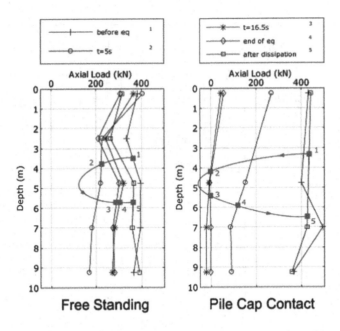

Figure 9.21 Changes in axial loads along the length of the pile during a liquefaction event.

right-hand side in Figure 9.21). With the onset of liquefaction, the axial loads reduce to near-zero values along the length of the pile (2) and remain very low throughout the earthquake (3). They start to increase as the earthquake ends (4) and recover to almost pre-earthquake level once the excess pore pressures dissipate (5). This suggests that the axial loads carried by the piles during the liquefaction phase are almost zero, and only in the post-earthquake phase, they recover. From an equilibrium of forces point of view, one should ask the question, if the piles in the pile group are not carrying the applied axial load (V), who is carrying this load? The answer to this question must be the pile cap. Revisiting Eq. 9.9, if the shaft resistance and end bearing of the piles disappear, then the applied vertical load to the pile group must be supported temporarily by the pile cap. In the post-earthquake period, the soil will regain its strength as the excess pore pressures dissipate, and the piles will start to carry the axial load again (both as shaft resistance and end bearing).

The above observations can be illustrated more clearly using Figure 9.22. Prior to any earthquake loading or liquefaction, the axial load is supported by shaft resistance and the end bearing as illustrated in Figure 9.22a. During liquefaction, the shaft friction and the end bearing reduce to near-zero values, and the axial load is borne by the pile cap by generated bearing pressures below the cap as illustrated in Figure 9.22b. Once the earthquake loading is completed, the excess pore pressures dissipate, and the soil will regain its strength. The shaft resistance and the end bearing of the piles are regained, and the axial load is once again supported by these. It is possible that due to settlement suffered by the pile group, some residual bearing pressures may persist below the pile cap, and the loads carried by the piles both in shaft friction and end bearing may reduce proportionally. This aspect however requires further research.

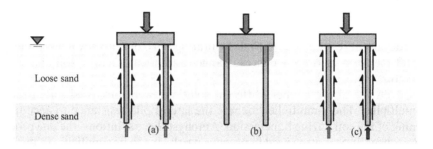

Figure 9.22 Load-carrying mechanisms in pile foundations: (a) pre-earthquake case, (b) during liquefaction case, and (c) post liquefaction case.

The observations made in the above paragraphs can have significant implications to the practising engineers. As mentioned before, in many practical designs, the pile cap bearing capacity is generally ignored. It turns out this is quite important during a liquefaction event, as the pile cap capacity is the only component that is supporting the vertical load during liquefaction. Having a larger pile cap may be advantageous when designing for pile groups in liquefiable soils. Also providing for drainage below the pile cap can be advantageous when dealing with such liquefiable sites.

9.8   LATERAL LOADING ON PILES IN LIQUEFIABLE SOILS

The response of sand as it approaches liquefaction is quite complex, and it gets even more involved due to the soil reaction close to the pile. Due to this complexity, numerical methods for pile-soil interaction during liquefaction have primarily been based on results of centrifuge tests or large-scale 1 g tests and extension of the simple engineering approach based on *p-y* curves. The following studies have indicated that pile-soil interaction is strongly dependent on the sand's relative density and pile movement relative to the soil. Chang & Hutchinson (2013) suggested the *p-y* relationships shown in Figure 9.23. In this figure, case (a) is for liquefied soil for large pile displacements and case (b) is for very loose sand after liquefaction (i.e., very little lateral resistance). Figure 9.23c shows the relative size/shape of the *p-y* curves in different liquefaction conditions (compared to a conventional *p-y* curve under no liquefaction condition).

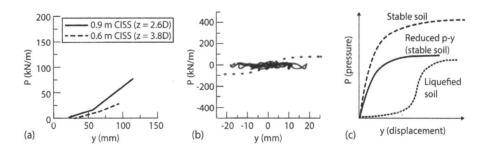

Figure 9.23 Example of *p-y* curves for different sand types, after Chang & Hutchinson (2013).

Tokimatsu et al. (2001) collected data from large-scale 1 g testing which cover the cases mentioned above. They plotted the sub-grade reaction against the relative displacement between the pile and the soil. Their results indicate that the liquefied soil offers very little lateral resistance to the piles vibrating horizontally due to earthquake loading, and the form of the p-y curves change dramatically as the soil approaches the liquefaction state.

Based on the above studies, different solutions were proposed, often referred to as p-multiplier. The p-multiplier reduces the lateral pile resistance to account for the softening of the soil during liquefaction. Amongst these solutions, the one proposed by Liu & Dobry (1995) is often used in practice partly due to its simplicity. In this solution, the p-multiplier is taken as $C_u = 1.0 - r_u$, where r_u is the excess pore water pressure ratio. One advantage of this method is that the data from a site response analysis can directly be used to compute the excess pore water pressure and hence the p-multiplier. Another approach is to represent the liquefied sand as an undrained material with a shear strength equal to the cyclic undrained or residual (S_r) shear strength (e.g., Brandenberg et al., 2007). There are several methods for estimating the residual shear strength of liquefied sand, such as CPT-based methods outlined earlier. Several models have also been proposed for reducing the stiffness of the p-y curves due to liquefaction. For example, the Japanese Geotechnical Society (JGS, 1998) has suggested a table for reduction of the reaction modulus as a function of FL, factor of safety against liquefaction, and STP N number. The reduction factor varies from zero for FL < 0.5 and N < 8 to 1.0 for FL approaching 1.0 and N > 20. In other words, the reaction modulus is severely reduced for FL < 0.5 and N < 8 and is almost unchanged for FL $\approx$ 1.0 and N > 20.

Goh & O'Rourke (1999) presented a method of estimating the lateral resistance to piles that are smooth or rough as shown in Figure 9.24. In this figure, the lateral resistance and the displacement are normalised. The resistance reduces with mobilisation of lateral displacement. Figure 9.24a shows the original curves proposed by Goh & O'Rourke (1999). Goh (2001) and Goh & O'Rourke (2008) have proposed further improvement to these curves as shown in Figure 9.24b that can be used for soils with shear strength degradation owing to liquefaction. For time domain dynamic analyses, it has been suggested to use strain-softening p-y curves (e.g., Goh & O'Rourke, 2008) in which the residual strength (or the ultimate lateral resistance) is reached at a relative pile-soil displacement of about $0.03D$ where D is the pile diameter. It should be noted that these curves have four constraints, i.e., y_p/D, y_{min}/D, k_L, and k_u which

Table 9.1 Reduction factor for modulus of subgrade reaction (JGS, 1998)

Factor of safety against liquefaction, FL	depth z below ground surface (m)	reduction factor r_k, applied to modulus of subgrade reaction k			
		$N < 8$	$8 < N < 14$	$14 < N < 20$	$N > 20$
<0.5	0 < z < 10	0	0	0.05	0.1
	10 < z < 20	0	0.05	0.1	0.2
0.5 < FL < 0.75	0 < z < 10	0	0.05	0.1	0.2
	10 < z < 20	0.05	0.1	0.2	0.5
0.75 < FL < 1.0	0 < z < 10	0.05	0.1	0.2	0.5
	10 < z < 20	0.1	0.2	0.5	1.0

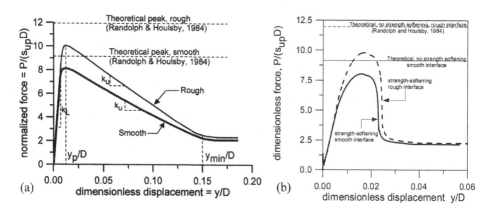

Figure 9.24 Normalised lateral resistance of piles (a) after Goh & O'Rourke (1999) and (b) after Goh (2001).

over-constrain the problem, i.e., there are more equations than unknowns. This would mean that such curves as shown in Figure 9.24 can fit any experimental data by choosing the four constraints listed above.

Another method was proposed by Rollins et al. (2005) based on experimental testing of piles. The experimental curves are reduced to an S-shaped back-bone curve as shown in Figure 9.25. In this figure, the API (1993) curves are also shown for reference. The lateral pressure increases at larger relative displacements in this figure, in contrast

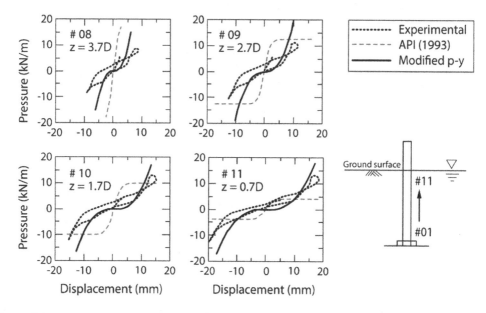

Figure 9.25 p-y curves for piles in liquefied soils (after Rollins et al. (2005)).

to the API (1993) curves. Further, the pressure exerted reduces towards the shallower depths of the pile.

More recently, Madabhushi (2018) has investigated the performance of dual row sheet pile walls tied at the top to protect coastal regions from multiple hazards of earthquake followed by a tsunami. Using centrifuge modelling, such wall configurations were tested to establish their performance when the foundation soil suffers liquefaction. In Figure 9.26, the shear stress-shear strain loops are presented for two locations A and B that are in the free-field and in the region between the sheet piles, respectively. In this figure, it can be seen that the free-field soil at location 'A' quickly liquefies and the stress-strain loops go nearly horizontal indicating very low shear stiffness. In contrast, the stress-strain loops at location 'B' show that the soil does not liquefy and show considerable shear stiffness. More details of this testing and the performance of dual row retaining walls can be found in Madabhushi & Haigh (2019, 2020).

These results show that the loss of stiffness in the free-field soil that liquefies may not be reflected in regions next to piles. The amount of stiffness that is retained in the regions adjacent to the piles may depend upon suction pressures that develop in the pore water due to shear-induced dilation and how long they are retained for, depending on the drainage paths available.

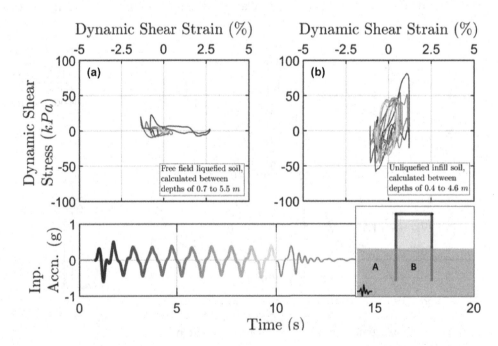

Figure 9.26 Shear stress-shear strain loops for liquefied and non-liquefied soils (after Madabhushi, 2018).

9.9 NUMERICAL MODELLING OF WIND TURBINES ON MONOPILES

With growth of offshore wind farms worldwide, the performance of their foundations in liquefiable soils needs to be established for certain locations where the seabed is likely to suffer liquefaction. Numerical modelling of such a problem is challenging as full 3D analyses in the time domain are required.

In recent years, several non-linear elastoplastic models have been developed for numerical simulation of site response, excess pore pressure, and accumulation of shear/volumetric strain in soils susceptible to liquefaction (e.g., Manzari & Dafalias, 1997; Elgamal et al., 2002; Andrianopoulos et al., 2010; Taiebat et al., 2010). Implementation of these models in finite element/finite difference tools has provided more robust methods for modelling pore-pressure generation and liquefaction in sand and their effects on piles (e.g., Cuéllar et al., 2014). Due to the strict performance criteria of offshore wind turbines, rigorous numerical methods have recently been applied for assessment of the response of these structures due to liquefaction. In most of these studies, pore pressure generation has been assessed only under wind or wave loads. Kementzetzidis et al. (2019) investigated the OWT-monopile-soil dynamic interaction and studied the pore pressure buildup, stress paths, and shear stress-strain responses in medium-dense/dense sand for different wind/wave cyclic loading conditions. Most recently, Kazemi Esfeh & Kaynia (2020) studied the earthquake response of a reference 5 MW OWT model in a liquefiable soil under earthquake shaking with and without the presence of a slowly varying lateral load representing the wind/wave loading. They used the SANISAND constitutive model (Dafalias & Manzari, 2004) implemented in FLAC3D for their numerical simulations. The soil profile is the one used by Ramirez et al. (2018) with soil parameters calibrated against centrifuge tests. The pile has a diameter of 6 m, founded 25 m in the liquefiable soil layer. The total mass of the reference wind turbine including the tower, blades, and in hub is about 700 tons. The model was shaken by the same acceleration time history used in the centrifuge experiment. A schematic view of the cross-section of the problem is shown in Figure 9.27. Figure 9.28 shows a representative set of results from this study. The figure shows the acceleration time history applied at the base of the soil at depth 34 m together with the time histories of the computed excess pore water pressures at five points in the soil at different depths. Points P1 and P2 are, respectively, 5 and 10 m from the pile edge normal to the direction of shaking. Point P3 is at the center of the pile, and Points P_b and P_f are at about 1.3 m from the pile edge in the back and in front of the pile in the direction of shaking (see Kazemi Esfeh & Kaynia, 2020, for details). The figure illustrates the development of pore pressure as the shaking continues and the soil approaches the state of liquefaction at most points. The results show that the point inside the monopile generates less pore pressure because of the kinematic confinement provided by the pile. Whilst such models with carefully calibrated soil parameters have until recently been considered mostly academic, they are now becoming more robust and acceptable for design purposes. Of course, the performance of these numerical simulations needs to be compared against centrifuge test data to increase the confidence of such analysis. This is the next logical step in further improving our understanding of the large diameter monopile behaviour in liquefied soils and in assessing the overall performance of the OWTs with the onset of liquefaction.

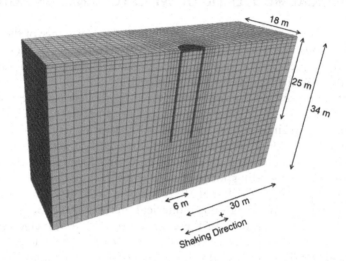

Figure 9.27 Schematic cross-section of 3-D monopile foundation for OWT (after Ka-zemi-Esfeh & Kaynia (2020)).

9.10 SUMMARY

In this chapter, the main focus has been liquefiable sites at which pile foundations have to be designed and constructed. The liquefaction potential of a site can be assessed using in-situ testing methods such as an SPT or CPT. However, designers need to be cautious when classifying a site as liquefiable or non-liquefiable based on a binary demarcation offered by liquefaction potential curves in existing literature. Such a distinction can be blurred as recent centrifuge test data show that significant excess pore pressures and consequent soil softening can occur even when a site falls towards the 'no-liquefaction' category.

Behaviour of pile foundations passing through liquefiable layers can fall into two categories based on the end conditions at the pile tip. A pile that is driven to bed rock (rock socketed), or if it has sufficient length into a non-liquefiable stiff stratum, can suffer failure due to instability or buckling, once the surrounding soil in shallow layer suffers liquefaction. This is true for single piles as well as pile groups. However, it is fairly straightforward for designers to check for instabilities using simple Euler's equation or Perry-Robertson equation that allows for load eccentricities or non-verticality of driven piles. Dynamic centrifuge test data could consistently predict the buckling failure of piles due to liquefaction.

For pile groups passing through a loose, liquefiable sand layer and passing into a denser sand or more competent underlain soil strata, settlement is the main issue. Again, dynamic centrifuge testing was used to establish the magnitude of these settlements. In addition, for sites that see repeated earthquakes and therefore can suffer liquefaction multiple times, it was shown that the pile group settlements show a continuous slope, i.e., the stress history of the soil from previous earthquakes is retained. It was also shown that pile groups that have a higher SSF require higher levels of liquefaction to induce settlements, whilst pile groups with lower SSF suffer settlements even under partial liquefaction conditions.

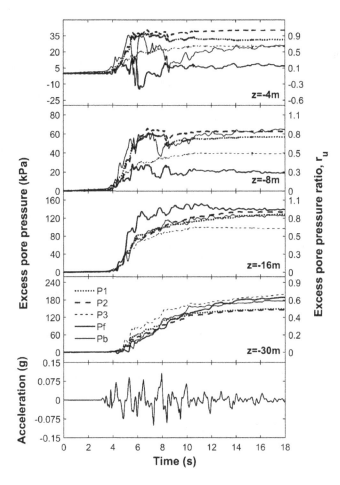

Figure 9.28 Time histories of excess pore pressures generated at different depths of soil profile due to acceleration time history shown at the bottom of the figure (after Kazemi-Esfeh & Kaynia (2020)).

In this chapter, we have also seen the interesting load transfer mechanisms identified through a series of centrifuge tests. It appears that the load carried by individual piles through shaft resistance and end bearing can disappear with the onset of liquefaction, when the pile cap is bearing on the ground surface. During this period, the pile cap bearing capacity provides the necessary equilibrium of forces to the applied loads on the pile group. In the post-earthquake period, the excess pore pressures dissipate, and the soil regains its strength. It was interesting to see that the shaft friction recovers in this period to almost the same level as the pre-earthquake values. In practical terms, this would mean that designers need to pay particular attention to the pile cap bearing capacity, when dealing with liquefiable sites. Use of a good drainage layer below the pile cap and using slightly larger pile caps can help ameliorate the behaviour of pile groups at such sites.

Finally, the existing methodologies in the literature to estimate the lateral loading on piles using the *p-y* concept are presented. Whilst these methodologies are not exhaustive, they can be used to estimate the reduction in the lateral loads due to the softening of the soil owing to liquefaction. Numerical methods, particularly the finite element analysis carried out in time domain, can now be used to carry out complex 3-D analyses of problems such as monopile foundations in liquefiable soils. A recent example of such a study is presented in this chapter. Of course, verification of the accuracy of these numerical simulations against dynamic centrifuge test data would be the next logical step that would increase the confidence in such simulations.

REFERENCES

American Petroleum Institute (1993), API recommended practice 2GEO/ISO 19901-4. Geotechnical and foundation design considerations, 1st Edition, April 2011, Addendum 1, October 2014.

Andrianopoulos, K.I., Papadimitriou, A.G. and Bouckovalas, G.D., (2010), Bounding surface plasticity model for the seismic liquefaction analysis of geostructures, *Soil Dynamics & Earthquake Engineering*, 30(10), pp. 895–911.

Berrill, J.B., Christensen, S.A., Keenan, R.P., Okada, W. and Pettinga, J.R., (2001), Case studies of lateral spreading forces on a piled foundation, *Geotechnique*, 51(6), pp. 501–517.

Bhattacharya, S., (2003), Pile instability during earthquake liquefaction, PhD Thesis, University of Cambridge, UK.

Bhattacharya, S., Bolton, M.D. and Madabhushi, S.P.G., (2005), A reconsideration of the safety of piled bridge foundations in liquefiable soils, *Soils and Foundations*, 45(4), pp. 13–26.

Bhattacharya, S., Madabhushi, S.P.G. and Bolton, M.D., (2004), An alternative mechanism of pile failure during seismic liquefaction, *Geotechnique*, 54(3), pp. 203–213.

Boulanger and Idriss, (2014), CPT and SPT based liquefaction triggering procedures, Report No: UCD/CGM-14/01, University of California, Davis.

Brandenberg, S.J., Boulanger, R.W., Kutter, B.L. and Chang, D., (2007), Static pushover analyses of pile groups in liquefied and laterally spreading ground in centrifuge tests, *Journal of Geotechnical & Geo-Environmental Engineering*, 133(9), pp. 1055–1066. Doi: 10.1061/(ASCE)1090-0241.

Chang, B.J. and Hutchinson, T.C., (2013), Experimental evaluation of *p-y* curves considering development of liquefaction, *Journal of Geotechnical & Geo-Environmental Engineering*, 139, pp. 577–586.

Cuéllar, P., Mira, P., Pastor, M., Fernández-Merodo, J.A., Baeßler, M. and Rücker, W., (2014), A numerical model for the transient analysis of offshore foundations under cyclic loading, *Computers & Geotechnics*, 59, pp. 75–86.

Dafalias, Y. F., & Manzari, M. T. (2004). Simple plasticity sand model accounting for fabric change effects. *Journal of Engineering Mechanics*, 130(6), 622–634.

Dobrisan, A., (2018), Liquefaction of saturated sands under cyclic loading, First Year PhD report, University of Cambridge, Cambridge, UK.

Elgamal, A., Yang, Z. and Parra, E., (2002), Computational modeling of cyclic mobility and post-liquefaction site response, *Soil Dynamics & Earthquake Engineering*, 22(4), pp. 259–271. https://doi.org/10.1016/S0267-7261(02)00022-2.

Eurocode 8 – Part 5, (2003), Design provisions for earthquake resistance of structures – foundations, retaining structures and geotechnical aspects, CEN European Committee for Standardisation, prEN 1998-5:2003.

Garala, T.K. and Madabhushi, S.P.G., (2018), Seismic behaviour of soft clay and its influence on the response of friction pile foundations, *Bulletin of Earthquake Engineering*, 17, pp. 1919–1939. https://doi.org/10.1007/s10518-018-0508-4.

Garala, T.K., Madabhushi, S.P.G. and Di Lora, R., (2020), Experimental investigation of kinematic pile bending in layered soils using dynamic centrifuge modelling, *Géotechnique*, https://doi.org/10.1680/jgeot.19.P.185.

Goh, S.H., (2001), Soil-pile interaction during liquefaction-induced lateral spread, PhD Dissertation, Cornell University, Ithaca, NY.

Goh, S.H. and O'Rourke, T.D., (1999), Limit state model for soil-pile interaction during lateral spread, *Proceedings 7th US-Japan Workshop on Earthquake Resistant Design of Lifeline Facilities and Countermeasures against Liquefaction*, MCEER, Buffalo, NY.

Goh, S.H. and O'Rourke, T.D., (2008), Soil-pile interaction during liquefaction-induced lateral spread, *Journal of Earthquake and Tsunami*, 2, pp. 53–85.

Haskell, J.J.M., Madabhushi, S.P.G., Cubrinovski, M. and Winkley, A., (2013), Lateral spreading-induced abutment rotation in the 2011 Christchurch earthquake: Observations and analysis, *Geotechnique*, 63(15), pp. 1310–1327.

Japanese Geotechnical Society, JGS, (1998), *Remedial Measures against Soil Liquefaction*. A.A. Balkema: Rotterdam.

Kazemi Esfeh, P. and Kaynia, A.M., (2020), Earthquake response of monopiles and caissons for Offshore Wind Turbines founded in liquefiable soil, *Soil Dynamics and Earthquake Engineering*, 136, p. 106213, http://www.elsevier.com/locate/soildyn.

Kementzetzidis, E., Corciulo, S.G., Versteijlen, W. and Pisanò, F., (2019), Geotechnical aspects of offshore wind turbine dynamics from 3D non-linear soil-structure simulations, *Soil Dynamics and Earthquake Engineering*, 120, pp. 181–199, https://doi.org/10.1016/j.soildyn.2019.01.037.

Knappett, J.A. (2006). Piled foundations in liquefiable soils: accounting for axial loads, PhD Thesis, University of Cambridge, UK.

Knappett, J.A. and Madabhushi, S.P.G., (2008), Liquefaction-induced settlement of pile groups in liquefiable and laterally-spreading soils, *ASCE Journal of Geotechnical & Geoenvironmental Engineering*, 134(11), pp. 1609–1618.

Knappett, J.A. and Madabhushi, S.P.G., (2009a), Influence of axial load on lateral pile response in liquefiable soils, Part I: Physical modelling, *Geotechnique*, 59(7), pp. 571–581.

Knappett, J.A. and Madabhushi, S.P.G., (2009b), Influence of axial load on lateral pile response in liquefiable soils, Part II: Numerical modelling, *Geotechnique*, 59(7), pp. 583–592.

Knappett, J.A. and Madabhushi, S.P.G., (2009c), Seismic bearing capacity of piles in liquefiable soils, *Soils & Foundations*, 49(4), pp. 525–536.

Kutter, B.L., Carey, T.J., Hashimoto, T., Zeghal, M., Abdoun, T., Kokkal, P., Madabhushi, S.P.G., Haigh, S.K., d'Arezzo, F.B., Madabhushi, S.S.C., Hung, W-Y., Lee, C-J., Cheng, H-C., Iai, S., Tobita, T., Ashino, T., Ren, J., Zhou, Y-G., Chan, Y-M., Sun, Z-B. and Manjari, M.T., (2018), LEAP-GWU-2015 experiment specifications, results, and comparisons, *International Journal of Soil Dynamics and Earthquake Engineering*, 113, pp. 616–628. https://doi.org/10.1016/j.soildyn.2017.05.018.

Liu, L. and Dobry, R., (1995), Effect of liquefaction on lateral response of piles by centrifuge model tests, *NCEER Bulletin*, 9(1), pp. 7–11.

Madabhushi, S.P.G., Knappett, J.A. and Haigh, S.K., (2009), *Design of Pile Foundations in Liquefiable Soils*. Imperial College Press, World Scientific: Singapore.

Madabhushi, S.P.G. and Schofield, A.N., (1993), Centrifuge modelling of tower structures subjected to earthquake perturbations, *Geotechnique*, 43(4), pp. 555–565.

Madabhushi, S.S.C. (2018), Multi-Hazard Modelling of Dual Row Retaining Walls, PhD Thesis, University of Cambridge, UK.

Madabhushi, S.S.C. and Haigh, S.K., (2019), Centrifuge testing of dual row retaining walls in dry sand: the influence of earthquake sequence and multiple flights. *International Journal of Soil Dynamics and Earthquake Engineering*, 125. https://doi.org/10.1016/j.soildyn.2019.05.029.

Madabhushi, S.S.C. and Haigh, S.K., (2020), Dual row retaining walls in dry sand: The influence of wall stiffness on the seismic response. *Canadian Geotechnical Journal*. https://doi.org/10.1139/cgj-2020-0538.

Manzari, M.T. and Dafalias, Y.F., (1997), A critical state two-surface plasticity model for sands, *Géotechnique*, 47(2), pp. 255–272. https://doi.org/10.1680/geot.1997.47.2.255.

Ramirez, J.R., Barrero, A., Chen, L., Dashti, S., Ghofrani, A., Taiebat, M. and Arduino, P., (2018), Site response in a layered liquefiable deposit: Evaluation of different numerical tools and methodologies with centrifuge experimental results, *Journal of Geotechnical & Geoenvironmental Engineering*, 144(10). https://doi.org/10.1061/(ASCE)GT.1943-5606.0001947.

Riks, E., (1972), The application of Newton's method to the problem of elastic stability, *Journal of Applied Mechanics*, 39, pp. 1060–1065.

Riks, E., (1979), An incremental approach to the solution of snapping and buckling problems, *International Journal of Solids & Structures*, 15, pp. 529–551.

Robertson, P.K. and Wride, C.E. (1998), Evaluating cyclic liquefaction potential using the cone penetration test, *Canadian Geotechnical Journal*, 35(3), pp. 442–459.

Rollins, K.M., Gerber, T.M., Lane, J.D. and Ashford, S.A., (2005), Lateral resistance of a full-scale pile group in liquefied sand, *Journal of Geotechnical & Geo-Environmental Engineering*, 131, pp. 115–125.

Seed, H.B. and Idriss, I.M., (1971), Simplified procedure for evaluating soil liquefaction potential, *Journal of Soil Mechanics and Foundations Division*, ASCE, 97(SM9), pp. 1249–1273.

Stringer, M.E., (2012), The axial behaviour of piled foundations in liquefiable soil, PhD Thesis, University of Cambridge, UK.

Stringer, M.E. and Madabhushi, S.P.G., (2012), Axial load transfer in liquefiable soils for free-standing piles, *Geotechnique*, 63(5), pp. 400–409. Doi: 10.1680/geot.11.P.078.

Stringer, M.E. and Madabhushi, S.P.G., (2013), Re-mobilisation of pile shaft friction after an earthquake, *Canadian Geotechnical Journal*, 50(9), pp. 979–988. Doi: 10.1139/cgj-2012-0261.

Taiebat, M., Jeremi´c, B., Dafalias, Y.F., Kaynia, A.M. and Cheng, Z. (2010). Propagation of seismic waves through liquefied Soils, *Soil Dynamics & Earthquake Engineering*, 30(4), 236–257.

Takahashi, A., Kuwano, J., Arai, Y., Yano, A., (2002), Lateral resistance of buried cylinder in liquefied sand, *Proceedings International Conference on Physical Modelling in Geotechnics: ICPMG '02*, pp. 477–482.

Timoshenko, S.P. and Gere, J.M., (1961), *Theory of Elastic Stability*. McGraw-Hill Book Company Inc.: New York.

Tokimatsu, K., Suzuki, H., Suzuki, Y. and Fujii, Y., (2001), Back-calculated p-y relation of liquefied soils from large shaking table tests, *4th International Conferences on Recent Advances in Geotechnical Earthquake Engineering and Soil Dynamics*, 6.24.

Towhata, I., Vargas-Monge, W., Orense, R.P. and Yao, M., (1999), Shaking table tests on subgrade reaction of pipe embedded in sandy liquefied subsoil, *Soil Dynamics & Earthquake Engineering*, 18, pp. 347–361.

Chapter 10

Analysis and characteristics of dynamic response of large pile groups

Amir M. Kaynia

Norwegian Geotechnical Institute (NGI)
Norwegian University of Science and Technology (NTNU)

CONTENTS

10.1 INTRODUCTION

Over the past four decades, a suite of models and solutions have been developed for dynamic analysis of piles and pile groups. These solutions range from analytical or approximate solutions mostly based on the Winkler medium or the concept of dynamic subgrade modulus (e.g. Novak, 1974; Dobry & Gazetas, 1988; Makris & Gazetas, 1992; Mylonakis & Gazetas, 1999) to rigorous analytical or numerical solutions. The latter include solutions based on Green's functions (e.g. Waas & Hartmann, 1981; Kaynia, 1982; Roesset, 1982; Miura et al. 1994), boundary element (BE) method (e.g. Mamoon et al., 1990), finite element (FE) method (e.g. Wolf et al., 1981; Taherzadeh et al., 2009; Latini et al., 2015), and coupled FE/BE methods (e.g. Padrón et al., 2008; Auersch, 2019). The Winkler-type solutions have been further extended to more complex soil and response conditions (e.g. Mylonakis et al., 1997; Di Laora & Rovithis, 2015; Dezi et al., 2016; Goit et al., 2016) and to nonlinear problems including development of gap at soil-pile interface and liquefaction (e.g. Boulanger et al., 1999; Varun et al., 2013). For an overview of the principles behind these methods, see Pender (1993). With the increased interest in renewable energy, the available methods have been applied or extended to the dynamic analysis of monopiles in offshore wind turbines in the past several years (e.g. Kjørlaug & Kaynia, 2015; Bayat et al., 2016; Shadlou & Bhattacharya, 2016; Álamo et al., 2018; Markou & Kaynia, 2018; Kaynia, 2018). These piles are characterized by large diameter steel pipe sections (diameters ranging from 5 to 10 m) and relatively short length. With such dimensions, monopiles approach embedded caissons where the vertical shear stresses on the pile due to horizontal loading has a nonnegligible role (e.g. He et al., 2019a). Some attention has also been given to the dynamic poroelastic response of the soil inside the pile (Kaynia, 1992; He et al., 2019b).

With the origin in the principles of dynamic soil-structure interaction (e.g. Kausel, 2010), the dynamic analysis of piles and pile groups has traditionally been based on linear viscoelastic response of the soil. Moreover, in the case of pile groups, it has often been assumed that the piles are connected to rigid caps. These assumptions have been motivated partly by the traditional soil-structure interaction solutions, such as the three-step method (Kausel et al., 1978). Most of the existing solutions reported on the impedances of pile groups have been performed for rigid pile cap conditions. Kaynia (2012) investigated the impact of this assumption by comparing the apparent average stiffness of flexible pile caps to the corresponding stiffness of rigid-cap foundation. He observed a minor effect on the vertical stiffness, but noticed a considerable effect on the horizontal stiffness that he attributed to the rotation of the pile heads in the case of flexible pile caps.

Some of the features of the dynamic response of pile groups, most notably the large stiffness and damping at certain frequency bands, have been attributed to the out-of-phase motions of soil particles between the neighboring piles (Kaynia, 1982). Therefore, there have been speculations about the applicability of these features to more realistic soil profiles than the idealized homogeneous ones and symmetric pile configurations. This chapter assesses these issues by considering the dynamic impedances of a large pile group comprising more than 200 piles. The cases considered include irregular pile configurations and heterogeneous soil profiles, both vertically and laterally. Moreover, a limited number of results are presented for highlighting the effect of soil nonlinearity.

10.2 METHODOLOGY AND MODEL FEATURES

The computational model used in the present study is PILES (Kaynia, 1982). The pile-soil tractions in this model are replaced by piecewise constant cylindrical loads (barrel loads) on the pile shafts and circular loads at the pile tips (Figure 10.1). By using analytically derived Green's functions for these loads in layered soil media, a soil stiffness matrix is established and is coupled to the dynamic stiffness matrix of the piles. The responses of the piles are then computed by imposition of the relevant boundary and traction conditions. The main advantage of using the Green's function is that the wave propagation is handled analytically without needing element mesh as in the FE method, and the infinite boundary conditions are satisfied rigorously (see Kaynia & Kausel, 1991) for the analytical derivation of the Green's functions and solution of pile-soil interaction. The analyses are carried out under steady-state harmonic vibration in the frequency domain. Therefore, both the foundation impedances and the seismic

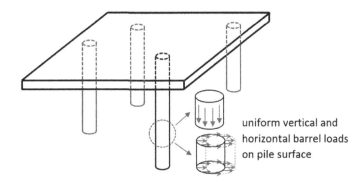

Figure 10.1 Pile group and representation of pile-soil tractions in horizontal and vertical directions.

response of the foundations are complex quantities. The seismic excitation is assumed to be due to vertically incident shear waves.

As stated in Section 10.1, the numerically observed large stiffness and damping at certain frequency bands have been attributed to the out-of-phase motions of the soil between the neighboring piles in a group. Therefore, there have been speculations that these features are primarily due to the idealization of soil profiles and pile configurations. The present investigation intends to evaluate the impact of these assumptions on the pile response for better understanding the dynamic behavior of pile foundations for practical conditions.

In all the cases considered here, the results of the real and imaginary parts of the horizontal and vertical impedances (i.e. stiffness and damping) of pile groups comprising 225 piles are compared with those of an idealized Base Case consisting of a square configuration of piles in a homogeneous half space with the following parameters:

- Circular piles with diameter, $d = 0.5$ m, length, $L = 20d$, elastic modulus, $E_p = 30$ GPa, unit mass, $\rho_p = 2500$ kg/m³ and spacing, $s = 5d$.
- Homogeneous viscoelastic half space with elastic modulus, $E_s = E_p/1000$, unit mass, $\rho_s = 0.75\,\rho_p$, Poisson's ratio, $v = 0.4$, and hysteretic soil damping, $\beta = 5\%$.

The cases that are considered with parameters different from the above Base Case include the following:

1 Vertically heterogeneous soil (linear and parabolic variation with depth)
2 Laterally heterogeneous soil
3 Circular pile layout
4 Nearly incompressible soil
5 Stiff soil with $E_p/E_s = 100$

In addition, the accuracy of the superposition method is investigated for large pile groups and the results of a limited FE numerical study are presented that highlight the effect of soil nonlinearity on pile-soil-pile interaction.

For comparison of the results, the real and imaginary parts of the horizontal and vertical impedances in all cases are normalized by 225 times the respective static stiffnesses of a single pile in the Base Case (i.e., homogeneous soil and $E_p/E_s = 1000$), that is, $225K_{0,BC}$ (0 stands for frequency, $\omega = 0.0$ rad/s). The values of $K_{0,BC}$ for the horizontal and vertical directions were computed as listed below.

- $K_{0,BC}^x = 64\,\text{MN/m}$

- $K_{0,BC}^z = 157\,\text{MN/m}$

The dynamic impedances in all cases are presented as functions of the normalized frequency, $a_0 = \omega d/V_{s,BC}$ where $V_{s,BC}$ is the shear wave velocity of the Base Case, which is equal to 77 m/s.

10.3   DYNAMIC IMPEDANCES

10.3.1   Group configuration

To investigate the influence of pile layout on the dynamic response of pile groups, a circular pile arrangement with the same number of piles (i.e. 225) and same footprint area as in the Base Case was constructed as shown in Figure 10.2a. For reference, the pile layout for the Base Case is also displayed in Figure 10.2b. Although the pile spacing varies in the circular configuration, the ratio s/d is generally close to 5 as in the rectangular group. This analysis was partly meant for studying how the regular pattern of piles, as has been used in most previous studies, affects the large peaks in the foundation impedances.

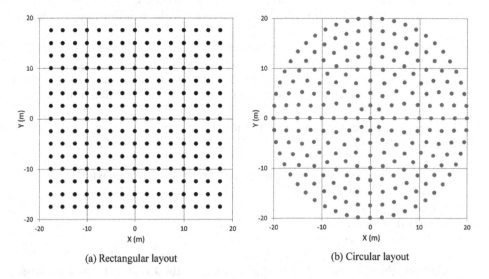

(a) Rectangular layout   (b) Circular layout

Figure 10.2 Pile layouts with the same footprint and the number of piles considered in this study: (a) rectangular (Base Case) and (b) circular.

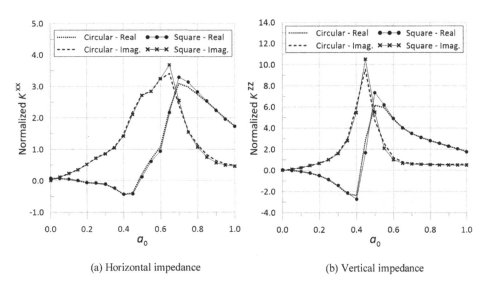

(a) Horizontal impedance   (b) Vertical impedance

Figure 10.3 Variations of normalized horizontal impedances (a) and normalized vertical impedances (b) of square and circular pile layouts. Solid lines are for the square group, and dashed lines are for the circular group.

Figure 10.3 presents the variations with frequency of the real and imaginary parts of the horizontal (K^{xx}) and vertical (K^{zz}) impedances for the two pile groups considered above. The similarities of the results, despite the different patterns and spacing of the piles, are striking. While the large peaks in the impedances are directly related to the out-of-phase motions of the soil at neighboring piles, the results clearly show that the slight variations in the pile spacing do not disturb this feature. The reduction in the real part of the impedances is an indication of the added soil mass, and the peaks in the imaginary parts indicate large radiation damping in a wide frequency band.

10.3.2   Vertically heterogeneous soil profiles

Two soil profiles were considered for the assessment of the effect of soil modulus variation with depth: (i) linear variation of shear modulus, representing typical normally-consolidated saturated clays, and (ii) parabolic variation, representing typically sandy sites. The small-strain shear modulus, G_{max}, for the linear (clayey) profile was taken as 1500 times a shear strength varying with depth equal to 0.25 σ_v' where σ_v' is the effective vertical stress. For the parabolic (sandy) profile, G_{max} was computed using the empirical formula by Seed and Idress (1970) for a sand with relative density $D_r = 60\%$. Using the suggested parameter by Seed and Idriss, one can establish the formula $G_{max} = 1140 \left(p_a \sigma_m' \right)^{0.5}$ where $p_a = 100$ kPa is the atmospheric pressure, and σ_m' is the average effective confining stress. Figure 10.4 displays these profiles together with that of the Base Case ($E_p/E_s = 1000$) and a homogeneous stiff profile with $E_p/E_s = 100$.

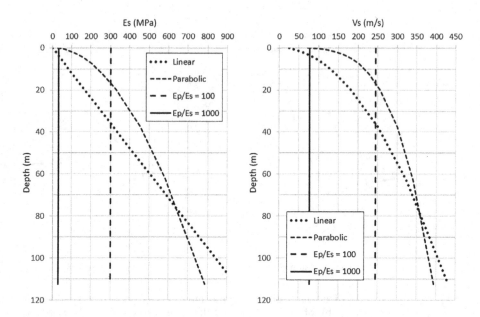

Figure 10.4 Variation of elastic moduli (left) and shear wave velocities (right) for considered homogeneous and heterogeneous soil profiles.

This case was considered in order to cover a wider range of shear moduli. The figure also shows the shear wave velocity profiles for the considered cases. In all these cases, the soil profiles below 110 m are continued as homogeneous half space.

Figure 10.5 presents the variations with frequency of the real and imaginary parts of the horizontal (K^{xx}) and vertical (K^{zz}) impedances for the two vertically heterogeneous soil profiles described above. For comparison, the corresponding impedances for the Base Case ($E_p/E_s = 1000$) and the homogeneous stiff soil with $E_p/E_s = 100$ are also shown in these plots. Note that, in all these cases, the results are presented as functions of the same frequency for direct comparison.

The results for the linear soil profile are consistent with those by Kaynia (1988) and show that the peaks in the impedances appear at lower frequencies representing softer soil. This is as expected considering the low shear moduli near the ground surface in the linear case. The impedances for the stiff soil profile are expected to demonstrate similar patterns with large peaks, as in the other cases, but at larger frequencies. The peaks being related to frequencies of out-of-phase soil motions between neighboring piles are expected to appear at a frequency about three times in the Base Case (related to the modulus being ten times larger).

10.3.3   Horizontally heterogeneous soil profiles

Geophysical methods like MASW, which map the soil parameters over a predetermined seismic line, result in average shear-wave velocities across a site. However, there is often lateral variability in the parameters. This variability is better captured by point-wise tests such as CPT sounding. A relevant question is then how this heterogeneity

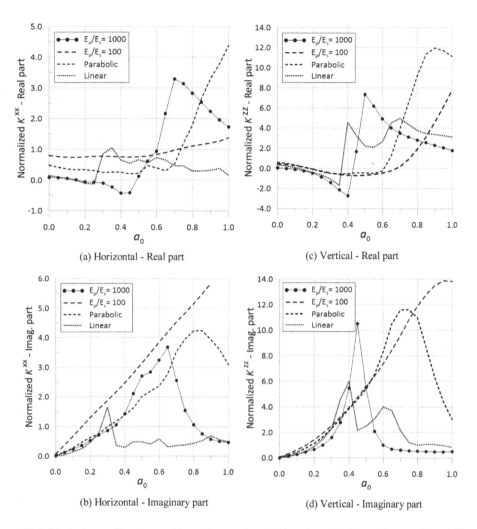

Figure 10.5 Variations of horizontal impedances (a and b) and vertical impedances (c and d) in homogenous and heterogeneous soil profiles shown in Figure 10.4 (styles of lines are consistent with those for profiles in Figure 10.4).

influences the foundation impedances of pile groups, especially in view of the general belief that the dynamic response of pile groups is dominated by pile-soil-pile interaction in which the spacing between adjacent piles plays a central role. To perform a rigorous study of the lateral heterogeneity, one should first generate a random soil, and next use general purpose FE tools that allow point-wise definition of the soil parameters. However, most such tools struggle with the classical limitations of FE codes, namely modelling of infinite domains and boundary conditions, especially as the frequency increases. To take advantage of the modelling features in PILES, an approximate, but practical, approach was adopted in this study. Instead of varying the soil profile across the foundation, the spacings among the piles were varied in a random manner. This simple scheme ensures different interactions between adjacent piles, while it preserves the general baseline average soil profile for interaction of piles at larger distances. Rigorous wave propagation analyses with random soil variations have indicated that the general features of the wave propagation are not dramatically different from those in a medium with the homogeneous average profile.

Figure 10.6 displays two realizations of pile layout for the square configuration considered in Figure 10.2a. The pile layout in Figure 10.6a was generated by assuming a modest random variation in the shear wave velocity about 15% of $V_{s,\,BC}$. The layout in Figure 10.6b was generated by imposing a relatively strong variation in the shear wave velocity about 30% of $V_{s,\,BC}$. The more erratic position of the piles in Figure 10.6b clearly demonstrates the large lateral heterogeneity for this case.

Figure 10.7 presents the variations with frequency of the real and imaginary parts of the horizontal (K^{xx}) and vertical (K^{zz}) impedances for the two lateral heterogeneous cases described above. These cases are identified as Heterogeneous 1 and Heterogeneous 2 in the figure. A few observations can be made from these figures. Firstly, these cases; namely, homogeneous profile, Heterogeneous 1 and Heterogeneous 2, practically display identical impedances over a large range of frequencies. This includes the static stiffnesses and the added soil masses, represented by the parabolic variations of the real parts. Secondly, the large peaks related to the pile spacing are still present, but they are "damped". Although this could be expected, it is surprising to observe that these peaks are not remarkably affected. These peaks appear at relatively large frequencies. For example, for the Base Case, the range is about 10–15 Hz. For a soil profile with about twice the shear wave velocity, the range would be about 20–30 Hz that are only important for machinery vibrations and have less consequence in earthquake designs.

These results have interesting design implications. They illustrate that, in characterizing a site, what matters most is to establish an average shear stiffness profile, determined either by techniques such as MASW or by a statistically sound averaging of point-wise soil parameters. In drawing such conclusions, one should also consider the limitations

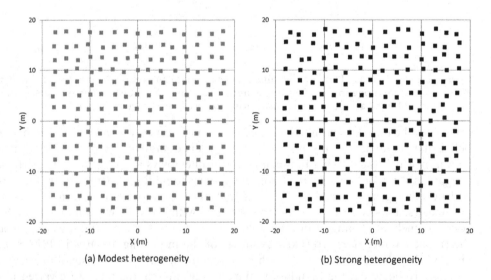

(a) Modest heterogeneity   (b) Strong heterogeneity

Figure 10.6 Two pile layouts considered for representing lateral soil heterogeneity in Base Case pile group and soil profile.

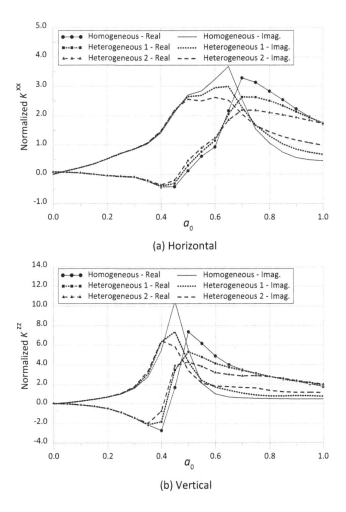

(a) Horizontal

(b) Vertical

Figure 10.7 Variations of real and imaginary parts of normalized horizontal impedances (a) and vertical impedances (b) in homogeneous and laterally heterogeneous soil profiles (solid lines are for homogeneous profiles, and dashed lines are for heterogeneous profiles in Figure 10.6).

of solutions based on linear-elastic soil response. As is shown in detail in Chapter 7 and in the next section through a limited study, consideration of soil nonlinearity could dramatically influence the strong pile-group effects presented here.

10.3.4 Effect of soil compressibility

In the preceding results, Poisson's ratio, v, was taken as 0.4. In saturated soil, v approaches 0.5, which represents an incompressible behavior. Poisson's ratio is generally known to have minor effects on horizontal foundation impedances. However, it does have a considerable effect on the vertical and rotational impedances. The analyses performed in this study confirmed this. Figure 10.8 compares the variation with

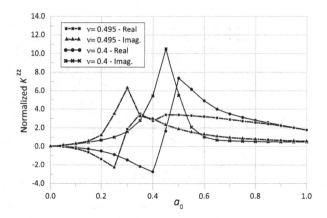

Figure 10.8 Effect of soil compressibility on vertical impedance in homogeneous soil.

frequency of the vertical impedance of the pile group in the Base Case with a case for $v = 0.495$ (close to incompressible soil). The clear features associated with incompressible soil include large added soil mass at low frequencies and shift of the response peaks to lower frequencies. The incompressibility appears to have a minor effect at high frequencies (above 18 Hz for the parameters considered in this study). This points to the importance of using representative values for Poisson's ratio, which is often overlooked in design.

10.3.5   Accuracy of the superposition method

The superposition method is based on the classical influence-function method used in structural analysis by considering two pints of a system at a time. Following the influence function method, one can compute the flexibility matrix of the pile group for the considered degrees of freedom at the pile heads and compute the pile group response (with or without a pile cap). In the superposition method, only two piles are considered at a time in computation of the influence functions. This reduces drastically the computational efforts, which can be significant in large pile groups. The superposition method was introduced by Poulos (1968) for static response of piles groups and was later shown to apply to dynamic problems by Kaynia (1982). To facilitate the analyses, Kaynia (1982) presented dynamic interaction factors as functions of frequency for different loads (horizontal and vertical), different pile spacings, and two pile-soil stiffness ratios. For two piles, an interaction factor gives the response of one of the piles to the load on the other normalized by the displacement of the loaded piles. The usefulness of the method has since motivated many researchers to develop simple expressions for the interaction factors (e.g. Dobry & Gazetas, 1988; Makris & Gazetas, 1992). More details about the method can be found in Chapters 5 to 7 of this book. Kaynia (1982) demonstrated the applicability of the superposition method for small pile groups (up to 4×4 groups). The following presents the results for the larger group studied in this chapter.

Figure 10.9 displays the horizontal and vertical impedances of the 225-pile group in the homogeneous profile for $s/d = 5$ (Figure 10.2a) computed using the superposition method and compared with the complete solution presented above and referred to as

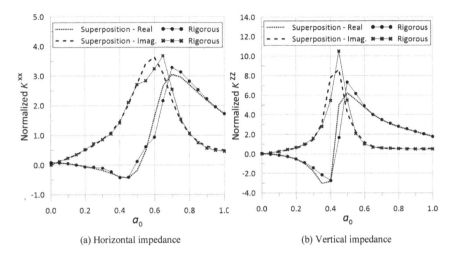

(a) Horizontal impedance   (b) Vertical impedance

Figure 10.9 Comparison of normalized impedances by rigorous and superposition methods for Base Case pile group and $s/d = 5$: (a) horizontal impedances and (b)
vertical impedances. Solid lines are for rigorous solution, and dashed lines are
for solution based on superposition.

rigorous solution in the figure. Other than minor local differences, the results reveal
the success of the superposition method even when used in large groups.

Because the superposition considers only two piles at a time, that is, it ignores the
neighboring piles in that computation, it is expected that its accuracy deteriorates as
piles get closer. This was investigated here by repeating the same analyses for $s/d = 3$.
The results for the superposition and rigorous analyses are presented in Figure 10.10.
While larger deviations from the rigorous solutions are observed as frequency increases
(especially in the vertical impedance), the trends are still captured reasonably well. The
largest deviations occur around the frequency with a wavelength about half the distance between the piles (out-of-phase motion of neighboring piles). The noted deviations are believed not to have any significant effect in earthquake responses. However,
for machine foundations operating at distinct high frequencies, one must be cautions
about the potential consequences of the superposition method if the frequencies are
close to the out-of-phase mode stated above.

10.4 EARTHQUAKE RESPONSE

It is generally known that typical pile-group foundations tend to follow the earthquake motions in the free field, but they filter out the higher frequency components
of the motions (Kaynia, 1982; Kaynia & Novak, 1992; Gazetas et al., 1993; for more
details see Chapter 8). This aspect of the response is often studied through the seismic
transfer function of the foundation, which is defined as the ratio between the absolute
value of the horizontal motion of the foundation and that on the ground surface in the
free field. For simplicity, this term is referred to as foundation transfer function in the
following discussion. The transfer function can be used to modify the earthquake's
acceleration response spectrum in seismic SSI analyses.

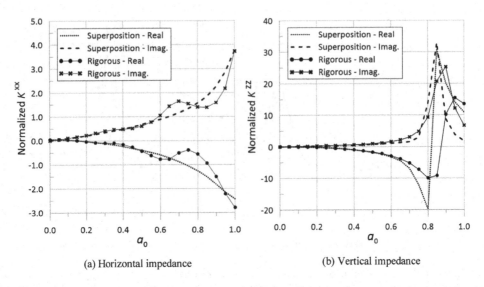

(a) Horizontal impedance   (b) Vertical impedance

Figure 10.10 Comparison of normalized impedances by rigorous and superposition methods for Base Case pile group and $s/d = 3$: (a) horizontal impedances and (b) vertical impedances. Solid lines are for rigorous solution, and dashed lines are for solution based on superposition.

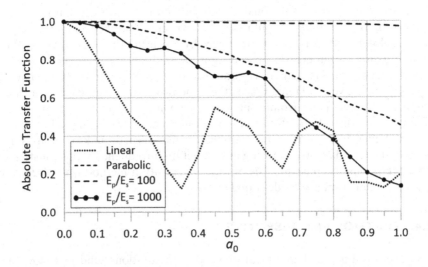

Figure 10.11 Absolute values of seismic transfer functions for homogeneous and heterogeneous soil profiles considered in this study.

Figure 10.11 displays the foundation transfer functions for the four soil profiles considered in Figure 10.7 (i.e. Base Case, stiff homogeneous soil, linear and parabolic soil profiles). As expected, the piles in the softest soil profile (linear with depth) start filtering out the earthquake motions at the lowest frequencies. Other than this soft soil case, the pile foundations tend to display high transfer functions in the frequency range of interest in typical earthquake analyses (up to nondimensional frequency of 0.2, corresponding to 5 Hz for the parameters considered in this study). Refer to Chapter 8 of this book for more in-depth discussion on the filtering effect of pile groups and other seismic design issues such as kinematic interaction forces in piles which arise from the passage of seismic waves through the soils and the piles.

10.5 SOIL NONLINEARITY

Dynamic analyses of piles and pile groups have traditionally been carried out assuming linear viscoelastic soil behavior. This assumption has been made partly due to the interest in the machine-foundation industry where machines often operate at small vibration amplitudes and high frequencies and partly because of the general practice of using primary soil nonlinearity, with the help of the equivalent linear method, in geotechnical earthquake engineering. If the inertial interaction in an SSI analysis involves large loads on the piles, then the secondary nonlinearity due to the nonlinear soil response near the piles should be accounted for. This is not only because of the larger induced stresses in the piles but also primarily because the dynamic SSI might change remarkably. As the pile is pushed against the soil, the dynamic response tends to be more dominated by the soil adjacent to the soil which in turn weakens the role of pile-soil-pile interaction.

Dynamic analyses of large pile groups with inelastic soil behavior can only be performed by using general FE solutions which are computationally intensive. An approximate solution is with the superposition method. The successful performance of the superposition in case of large group presented above motivate the study presented in this section. A limited set of analyses were performed in this study to shed some light on the significance of soil nonlinearity on the pile group response. More in-depth discussion on this topic, including dynamic effects, can be found in Chapter 7.

In developing simple methods for the analyses of pile groups, Dobry & Gazetas (1988) observed that pile-soil-pile interaction could be captured by simply ignoring the passive pile when the active pile is subjected to a harmonic load. In view of this observation,

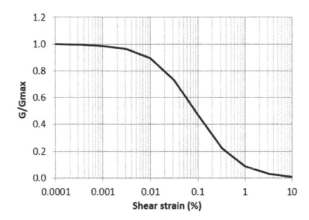

Figure 10.12 Variation of shear modulus with shear strain considered in nonlinear pile-soil analyses.

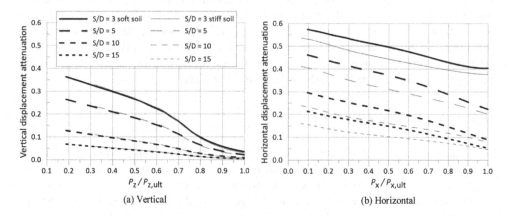

Figure 10.13 Attenuation of soil displacements as a function of pile capacity mobilization for different distances from pile center. Thick lines are for
E_p/E_s = 1000 (soft soil), and thin lines are for E_p/E_s = 100 (stiff soil).

several nonlinear FE analyses were performed on a single pile using Plaxis3D (Plaxis,
2018) to investigate the effect of soil nonlinearity on the interaction factors. The two homogeneous soil cases with E_p/E_s = 100 and 1000 were considered, and the soil was taken
to behave nonlinearly following the shear stress-shear strain response corresponding to
the G/G_{max} curve shown in Figure 10.12. This curve corresponds to a clay with plasticity
index Ip = 25% and OCR = 1 according to Darendeli (2001). The shear strengths for the
two soil profiles were taken as $G_{max}/1000$, which is representative for the assumed Ip.

Figure 10.13 plots the attenuation of the soil displacements with distance from the
loaded pile as a function of the load normalized by the corresponding pile capacity.
Figure 10.13a shows the vertical displacements for normalized distances (pile spacing)
s/d = 3, 5, 10, and 15. Figure 10.13b presents the corresponding results for horizontal
loading. In the latter case, the pile head was restrained from rotation, as this is the case
of loading of pile groups with piles rigidly connected to the pile cap.

The figures clearly indicate the remarkable reduction of the soil displacement with
distance as the loads approach the pile capacity. It is interesting to note that the results
for the soft and stiff soil cases are close (practically identical for the vertical loading).
This is a useful observation when working with different soil profiles.

For a given distance, one can obtain a factor defining the ratio between the nonlinear and linear interaction factors as a function of pile load by dividing the soil
displacements (Figure 10.13) by the displacement at small loads (elastic condition).
Figure 10.14 shows this factor for the Base Case (s/d = 5 and E_p/E_s = 1000) for the vertical and horizontal loadings. The advantage of using the interaction reduction factor is
that one could extend the simple formulas for elastic interaction factors (as described
in detail in Chapters 5 and 6) to inelastic soil response. The results in Figure 10.14
indicate that the reduction in interaction factor due to soil nonlinearity is considerably larger in the vertical direction than in the horizontal direction. For example, for
a capacity mobilization of 50%, the nonlinear interaction factors for the vertical and
horizontal loadings are, respectively, about 60% and 80% of their elastic counterparts.

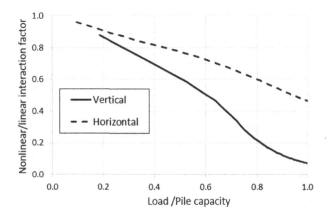

Figure 10.14 Ratio of nonlinear interaction factor to linear interaction factor as a function of pile load relative to pile capacity.

10.6 CONCLUSIONS

This chapter presented results of dynamic impedances of a large group comprising 225 piles for different pile configurations and soil conditions. The intention was to study how the pile layout and soil heterogeneity, both horizontally and vertically, affect the well-known features observed in dynamic response of pile groups for idealized soil conditions. The analyses have revealed the following.

a   The pile group layout has little impact on the impedances of large pile groups if the pile spacing is not very different. In the case considered in this study, the two group configurations have the same number of piles and about the same footprint.

b   Vertical soil heterogeneity affects the impedances to the extent the soil profiles have different equivalent moduli. However, the general features of pile group responses in terms of added soil mass and strong influence of pile-soil-pile interaction are preserved.

c   Lateral soil heterogeneity, defined in terms of variation of soil moduli relative to an average profile across the site, has a minor effect on the overall dynamic properties of large pile groups.

d   The kinematic interaction response of large pile groups confirms the general belief that pile groups generally follow the free-field earthquake motions; however, they filter out the high frequency components.

e   Limited nonlinear FE analyses have indicated the expected results that pile-soil-pile interaction is reduced as the loads approach the pile capacity. For example, for the case considered in this study, for the loads reaching 50% of the pile capacities, the elastic interaction factors for the vertical and horizontal loading are reduced by about 40% and 20%, respectively.

REFERENCES

Álamo, G.M., Aznarez, J.J., Padron, L.A., Martínez-Castro, A.E., Gallego, R. & Maeso, O. (2018). Dynamic soil-structure interaction in offshore wind turbines on monopiles in layered seabed based on real data. *Ocean Engineering*, 156: 14–24.

Auersch, L. (2019). Compliance and damping of piles for wind tower foundation in nonhomogeneous soils by the finite-element boundary-element method. *Soil Dynamics and Earthquake Engineering*, 120: 228–244.

Bayat, M., Andersen, L.V. & Ibsen, L.B. (2016). p-y-ẏ curves for dynamic analysis of offshore wind turbine monopile foundations. *Soil Dynamics and Earthquake Engineering*, 90: 38–51.

Boulanger, R.W., Curras, C.J., Kutter, B.L., Wilson, D.W. & Abghari, A. (1999). Seismic soil-pile-structure interaction experiments and analyses. *Journal of Geotechnical and Geoenvironmental Engineering*, 125(9): 750–759.

Darendeli, M. (2001). Development of a new family of normalized modulus reduction and material damping curves. Ph.D. Dissertation, University of Texas, Austin, TX.

Dezi, F., Carbonari, S. & Morici, M. (2016). A numerical model for the dynamic analysis of inclined pile groups. *Earthquake Engineering & Structural Dynamics*, 45: 45–68.

Di Laora, R. & Rovithis, E. (2015). Kinematic bending of fixed-head piles in nonhomogeneous soil. *Journal of Geotechnical and Geoenvironmental Engineering*, 141(4): 04014126.

Dobry, R. & Gazetas, G. (1988). Simple method for dynamic stiffness and damping of floating pile groups. *Geotechnique*, 38(4): 557–574.

Gazetas, G., Fan, K. & Kaynia, A.M. (1993). Dynamic response of pile groups with different configurations. *Soil Dynamics and Earthquake Engineering*, 12: 239–257.

Goit, C.S., Saitoh, M. & Mylonakis, G. (2016). Principle of superposition for assessing horizontal dynamic response of pile groups encompassing soil nonlinearity. *Soil Dynamics and Earthquake Engineering*, 82: 73–83.

He, R., Kaynia, A.M. & Zhang, J. (2019b). A Poroelastic solution for dynamics of laterally loaded offshore monopiles. *Ocean Engineering*, 179(2019): 337–350.

He, R., Kaynia, A.M., Zhang, J., Chen, W. & Guo, Z. (2019a). Influence of vertical shear stresses due to pile-soil interaction on lateral dynamic responses for offshore monopiles. *Marine Structures*, 64: 341–359.

Kausel, E. (2010). Early history of soil-structure interaction. *Soil Dynamics and Earthquake Engineering*, 30: 822–832.

Kausel, E., Whitman, R.V., Morray, J.P. & Elsabee, F. (1978). The spring method for embedded foundations. *Nuclear Engineering and Design*, 48: 377–392.

Kaynia, A.M. (1982). Dynamic stiffness and seismic response of pile groups. Research Report R82–03, Dept. Civil Eng., M.I.T. Cambridge, MA.

Kaynia, A.M. (1988). Characteristics of the dynamic response of pile groups in homogeneous and nonhomogeneous media. *Proc. 9th World Conf. Earthquake Eng*, Tokyo-Kyoto, Japan, 3: 575–580.

Kaynia, A.M. (1992). Transient Green's functions of fluid-saturated porous media. *Computers and Structures*, 44(1–2), 19–27.

Kaynia, A.M. (2012). Dynamic response of pile foundations with flexible slabs. *International of Journal of Earthquakes and Structures*, 3(3–4): 495–506.

Kaynia, A.M. (2018). Seismic considerations in design of offshore wind turbines. Seismic considerations in design of offshore wind turbines. *Soil Dynamics and Earthquake Engineering*. https://doi.org/10.1016/j.soildyn.2018.04.038.

Kaynia, A.M. & Kausel, E. (1991). Dynamics of piles and pile groups in layered soil media. *Soil Dynamics and Earthquake Engineering*, 10(8): 386–401.

Kaynia, A.M. & Novak, M. (1992). Response of pile foundations to Rayleigh waves and obliquely incident body waves. *Earthquake Engineering & Structural Dynamics*, 21(4): 303–318.

Kjørlaug, R.A. & Kaynia, A.M. (2015). Vertical earthquake response of megawatt-sized wind turbine with soil-structure interaction effects. *Earthquake Engineering and Structural Dynamics*, 44: 2341–2358.

Latini, C., Zania, V. & Johannesson, B. (2015). Dynamic stiffness and damping of foundations for jacket structures. *Proc. 6th International Conference Earthq Geotech Eng*, Christchurch, New Zealand.

Makris, N. & Gazetas, G. (1992). Dynamic pile-soil-pile interaction. Part II: Lateral and seismic response. *Earthquake Engineering and Structural Dynamics*, 21(2): 45–162.

Mamoon, S., Kaynia, A.M. & Banerjee, P.K. (1990). Frequency domain analysis of piles and pile groups. *Journal of the Engineering Mechanics Division ASCE*, 116: 2237–2257.

Markou, A.A. & Kaynia, A.M. (2018). Nonlinear soil-pile interaction for offshore wind turbines. *Wind Energy*, 21: 558–574.

Miura, K., Kaynia, A.M., Masuda, K., Kitamura, E. & Seto, Y. (1994). Dynamic behaviour of pile foundations in homogeneous and nonhomogeneous media. *Earthquake Engineering and Structural Dynamics*, 23(2): 183–192.

Mylonakis, G.E. & Gazetas, G. (1999). Lateral vibration of internal forces of grouped piles in layered soil. *Journal of Geotechnical and Geoenvironmental Engineering*, 125: 16–25.

Mylonakis, G., Nikolaou, A. & Gazetas, G. (1997). Soil-pile-bridge seismic interaction: Kinematic and inertial effects. Part I: Soft Soil. *Earthquake Engineering & Structural Dynamics*, 26(3): 337–359.

Novak, M. (1974). Dynamic stiffness and damping of piles. *Canadian Geotechnical Journal*, 11(4): 574–598.

Padrón, L.A., Aznárez, J.J. & Maeso, O. (2008). Dynamic analysis of piled foundations in stratified soils by a BEM-FEM model. *Soil Dynamics and Earthquake Engineering*, 28: 333–346.

Pender, M. (1993). Aseismic pile foundation design analysis. *Bulletin of the New Zealand National Society for Earthquake Engineering*, 26(1): 49–160.

PLAXIS 3D (2018). *PLAXIS BV.* Delft, The Netherlands.

Poulos, H.G. (1968) Analysis of the settlement of pile groups. *Geotechnique*, 18(4): 449–471.

Roesset, J.M. (1982). *Dynamic Stiffness of Pile Groups, in Pile Foundations*, ASCE, New York, 263–286.

Seed, H.B. & Idriss, I.M. (1970). Soil moduli and damping factors for dynamic response analyses. Report EERC 70-10, EERC, Univ. of California, Berkeley, CA.

Shadlou, M. & Bhattacharya, S. (2016). Dynamic stiffness of monopiles supporting offshore wind turbine generators. *Soil Dynamics and Earthquake Engineering*, 88: 15–32.

Taherzadeh, R., Clouteau, D. & Cottereau, R. (2009). Simple formulas for the dynamic stiffness of pile groups. *Earthquake Engineering & Structural Dynamics*, 38: 1665–1685.

Varun, V., Assimaki, D. & Shafieezadeh, A. (2013). Soil-pile-structure interaction simulations in liquefiable soils via dynamic macroelements: Formulation and validation. *Soil Dynamics and Earthquake Engineering*, 47: 92–107.

Waas, G. & Hartmann, H.G. (1984). Seismic analysis of pile foundations including pile-soil-pile interaction. *Proc. 8th World Conf. Earthquake Eng*, San Francisco, CA, 5: 55–62.

Wolf, J.P., von Arx, G.A., de Barros, F.C.P. & Kakubo, M. (1981). Seismic analysis of the pile foundation of the reactor building of the NPP Angra 2. *Nuclear Engineering and Design*, North Holland, Amsterdam, 65: 329–341.

Index